Medical and Laboratory Examinations for Neuropsychologists

Medical and Laboratory Examinations for Neuropsychologists

A Clinical Guide

Edited by

THOMAS J. FARRER
R. BROCK FROST

OXFORD
UNIVERSITY PRESS

OXFORD
UNIVERSITY PRESS

Oxford University Press is a department of the University of Oxford. It furthers the University's objective of excellence in research, scholarship, and education by publishing worldwide. Oxford is a registered trade mark of Oxford University Press in the UK and in certain other countries.

Published in the United States of America by Oxford University Press
198 Madison Avenue, New York, NY 10016, United States of America.

CIP data is on file at the Library of Congress.

ISBN 9780197619506

DOI: 10.1093/9780197619537.001.0001

Printed by Integrated Books International, United States of America

MIX
Paper
FSC FSC® C183721

The manufacturer's authorized representative in the EU for product safety is Oxford University Press España S.A. of Parque Empresarial San Fernando de Henares, Avenida de Castilla, 2 – 28830 Madrid (www.oup.es/en or product.safety@oup.com). OUP España S.A. also acts as importer into Spain of products made by the manufacturer.

Contents

List of Contributors

C.J. Brush PhD is an Assistant Professor, School of Kinesiology, Auburn University, Auburn, AL, US.

Carol Kave Chamoff, OTR/L, Nassau University Medical Center, East Meadow, NY, US.

Paula-Ann E. Chin M.A. Department of Neurology, Duke University School of Medicine, Durham, NC, US.

Sarah E. Cook PhD, ABPP, is an Associate Professor in Psychiatry and Behavioral Sciences and in Neurology, Duke University, Durham, NC, US.

Stephen Correia PhD, ABPP, is an Associate Professor, Institute of Gerontology, University of Georgia, Athens, GA, US.

Jorg Dietrich MD, PhD, is an Associate Professor of Neurology, Mass General Brigham Cancer Institute, Boston, MA, US.

Thomas J. Farrer, PhD, MHS, ABPP, is an Assistant Professor, School of Health and Medical Professions, University of Idaho, Moscow, ID, US.

R. Brock Frost PhD, MSCP, ABPP, is associate faculty in the University of New Mexico School of Medicine's Rural Family Medicine Residency (Santa Fe) and practices at Christus St. Vincent.

Amanda Gooding PhD, ABPP is Associate Professor of Psychiatry, University of California San Diego, San Diego, CA, US.

Marc W. Haut, PhD, ABPP, is a Professor, Department of Behavioral Medicine/Psychiatry, Neurology and Radiology, Director, Memory Health Clinic, Rockefeller Neuroscience Institute, West Virginia University, Morgantown, WV, US.

Rebecca C. Hendrickson, MD, PhD, Department of Psychiatry and Behavioral Sciences, University of Washington School of Medicine, Seattle, WA, US.

Nada El Husseini MD, MHSc., is an Associate Professor of Neurology, Duke University, Durham, NC, US.

Cierra M. Keith PhD, ABPP, is an Associate Professor, Rockefeller Neuroscience Institute, West Virginia University, Morgantown, WV, US and Professor of Psychiatry, Neurology, and Radiology, Rockefeller Neuroscience Institute West Virginia University, Morgantown, WV, US.

Gabriel M. MacAlevy BS School of Medicine, University of Washington, Seattle, WA, US.

Seth A. Margolis PhD, ABPP, is an Assistant Professor, Department of Psychiatry and Human Behavior, The Warren Alpert Medical School of Brown University, Providence, RI, US.

Thomas A. Nelson MD,Division of Neuro-oncology, Department of Neurosurgery, University of California, San Francisco, San Francisco, CA, US.

Kathleen F. Pagulayan PhD is a Professor, Department of Rehabilitation Medicine, University of Washington, Seattle, WA, US.

Brett Parmenter PhD, ABPP is a Clinical Professor, University of Washington School of Medicine, Seattle, WA, US and Clinical Neuropsychologist, Mt Tahoma Psychological Services, Tacoma, WA, US.

Michael Parsons PhD, ABPP is an Associate Professor, Neuropsychology & Neuro-oncology, Mass General Brigham/Harvard Medical School, Boston, MA, US.

Howard D. Pomeranz MD, PhD, is a Professor, Northwell Health, Great Neck, NY, US.

Heidi Sarles-Whittlesey, MA, Department of Psychology, Franklin College of Arts and Sciences, University of Georgia, Athens, GA, US.

Lynn A. Schaefer PhD, ABPP is Director of Neuropsychology Services & Training, Department of Psychiatry & Behavioral Sciences, Nassau University Medical Center, East Meadow, NY, US.

David P. Sheppard PhD is an Assistant Professor, Department of Rehabilitation Medicine, University of Washington, Seattle, WA, US

Miran Skrap, MD, Department of Neurosurgery, Azienda Ospedaliero-Universitaria di Udine (Santa Maria della Misericordia University Hospital), Udine, Italy.

Skyler J. Smith, B.S., School of Medicine, University of Washington, Seattle, WA, US.

Marjorie Soltis MD is an Assistant Professor, Duke University, Durham, NC, US.

Jill Z. Stuart PhD, ABPP, is Assistant Professor, Psychiatry and Behavioral Sciences, Duke University School of Medicine, Durham, NC, US.

M. Omar Subei MD, Neurology Department, Duke University Medical Center, Durham, NC, US.

Lawrence H. Sweet, PhD, Professor in Clinical Psychology, Department of Psychology, Franklin College of Arts and Sciences, University of Georgia, Athens, GA, US.

Barbara Tomasino PhD is Senior Researcher, IRCCS Eugenio Medea, Pasian di Prato (Udine), Italy, and Senior Researcher, IRCCS Eugenio Medea, Pasian di Prato (Udine), Italy.

Melanie Ward, MD is an Associate Professor, Department of Neurology, Memory Health Clinic, Rockefeller Neuroscience Institute, West Virginia University, Morgantown, WV, US.

Kirk Wilhelmsen, MD, PhD is a Professor, Rockefeller Neuroscience Institute, West Virginia University, Morgantown, WV, US.

Introduction

Thomas J. Farrer

The field of neuropsychology has evolved substantially over time, but at its roots, neuropsychology training focuses on understanding functional neuroanatomy and psychometric correlates with disease and injury, and neuropsychologists develop strong clinical skills in lesion localization based on a keen knowledge of brain-behavioral associations. Neuropsychology has also evolved alongside other disciplines in clinical medicine (psychiatry, neurology, physiatry) and allied professions (physical therapy, speech therapy, and occupational therapy). These advances have come with a plethora of neurodiagnostic tools. Neurodiagnostics include a broad set of techniques that evaluate the structure, function, chemistry, or electrical activity of the central nervous system. Neurodiagnostics may include a variety of imaging modalities, electrophysiological assessments, laboratory markers of blood or CSF, or other assessments that offer critical insight into the integrity of the central nervous system. Such tools, which are discussed throughout this text, have aided in assessment, diagnosis, and treatment monitoring.

This volume seeks to bridge a divide between traditional neuropsychological assessment that focuses on psychometric and neurobehavioral assessment, with a wide array of diagnostic tools that could aid clinical neuropsychologists in understanding patient conditions. This book provides a comprehensive overview of neurodiagnostic tools that a neuropsychologist will frequently encounter in clinical practice while performing record review or while consulting with other professions in clinical medicine and allied health professionals. The chapters focus on a review of the relevant procedures, their clinical utility and limitations, relevant lab values where appropriate, and case illustrations when possible.

The chapters are organized by modality and written with an emphasis on clinical utility, interpretive guidance, and relevance to neuropsychological evaluation. The following is a roadmap to the chapters:

Chapter 1 extensively reviews the neurological examination. This exceptional chapter provides a comprehensive review of cranial nerve function and examination. Additionally, the authors detail how physicians conduct motor and

Thomas J. Farrer, *Introduction*. In: *Medical and Laboratory Examinations for Neuropsychologists*.
Edited by: Thomas J. Farrer and R. Brock Frost, Oxford University Press.
DOI: 10.1093/9780197619537.003.0001

sensory examination, and examination of cerebellar functions and gait. The findings of such exams can distinguish between central versus peripheral and cortical versus subcortical disorders.

Chapters 2–3 review structural and functional imaging and vascular imaging, respectively, and provide a comprehensive review of the clinical use of computed tomography (CT), magnetic resonance imaging (MRI), positron emission tomography (PET) imaging, magnetic resonance angiography (MRA), computed tomography angiography (CTA), digital subtraction angiography (DSA), single-photon emission computed tomography (SPECT), and neurovascular ultrasound. These chapters demonstrate how neuropsychologists can interpret findings in the context of cognitive and behavioral data.

Chapter 4 uniquely discusses how body imaging, particularly PET, can uncover malignancies or paraneoplastic syndromes and discusses when such scans are indicated. A detailed case study illustrates how integrating systemic imaging results can enhance differential diagnosis, guide treatment planning, and support interdisciplinary collaboration in neuro-oncology and behavioral neurology. Chapter 5 introduces the reader to measures of electrophysiology, including EEG, EMG, and nerve conduction studies, a valuable chapter for neuropsychologists working in epilepsy, movement disorders, and neuroimmunology clinics.

Chapter 6 provides a dynamic and comprehensive review of intraoperative mapping and the role of the neuropsychologists in awake brain surgeries and the use of real-time cognitive testing during resection. The case illustrates how results help surgeons avoid harming brain areas that support key functions. It also discusses how such assessments improve recovery and shows the value of neuropsychologists in surgical teams. Similarly, Chapter 7 is an exceptional chapter on the well-known Wada procedure. This chapter explains how the Wada procedure works, what information it provides, and how neuropsychologists help interpret the results. These two chapters (6 and 7) highlight the importance of using such procedures correctly to guide surgical planning and protect important brain functions.

In Chapter 8, the authors expertly detail the nature and utility of the polysomnography and how it is used to diagnose sleep disorders like sleep apnea, narcolepsy, and REM sleep behavior disorder. The chapter also explores how sleep problems can affect cognition, and why assessment of sleep is an important part of a neuropsychological evaluation. A case example shows how understanding a patient's sleep patterns can improve diagnosis and treatment planning.

In reviewing what are likely the most common labs almost all patients have in their record, Chapter 9 discusses blood and urine panels and lab values related to organ function, vitamin deficiencies, infections, medication effects, and substance use. The authors explain how to interpret lab results and use them to explain changes in cognition, mood, or behavior. Case examples show how lab work can improve diagnosis and guide treatment, especially when medical issues may be contributing to cognitive problems.

Chapter 10 details the ophthalmology exam and common types of visual impairments, such as disorders of eye movement, visual fields, and visual processing, and the tools used to assess these. The chapter also explains how vision issues can overlap with or mimic cognitive problems. It includes a unique section on how occupational therapists assess vision in hospital and rehabilitation settings. A case example shows how combining neuropsychological and visual testing can improve patient care.

Lastly, Chapter 11 is a highly dynamic, yet critical chapter, which reviews cerebrospinal fluid (CSF) analysis, tissue biopsy, and genetic testing related to neurological diseases. Such assessment tools can identify causes of rapid or unusual cognitive decline and help differentiate between types of neurodegenerative diseases. The chapter discusses when these tests are used, what they measure, and how the results can guide diagnosis and treatment.

Conclusion

This volume was created to address a key gap in clinical training and practice: the lack of formal instruction on how to interpret neurodiagnostic tests in neuropsychological contexts. While neuropsychologists are rarely the ordering provider, they routinely encounter findings from laboratory panels, neuroimaging, electrophysiology, and other diagnostic modalities in their work. This book offers a structured, practical guide to these tools, their indications, and their implications for cognitive functioning. By covering these diagnostic domains, the book equips readers with clinically actionable knowledge that strengthens interdisciplinary collaboration and improves diagnostic accuracy in clinical neuropsychological assessment. Whether used as a reference for practicing professionals or as a supplemental text for training programs, this volume is intended to support a more integrated and medically informed model of neuropsychological assessment.

1
The Neurological Examination

Melanie Ward, Kirk Wilhelmsen, Cierra M. Keith, and Marc W. Haut

Introduction

The neurological examination is an important complement to the neuropsychological record review, and integrating neurological examination findings can enhance the accuracy, results, and conclusions of neuropsychological assessment. The importance of understanding neurological findings is amplified given the number of referrals neuropsychologists receive from neurology. In this chapter, we will introduce the most pertinent aspects of the neurological examination for practicing neuropsychologists with a focus on distinguishing between central and peripheral nervous system findings and cranial versus spinal findings. We will not review all aspects of the examination and readers are referred to more comprehensive discussions (Campbell, 2005). We will also describe a multidisciplinary approach, during which a neurological examination and a neurobehavioral examination are both conducted (see Kiefer et al., 2014, for details of the neurobehavioral examination). Finally, we will leave the reader with a list of "gems" that we believe are useful for neuropsychologists.

Cranial Nerves

Cranial nerve (CN) nuclei are located in the brainstem and contribute to numerous motor functions in the head and neck, sensory functions including taste and smell, and bulbar functions such as swallowing and speech. Many injuries and neurodegenerative disorders present with symptoms involving the cranial nerves. Evaluating the cranial nerves can provide important clues regarding the localization and etiology of neurodegenerative processes.

Cranial Nerve I (Olfactory)

The olfactory nerve is a sensory nerve responsible for sense of smell, and patients with neurodegenerative disorders often report anosmia. Patients with frontal

Melanie Ward et al., *The Neurological Examination*. In: *Medical and Laboratory Examinations for Neuropsychologists*. Edited by: Thomas J. Farrer and R. Brock Frost, Oxford University Press.
 DOI: 10.1093/9780197619537.003.0002

lobe tumors or structural lesions involving the skull base may also have decreased sense of smell. This occurs because of the anatomical location of the olfactory bulbs within the anterior cranial fossa and their abutment to the cribriform plate. During head injury, the brain will move within the cranial vault and the small olfactory nerve fibers coursing through the cribriform plate may be sheared, thus losing the sensory nerve connection. In addition to head injury, other clinical populations that may have reduced olfaction include those with Parkinson's disease, Alzheimer's disease, and schizophrenia. However, there are several other reasons why someone may present with reduced smell and neuropsychologists should not always assume a primary CNS abnormality. For instance, normal aging is associated with a reduced sense of smell. Reduced olfaction is also common in individuals with nasal or paranasal sinus diseases, upper respiratory infections, SARS-CoV-2 infection (COVID), smoking history, or cocaine use history.

Smell can be evaluated by introducing a strong smell, such as coffee or an alcohol swab, to the patient (with their eyes closed) and asking them to identify the smell. In most cases, the first cranial nerve is not formally tested. More applicable to general clinical practice, patients can be asked if they have noticed a change in their sense of smell. Decreased sense of smell typically localizes to the central nervous system in the absence of other non-neurologic processes such as sinus issues or infections. Anosmia is the clinical term for the absence of olfaction. Other less common terms may include hyposmia (diminished olfaction) and parosmia (atypical sense of smell).

Cranial Nerve II (Optic)

The optic nerve is a sensory nerve and conducts visual signals from the eye to the brain. Vision can be evaluated by assessing acuity with a Snellen chart and by evaluating pupillary responses with a flashlight. At least 30% of individuals with mild to moderate dementia have vision impairment, which is associated with worse quality of life (Dawes et al., 2018; Leroi et al., 2020). Beyond CN II, it is important to remember that there are numerous (and more common) other etiologies of vision impairment that localize to the eye itself such as cataracts, glaucoma, macular degeneration, and refractive error. Visual impairment may affect some cognitive tests and lead to misdiagnosis of mild cognitive impairment or dementia (Macnamara et al., 2022).

Visual fields can be rapidly assessed in the clinic by having patients cover one eye and count fingers in all four visual quadrants, then repeating with the other eye. Anatomically, retina ganglion cells exit the posterior aspect of the eye at the optic disc, forming the optic nerve. The bilateral optic nerves then

Diagnostic Approach to Vision Loss

Newman. Nancy: Biousse, Valérie
CONTINUUM: Lifelong Learning in
Neurology 20(4): 785–815, August 2014.
doi: 10.1212/01.CON.0000453317.67637.46

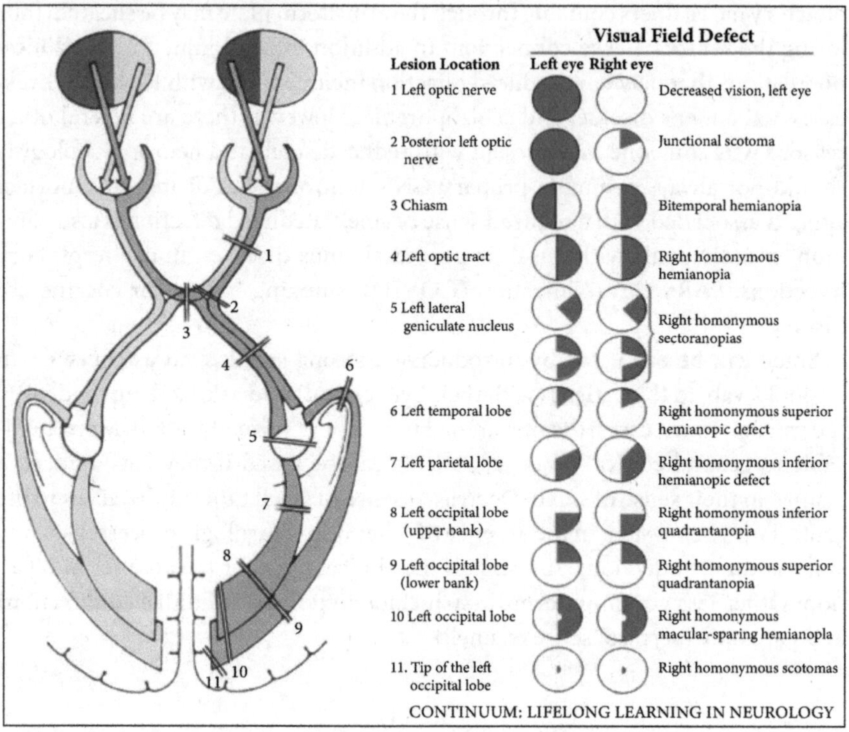

Diagram showing the effects on the field of vision produced by lesions at various points along the visual pathways. Lesions anterior to the chiasm produce ipsilateral, monocular visual loss, while lesions at the level of, or behind, the chiasm produce bilateral visual field defects with respect of the vertical meridian. Reprinted with permission from Biousse V, Newman NJ, Thieme.1 © 2009 Thieme Medical Publishers, Inc.

Figure 1.1 Schematic of visual pathways

come together, forming the optic chiasm, which bifurcates to become the optic tracts. Each tract synapses at the lateral geniculate nucleus of the thalamus. From here, the path projects back to the primary visual cortex in the occipital lobe via the superior and inferior optic radiations. Interruptions at any point in the pathway can result in visual field deficits. Visual field deficits (i.e., visual anopsia) in one eye only are due to peripheral lesions anterior to the optic chiasm. Lesions of the optic chiasm cause bitemporal field deficits and often portend an underlying pituitary growth pressing on the chiasm. Lesions of the optic tracts cause contralateral homonymous hemianopia. Superior and inferior quadrantanopias are due to inferior and superior lesions of the optic radiations or visual cortex, respectively. The inferior radiations from the lateral geniculate body to the occipital cortex loop back into the posterior temporal lobe. Thus, a superior quadrantanopia can also be due to a temporal lobe lesion. Figure 1.1 schematizes the visual pathways and deficits produced by interruptions of the pathways at their varying points.

Visual neglect/inattention is a subtler deficit often not recognized by patients. Neglect is not caused by vision impairment or a true visual field cut but represents inattention to the opposite side of a lesion in the setting of an otherwise intact visual pathway. It is important to distinguish neglect from the above-described visual field deficits, as localization is different and has important clinical implications. The most common cause of visual neglect is a stroke in the right middle cerebral artery distribution. Frequently, strokes with visual neglect also cause hemibody neglect. Visual neglect may improve after a stroke. Residual neglect can still be detected by testing for simultaneous detection of movements in a patient's peripheral vision or asking a patient to bisect lines on a piece of paper.

Common visual symptoms in Alzheimer's disease can include decreased acuity (possibly related to tissue loss in the retinal ganglion cells in the eye), hemianopia, and visual agnosia (Kaeser et al., 2015).

Cranial Nerves III (Oculomotor), IV (Trochlear), and VI (Abducens)

Cranial nerves III, IV, and VI are motor nerves grouped together because they all impact extraocular movements and contribute to conjugate eye movement (see also Chapter 10). The long course of the nerves that control eye movements from the midbrain to eye muscles makes them sensitive to damage by many processes. Screening evaluation of extraocular movements includes the "H" test, in which a patient is asked to visually track the examiner's finger in an "H" shape drawn in the air. Innervation for CN III includes a superior division: levator palpebrae superioris, superior rectus; and an inferior division: medial rectus, inferior rectus, inferior oblique. CN III allows the eye to look up, down, and medially (adduction), elevates the eyelid, and carries the nerves that constrict pupils and aid in accommodation of the lens. Patients with a CN III lesion have a deviation of the eye down and out. Pupillary fibers of CN III are located on the outer edge of the nerve. Thus, compressive lesions (such as posterior communicating artery aneurysms) typically cause pupil dilation, whereas microvascular infarcts of the nerve typically spare the pupil. CN IV allows the eye to rotate outward slightly with superior oblique muscles. Lesions of CN IV cause double vision in primary gaze such that affected patients tend to tilt their heads to align the vision of the affected eye with the unaffected eye. CN VI nerve lesions result in a reduction of lateral gaze in the affected eye and influences the lateral rectus muscle. Affected patients often rotate their heads to reduce diplopia.

A dysconjugate gaze can be detected in many patients who deny double vision. Shining a light toward the patient's eyes from a distance and examining its reflection on the pupils can be used to detect a dysconjugate gaze.

Early in life, many patients develop a dominant eye such that they can see with either eye but suppress the vision in one eye, a condition referred to as amblyopia. Amblyopia is a benign condition that can be confirmed by watching the movement of the eyes when they are covered, uncovered, or alternately covered. The affected eye will move to fixation when the unaffected eye is covered.

Following the examiner's finger, smooth pursuit requires the paramedian pontine reticular formation, multiple brain regions, the striatum, and the cerebellum. Rapid movements of the patient's eyes from one remembered position to another, called saccades, require, in addition to brain regions required for smooth pursuit, the superior colliculus and frontal eye fields. Because smooth pursuit and saccades require the normal function of many regions, they are sensitive to many neurologic conditions.

While vertical gaze can change to a degree with normal aging, impaired upgaze is often present in Parkinson's disease and atypical parkinsonism, including corticobasal degeneration, dementia with Lewy bodies (DLB), progressive supranuclear palsy, and multiple system atrophy. Progressive supranuclear palsy gets its name because of the loss of vertical eye movements.

In intranuclear ophthalmoplegia (INO), when testing horizontal eye movements, the ipsilateral eye cannot fully adduct and the contralateral eye has horizontal nystagmus. INO localizes to the medial longitudinal fasciculus and is most often seen in multiple sclerosis (MS), but can also occur in brainstem stroke.

Cranial Nerve V (Trigeminal)

As one of the largest and multifaceted cranial nerves, CN V has both sensory and motor functions. There are three main divisions of this nerve (V_1 = ophthalmic, V_2 = maxillary, V_3 = mandibular). Overall, CN V provides sensory innervation to the face, sinuses, and anterior two-thirds of the tongue and includes somatic sensations of pain, temperature, vibration, and dental pressure. CN V also relays proprioceptive information of the face. The V_1 branch provides information from the upper face, forehead, anterior scalp, cornea, and nasal cavity.

A portion of V_1 includes the lacrimal nerve and provides sensory information related to tear production. The V_2 segment provides information about the middle portion of the face, including the cheeks, lower eyelid, upper lip, nasal septum, hard palate, lateral zygomatic region, and upper teeth and gums. Lastly, division V_3 has both motor and sensory functions. The motor function of CN V_3 is related to its innervation to muscles for mastication. The sensory function relates the bottom portion of the face, including lower teeth, lower lip, chin, portions of the tongue, and lower gums.

Trigeminal Neuralgia

The complex syndrome of trigeminal neuralgia (TN) is often reported in clinical settings and is of clinical relevance to neuropsychologists. TN most often occurs after age 35 and frequently co-occurs with MS. There is mixed literature on the neurocognitive impairment in TN, although few studies have adequately addressed this question. One study suggested that TN may attenuate cognitive function, but also acknowledged that most study participants had medication treatments for TN that can impact cognition, such as opioid pain medications or other CNS (GABAergic) suppressive medications (i.e., benzodiazepines or anticonvulsants), which could confound study results (Meskal et al., 2014). Conversely, it is well-documented that chronic pain syndromes can negatively impact cognition and TN can result in paroxysms of severe pain that can occur for minutes at a time, dozens of times per day. TN is also associated with depression and anxiety (Zakrzewska & Linskey, 2014). Overall, it may be important for neuropsychologists to consider the degree and frequency of pain experienced by their patients with TN, as well as medication side effects and the impact of TN on emotional health.

Cranial Nerve VII (Facial)

The facial nerve has sensory and motor functions. It is responsible for most facial movements and facial expressions (smiling, frowning, closing eyes, raising eyebrows), as well as taste to the anterior two-thirds of the tongue and several parasympathetic functions. Peripheral CN VII nerve lesions, such as Bell's palsy, affect the whole face on one side. Central lesions affecting CN VII, most commonly stroke, spare the ability to wrinkle the forehead due to bilateral innervation of the nerve. Most patients with CN VII lesions have facial asymmetry. Distinguishing between a mild CN VII nerve lesion and long-standing asymmetry is difficult. Asking the patient or informant if their face has changed or looking at old pictures or driver's licenses can be informative. CN VII also plays a role in parasympathetic function, including lacrimation (tearing of eyes), salivation, and mucosal function in nasal cavity and part of the oral cavity.

Certain neurological diseases can result in reduced facial expression, but are not directly related to CN VII. For example, Parkinson's disease may present with masked facies (decreased facial expressions; also known as hypomimia), but this is due to basal ganglia dysfunction rather than damage to CN VII.

Cranial Nerve VIII (Vestibulocochlear)

The vestibulocochlear nerve is a sensory nerve responsible for hearing as well as sensory input regarding balance and equilibrium. Hearing loss is associated with dementia risk and social isolation (Lin & Albert, 2014). The Lancet 2020 commission on dementia recommended using hearing aids as a modifiable risk factor for dementia (Livingston et al., 2020). Hearing can be assessed by asking the patient and their family members if they have noticed hearing changes and by assessing hearing comprehension in conversation. The "finger-rub" test, in which the patient closes their eyes and is asked to identify whether they can hear fingers rubbing together approximately 3–4 inches from their ears, is also a simple screening test for hearing loss. Referral to audiology for hearing aid evaluation should be considered if there is evidence or report of hearing loss.

Many older patients report dizziness. It is important to obtain the history of whether the dizziness is due to a change in vestibular function, reduced blood pressure/autonomic dysfunction, or a reversible reason (i.e., medication side effects or dehydration). Acute changes in vestibular function cause vertigo, which manifests as a sensation of movement (often spinning or dysequilibrium). Vertigo can localize to cerebellar dysfunction or inner ear pathology. A Dix-Hallpike maneuver can aid in localizing the lesion and is typically positive in a peripheral (inner ear) process. The patient begins sitting up with their head turned 45 degrees toward the ear being tested, and then is lowered horizontally quickly until their neck is 20 degrees below the head of the bed. With a peripheral etiology, nystagmus (typically torsional and upbeating) will occur after a few seconds. Given the relationship between balance and vestibulocochlear function, the cranial nerve examination can provide important clues about the localization of impaired balance.

Lightheadedness is made worse by standing in orthostatic hypotension, which may indicate cardiac disease, autonomic dysfunction, or atherosclerosis. Autonomic dysfunction is seen in many neurodegenerative diseases, and atherosclerosis can contribute to strokes and vascular dementia. Poorly controlled diabetes can also cause autonomic dysfunction leading to orthostatic hypotension.

Cranial Nerve XI (Spinal-Accessory)

The spinal accessory nerve is a motor nerve that supplies the sternomastoid and upper trapezius muscles. The sternomastoid muscles turn the head side to side (left muscle turns head to the right; right muscle turns head to the left), and the upper trapezius allows for shoulder shrug. CN XI is typically tested by placing a hand on the side of the chin and having the patient turn their head in the direction of the physician's hand as the physician applies mild resistance. Additionally, the physician will place their hands on the patients' shoulders and have them shrug their shoulders to test for weakness.

Cranial Nerves IX (Glossopharyngeal), X (Vagus), and XII (Hypoglossal)

The glossopharyngeal and vagus nerves have both sensory and motor function, while the hypoglossal nerve is motor-only. These cranial nerves help control swallowing and speech, with IX and X also contributing to taste. Dysarthria is typically assessed in conversation and can be present in many neurologic diseases including (but certainly not limited to) stroke, myasthenia gravis, cerebral palsy, and other diseases discussed below. Bulbar (the nerves of the medulla, which looks like a bulb) weakness including dysphagia and dysarthria can be prominent in atypical parkinsonisms and frontotemporal dementia–amyotrophic lateral sclerosis (FTD-ALS). Bulbar symptoms are common in more advanced stages of other dementias as well and can also occur in stroke and neuroinflammatory disorders including MS. Other distinguishing features of bulbar weakness include tongue and palate deviation, drooling, and decreased gag reflex. Tongue fasciculations and atrophy are also frequently seen in ALS, and may especially be early findings in bulbar-onset ALS.

A summary of the cranial nerves and their functions is provided in Table 1.1.

Examination of Sensory Function

Neurologists routinely perform sensory examinations as part of a standard workup. The pattern of sensory loss corresponds to the human dermatome, and loss of sensation in these patterns allows the physician to isolate where in the central or peripheral nervous system a lesion may have occurred. Often patients are aware of their sensory deficits. Patients are not aware if they have sensory neglect, or not fully aware if they have inattention. The presence of sensory neglect is assessed by checking light touch in one extremity, then the

Table 1.1 Synopsis of the cranial nerves

Nerve	Name	Functional Type	Clinical Function	Findings in Deficit
CN I	Olfactory	Sensory	Smell	Anosmia
CN II	Optic	Sensory	Visual signal from eye	Unilateral vision loss or visual field cut
CN III	Oculomotor	Motor	Innervates multiple muscles involved in eye movement, eyelid elevation, & pupillary constriction	Diplopia with ipsilateral impaired adduction, infraduction, & supraduction (eye "down and out"), ptosis, variable pupillary involvement
CN IV	Trochlear	Motor	Eye movement, innervates superior oblique muscle which rotates eye forward when eye is adducted	Diplopia with ipsilateral hypertropia & excyclotorsion (eye "up & out"), patients often have head tilt
CN V	Trigeminal	Sensory & Motor	Sensations from face and motor function of mastication	Sensory impairment and/or jaw deviation or dysarthria
CN VI	Abducens	Motor	Eye movement, innervates lateral rectus, which abducts eye	Diplopia with ipsilateral impaired abduction
CN VII	Facial	Sensory & Motor	Facial expression	Forehead-sparing facial weakness, impaired taste
CN VIII	Vestibulocochlear	Sensory	Auditory and vestibular function	Hearing loss, vertigo
CN IX	Glossopharyngeal	Sensory & Motor	Sensation of touch, pain, & temperature in pharynx, middle, ear, & posterior tongue; motor function related to elevation of pharynx during speech and swallowing	Impaired taste and dysphagia

Nerve	Name	Functional Type	Clinical Function	Findings in Deficit
CNX	Vagus	Sensory & Motor	Parasympathetic innervation to heart, lungs, & digestive system; motor control of pharynx, larynx, and upper esophagus	Autonomic dysfunction, dysphagia, absent gag reflex, dysphonia, impaired taste
CN XI	Spinal-Accessory	Motor	Motor control of sternocleidomastoid and trapezius	Weakness in neck movement & shoulder shrug
CN XII	Hypoglossal	Motor	Tongue movement	Tongue deviation to the ipsilateral side of lesion, dysarthria, dysphagia

opposite extremity, then both simultaneously—patients with extinction will be able to identify touch in each individual extremity but not when both are tested simultaneously. Sensory extinction is typically representative of a central process involving the parietal lobe.

Without testing, patients would not be aware if they have problems with graphesthesia (inability to identify numbers or letters traced on the palm of the hand) or two-point discrimination (inability to recognize simultaneous stimulation by one or two points), which are due to cortical and subcortical lesions. Patients are also unaware of problems with proprioception because it is not a sensation that reaches our consciousness.

A formal typical screening sensory exam consists of assessing for intact sensation to light touch in the upper and lower extremities as well as vibration and proprioception. Vibration and proprioception are typically assessed at the great toes or distal finger joints because the peripheral nerves from these positions travel the furthest from the tendons innervated to the posterior column of the spinal cord and thus are the most sensitive to damage. Vibration sense, which can be used as surrogate for proprioception, is tested using a tuning fork placed on a joint. Vibratory sense can be quantified by comparing when the patient stops feeling vibration relative to the examiner holding the tuning fork. Proprioception is evaluated by stabilizing a proximal joint while the patient's eyes are closed and moving the more distal joint slightly up or down, asking the patient if they can accurately determine if the limb moved and in which direction. Vibratory sense loss due to peripheral lesions can be distinguished from a posterior column lesion by testing for vibratory sensation along the spine. The length of the peripheral nerve for vibratory sense is the same for any position along the spine. The length of the nerve in the posterior column is longer and thus more sensitive to damage for nerves from the lower thoracic region than the cervical region. Neuropathies can be present in nutritional deficiencies that can cause cognitive symptoms, including vitamin B12 (cobalamin) and B1 (thiamine) deficiencies. Peripheral neuropathy can also occur in thyroid disorders, which can be associated with cognitive symptoms, and diabetes, which can cause vascular brain lesions that can lead to cognitive dysfunction. Unilateral sensory loss most frequently occurs in stroke and MS.

Motor

Tone

Muscle tone is assessed by passively moving a patient's extremities, most often by passively moving the patient's hand at the wrist or forearm. Spasticity (velocity-dependent increase in muscle tone) is commonly seen in MS, post-stroke, and in motor neuron disease and is more noticeable with fast movements. The modified Ashworth scale is commonly used to grade spasticity (Table 1.2). Rigidity is increased tone that is not velocity dependent. Cogwheeling rigidity, characterized by jerking movements when flexing and extending a patient's limb, is a tremor superimposed on rigidity and is commonly seen with parkinsonism. Other prominent features in Parkinson's disease and atypical parkinsonism can include bradykinesia (slow movements), tremor, decrement with finger tapping, and shuffling gait. Tremor in Parkinson's disease is typically high amplitude and low frequency, and typically occurs at rest. Vascular events and chronic ischemic changes can also cause parkinsonism, which may be asymmetric.

Essential tremor generally presents as low-amplitude, high-frequency tremor worse with posture and action, and is not necessarily associated with significant neurocognitive diseases.

Paratonia refers to increased tone with involuntary variable resistance during passive movement. Frequently patients with neurocognitive disorders assist and try to anticipate the examiner's movement during tone testing. Asking the patient to pat their knee with the contralateral hand can often unmask paratonia and rigidity.

Unilateral weakness on strength testing can occur from many etiologies including stroke or MS (although bilateral weakness can also occur in both etiologies depending on the location of the lesion) and can also occur in neurodegenerative diseases such as ALS. Pronator drift, in which one arm drifts downward and/or supinates when both hands are held outstretched with palms upward and the patient closes their eyes, can be a subtle indicator of weakness, and having the patient rotate their head at the same time can also bring out more subtle deficits. Strength is scored as shown in Table 1.3.

Table 1.2 Modified Ashworth Scale for grading spasticity

Modified Ashworth Scale	
0	No increase in muscle tone
1	Slight increase in muscle tone, with a catch and release or minimal resistance at the end of the range of motion when an affected part(s) is moved in flexion or extension
1+	Slight increase in muscle tone, manifested as a catch, followed by minimal resistance through the remainder (less than half) of the range of motion
2	Marked increase in muscle tone throughout most of the range of motion, but affected part(s) are still easily moved
3	Considerable increase in muscle tone, passive movement difficult
4	Affected part(s) rigid in flexion or extension

Table 1.3 Muscle strength grading

Muscle Strength Grading	
0	No movement
1	Muscle contraction can be seen, but strength is not sufficient to produce joint movement.
2	Muscle able to move in full range of motion when gravity is removed (e.g., for hip flexion or extension patient would need to be placed on their side to remove the effects of gravity).
3	Muscle can move in full range of motion with gravity, but not with resistance.
4	Muscle can move through full range of motion with some, but not full resistance from examiner.
5	Muscle can move through full range of motion with maximal resistance from examiner.

Abnormal Movements

Chorea is involuntary muscle contractions seen primarily in Huntington's disease, but it can also be drug induced from antipsychotic medications and levodopa (in addition to other causes). Chorea involves dancing movements that may be smooth or jerking. Athetotic movements are more writhing and slower than choreiform movements. Athetosis can be seen in thalamic stroke among other etiologies. Both chorea and athetosis localize to the basal ganglia.

Dyskinesia, abnormal involuntary movement that is often writhing or restless appearing, can occur with long-term use of levodopa in treatment of Parkinson's disease. Tardive dyskinesia can occur in the setting of antipsychotic use and consists of repetitive muscle movements in the face, neck, arms, and legs.

Pseudoathetosis is caused by severe loss of proprioception and often appears as piano-like movements with outstretched hands. Pseudoathetosis can be either a central or peripheral process and can occur in nutritional deficiencies (for example vitamin B12), MS, and spinal pathologies in addition to other processes that impair proprioception. Myoclonus is a brief, involuntary muscle jerk that can be seen in metabolic disturbances and in Creutzfeldt-Jakob disease. Asterixis, also known as negative myoclonus, is a sudden loss of muscle tone when attempting to maintain a posture (such as outstretched arms) commonly seen in liver disease (hepatic encephalopathy) and other metabolic disturbances. Fasciculations are brief, small muscle contractions that can be seen in ALS, and are typically identified through visualization (patients often describe this as muscle twitching). Finally, clonus refers to involuntary, rapid muscle contractions most commonly assessed by rapidly dorsiflexing the ankle and are indicative of a central lesion involving either the brain or spinal cord.

Reflexes

Reflexes are assessed by tapping a muscle tendon with a reflex hammer. Common locations to test tendon reflexes in clinical practice include the biceps, triceps, brachioradialis, quadriceps (knee), and Achilles. Reflexes are graded as shown in Table 1.4.

Increased reflexes (3–4+) are seen in upper motor neuron diseases including stroke, MS, and ALS. Vitamin B12 deficiency also typically causes hyperreflexia. A reflex can "spread" from the tendon being tested to an adjacent joint. For example, contraction of the biceps when testing the brachioradialis reflex may indicate hyperreflexia. Decreased reflexes (0–1+), especially distally, are commonly seen in peripheral neuropathy.

Table 1.4 Reflex grading

Reflex Grading	
0	Absent
1+	Present, but decreased
2+	Normal
3+	Increased, but without clonus and not necessarily pathologic
4+	Significantly increased, often with associated clonus

The Babinski reflex is assessed by moving a pointed object (often the sharper end of a reflex hammer) along the lateral plantar surface of the foot anteriorly. When the Babinski reflex is present the great toe extends (moves upward), as opposed to a normal response in which the toe flexes (moves downward). The Babinski reflex is a pathologic reflex indicating upper motor neuron dysfunction and typically localizes to the spinal cord or brain. Babinski reflex may be present in stroke (including advanced small vessel disease), MS, nutritional deficiencies that cause a myelopathy including B12 deficiency, and numerous other central etiologies.

Frontal Release Reflexes

Frontal release signs indicate frontal lobe dysfunction and are commonly seen in neurodegenerative disorders. These primitive reflexes from infancy naturally abate during healthy maturation. However, when injury to frontal regions impacts premotor or supplementary motor regions, these reflexes re-emerge. Frontal release signs include grasp reflex (involuntary grasping of the hand when an object or fingers are placed in the hand), snout reflex (puckering of the lips when tapped), palmomental reflex (contraction of the ipsilateral mentalis muscle with stimulation of the thenar surface of the hand), glabellar (persistent blinking with tapping of the forehead beyond 6 taps), and rooting/suck reflexes (sucking movements of the lips when touched). There is some, but limited support, on the relationship between frontal release signs and standard cognitive measures (Arango et al., 1999; Viewing et al., 1993).

Cerebellar

Historically, the cerebellum was thought to be primarily involved in motor processes, but the role of the cerebellum in cognitive functioning has been increasingly recognized (Schmahmann, 2019). Exam findings indicating cerebellar dysfunction include nystagmus (involuntary rapid eye movements either horizontally or vertically), dysarthria, dysdiadochokinesia (inability to perform rapid alternating movements), and ataxia/dysmetria (difficulty with fine motor movements and gait/balance). Of these, nystagmus and ataxia, dysmetria, and dysdiadochokinesia are the easiest signs to evaluate clinically as they are more likely to be representative of cerebellar dysfunction, while dysarthria can be present in multiple other neurologic disorders as previously discussed. Language impairments seen after cerebellar strokes can include dysarthria, loss of normal cadence, decreased fluency, and agrammatism (Schmahmann, 2019). Chronic alcohol abuse can cause cerebellar dysfunction in addition to cognitive impairment. Cerebellar dysfunction is also commonly seen in MS, and cerebellar signs in MS have been correlated with worse performance on the symbol digit modalities test and paced auditory serial addition test (Weier et al., 2014).

Gait

Gait is an important part of the neurologic exam and often can be informally assessed while a patient is walking to or from the exam room. Peripheral neuropathy can cause sensory ataxia, indicated by a high-stepping gait that occurs because the patient does not know where their feet are in space. Ataxia that is cerebellar in origin is typically wide-based and uncoordinated. Unilateral spastic gait with circumduction of the affected leg is typically seen in stroke or MS; bilateral spasticity can also occur in MS and stroke (among other etiologies), depending on the localization of the lesions. Foot drop is often seen in MS or lumbar radiculopathies. Shuffling gait, decreased arm swing, and postural instability are often hallmarks of Parkinson's disease or atypical parkinsonism syndromes. "Magnetic" gait, or inability to lift the feet off the floor, is part of the classic triad of normal pressure hydrocephalus (the remainder of the triad are urinary incontinence and cognitive impairment).

Integration in Clinical Practice

Both the neurological examination and neurobehavioral examination are crucial and complementary aspects of the evaluations completed in our multidisciplinary memory clinic. The neurologist and neuropsychologist may work in tandem through shared visits with patients to gather pertinent history and perform different aspects of their physical and behavioral examination. While certain aspects of the examination are better suited to be completed by the physician (e.g., testing of motor reflexes, cranial nerves, vibratory sensation, and proprioception), there are also some overlapping features of standard neurological and neurobehavioral exams that can be completed by either provider. Within our clinic, neuropsychologists evaluate for frontal signs through motor sequencing, go/no-go, copy of repetitive drawings, and certain frontal release signs. They also briefly assess repetition and comprehension of language, and basic visuospatial perception, and look for signs of apraxia. In our clinic, cognitive abilities are further assessed in a 1-hour neuropsychological testing session immediately following the clinical interview and initial evaluation with both providers.

There are many subtle findings observed during the neurobehavioral examination that raise concern for brain dysfunction, particularly when taken in combination with neurological examination findings. For instance, neuropsychologists may note stimulus-bound and/or perseverative responding, verbal and/or motor intrusions, or impulsivity during the neurobehavioral examination, which in some cases may present prior to evidencing more frank cognitive impairment. Increased understanding of the neurological examination with incorporation into neurobehavioral and neuropsychological assessment results can maximize diagnostic accuracy and lead to more nuanced interpretations. For example, interpretation of observed visuospatial impairments can assist in evaluation for cortical dysfunction, the location of which can inform diagnosis (i.e., visuospatial impairment in DLB or stroke). The presence of peripheral neuropathy may assist in determining the etiology of balance difficulties and certain vitamin deficiencies and metabolic causes to the differential of cognitive impairment. Careful motor examination can reveal parkinsonian features that may impact the differential, or reveal more subtle signs of motor neuron dysfunction that can be present in ALS. Presence of observed motor abnormalities such as cogwheel rigidity and paratonia can also inform both conceptualization and recommendations. For example, paratonia is commonly observed in individuals with dementia and can be associated with increased difficulties completing basic activities of daily living (e.g., washing, feeding, dressing) and increased caregiver burden (see Drenth et al., 2020), and cogwheeling raises concern for a parkinsonian disorder. Delineating other potential non-neurologic contributions to a clinical picture such as arthritis and joint replacements (for gait impairment) and cataracts or macular degeneration (for vision impairment) can also be beneficial. This real-time live feedback between neurologist and neuropsychologist is very efficient and helpful with developing a differential diagnosis.

Finally, below is an abbreviated list of "take-home" examination findings we believe are easiest to assess clinically and can be valuable in differentiating between varying etiologies of cognitive dysfunction.

- Distal loss of vibration or proprioception are often seen in peripheral neuropathy, which can indicate underlying metabolic diseases such as diabetes or hypothyroidism, or nutritional deficiencies (B12, thiamine).
- Restricted eye movements (especially restricted upgaze), masked facies, slowed movements, and shuffling gait with decreased arm swing are parkinsonian features that can be present in both Parkinson's disease as well as multiple other atypical parkinsonian syndromes, all of which are commonly associated with cognitive impairments.
- Weakness, fasciculations, hyperreflexia, bulbar signs, and cognitive/behavioral changes should raise concern for FTD-ALS.
- Unilateral weakness with motor or sensory hemineglect is often seen post-stroke.
- Hearing and vision changes may also contribute to cognitive impairment and performance on neuropsychological examination, and evaluating these etiologies prior to full assessment may be helpful.
- Brisk reflexes in conjunction with frontal release signs are pathognomonic markers for CNS disease/dysfunction.
- Identifying overlap syndromes causing both cognitive impairment and other deficits is important in the management of the non-cognitive symptoms that often co-occur in these conditions. For example, the occurrence of cognitive and behavioral dysfunction in FTD-ALS can negatively impact the care of individuals with significant dependence from the motor weakness in ALS (breathing, swallowing, and walking).

In summary, there is significant overlap between the neurologic exam, the neurobehavioral and the neuropsychological examination, and taking information from all in context with each other increases diagnostic yield, helps ensure that the clinical picture is viewed as a whole, and may allow for more thorough evaluation of potentially reversible causes of cognitive symptoms.

Case

A 65-year-old woman presents for evaluation of cognitive impairment. She feels that her memory is fine and is unhappy that her family "dragged me to this appointment so they can get my money." Her family, however, reports that 5 years ago she developed depression for the first time in her life and has gradually become more socially withdrawn. More recently, she has been speaking inappropriately to people at her church and has been much more irritable than usual. In the past year her family has noticed that she has difficulty finding her words, repeats herself, and has been neglecting her hygiene. She was once an avid cook but has not been able to follow recipes. She will start a task and wander off to something else before it is completed. She used to be much more meticulous and organized. Finally, she seems to be moving more slowly and has fallen several times, which is not normal for her.

On neurobehavioral examination she showed a grasp reflex, a glabellar tap, impaired learning of a new motor sequence, and continuous stimulus bound responding on a go/no-go task. On formal neuropsychological exam, she demonstrated impaired delayed recall on a word list task; she improved but not fully with yes/no recognition cues. She also showed moderate word-finding problems on confrontation naming, reduced fluency for phonemes relative to category fluency, moderate difficulty with verbal abstract reasoning, and impaired cognitive flexibility on Trail Making Test part B beyond slowed graphomotor processing speed observed during part A. Basic visuospatial skills were relatively preserved.

On neurologic exam, she has a snout reflex, masked facies, cogwheeling rigidity, and shuffling gait. In addition, she has decreased proprioception and loss of vibration sense in her bilateral great toes.

Discussion

In this case, her clinical history indicates mood and personality changes prior to onset of her cognitive symptoms, which primarily involve language and planning. She has parkinsonian features on exam as well as evidence of a peripheral neuropathy. Taken together, her history and exam are most consistent with behavioral variant frontotemporal dementia (bvFTD), which typically presents with mood and personality changes prior to onset of cognitive symptoms. FTD can present as an overlap syndrome with parkinsonism. The presence of a concomitant neuropathy should prompt screening for additional multifactorial causes of her deficits such as vitamin B12 deficiency, hypothyroidism, and diabetes. Impaired proprioception from peripheral neuropathy as well as parkinsonian gait are likely both contributing to her falls.

It is also important to be mindful that in some cases Alzheimer's disease can present with predominantly frontal/behavioral symptomatology early in the course, and this must be considered in differential diagnosis. Neuroimaging can be a helpful tool to further clarify etiology. See Figure 1.2 for brain magnetic resonance imaging (MRI) showing frontal and midbrain atrophy.

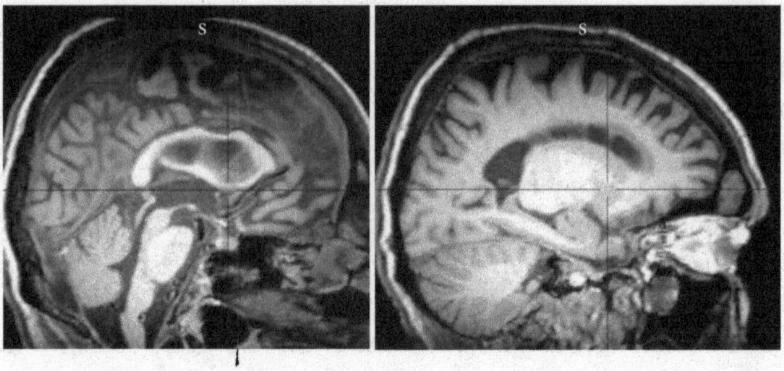

Figure 1.2 Sagittal MRI demonstrating midbrain signs and frontal atrophy

References

Arango, C., Bartko, J. J., Gold, J. M., & Buchanan, R. W. (1999). Prediction of neuropsychological performance by neurological signs in schizophrenia. *American Journal of Psychiatry, 156*(9), 1349–1357.

Campbell, W. (2005). *DeJong's The neurologic examination.* Lippincott Williams & Wilkins.

Dawes, P., Wolski, L., Himmelsbach, I., Regan, J., & Leroi, I. (2018). Interventions for hearing and vision impairment to improve outcomes for people with dementia: A scoping review. *International Psychogeriatrics, 31*(2), 203–221.

Drenth, H., Zuidema, S., Bautmans, I., Marinelli, L., Kleiner, G., & Hobbelen, H. (2020). Paratonia in dementia: A systematic review. *Journal of Alzheimer's Disease, 78*(4), 1615–1637.

Kaeser, P., Ghika, J., & Bourruat, F. X., (2015). Visual signs and symptoms in patients with the visual variant of Alzheimer disease. *BMC Ophthalmology, 15*(65). https://doi.org/10.1186/s12886-015-0060-9

Keifer, E., & Haut, M. W. (2014). Neurobehavioral examination. In M. W. Parsons & T. Hammeke (Eds.), *Clinical neuropsychology: A pocket handbook for assessment* (3rd ed., pp. 31–52). APA.

Leroi, I., Constantitidou, F., Langenbahn, D., Heyn, P., Yeung, K., & Dawes, P. (2020). Hearing and vision impairment in people with dementia: A guide for clinicians. *Archives of Physical Medicine and Rehabilitation, 101*(9), 1667–1670.

Lin, F. R., & Albert, M. (2014). Hearing loss and dementia—Who is listening? *Aging & Mental Health, 18*(6), 671–673.

Livingston, G., Huntley, J. Sommerlad, A., Ames, D., Ballard, C., Banerjee, S., Brayne, C., Burns, A., Cohen-Mansfield, J., Cooper, C., Costafreda, S., Dias, A., Fox, N., Gitlin, L., Howard, R., Kales, H., Kivimaki, M., Larson, E, Ogunniyi, A., . . . Mukadam, N. (2020). Dementia prevention, intervention, and care: 2020 report of the Lancet commission. *The Lancet Commissions, 3096*(10248), 413–446.

Macnamara, A., Schinazi, V. R., Chen, C., Coussens, S., & Loetscher, T. (2022). The effect of age-related macular degeneration on cognitive test performance. *Science Reports, 12*(4033). https://doi.org/10.1038/s41598-022-07924-8

Meskal, I., Rutten, G. J. M., Beute, G. N., Salden, M. E., & Sitskoorn, M. M. (2014). Cognitive deficits in patients with trigeminal neuralgia: Opportunities to improve care and quality of life. *Acta Neurochirurgica, 156*, 1565–1566.

Schmahmann, J.D. (2019). The cerebellum and cognition. *Neuroscience Letters, 688*, 62–75.

Vieweg, W. V. R., Brashear, H. R., Sautter, S. W., & Tabscott, J. A. (1993). Frontal release signs and their relationship with the initiation and perseveration component of the Mattis Dementia Rating Scale in persons with dementia. *American Journal of Geriatric Psychiatry, 1*(4), 310–315.

Weier, K., Pennert, I. K., Magon, S., Ammann, M., Naegelin, Y., Andelova, M., Derfuss, T., Stippich, C., Radue, E. W., Kappos, L., & Sprenger, T. (2014). Cerebellar abnormalities contribute to disability including cognitive impairment in multiple sclerosis. *PLos One, 22*(9): e86916. https://doi.org/10.1371/journal.pone.0086916

Zakrzewska, J. M., & Linskey, M. E. (2014). Trigeminal neuralgia. *BMJ Clinical Evidence, 2014*, 1207.

2

Structural and Functional Imaging

Computed Tomography, Magnetic Resonance Imaging, and Positron Emission Tomography in Clinical Neuropsychology

Stephen Correia, Heidi Sarles-Whittlesey, and Lawrence H. Sweet

Introduction

This chapter provides neuropsychologists with working knowledge about the clinical utility of computed tomography (CT), magnetic resonance imaging (MRI), and positron emission tomography (PET) of the brain. It also provides the reader with a brief background on the technical information about these methods to aid understanding of the clinical significance of imaging findings. A full detailed discussion of these techniques is beyond the scope of this chapter.

Incorporating neuroimaging findings into neuropsychological assessment, diagnosis, and intervention has advantages and potential pitfalls. From the perspective of neuropsychology, perhaps the greatest advantage of brain imaging is its ability to identify structural, functional, molecular, and metabolic pathology (using NMR spectroscopy, which not covered in this chapter) that can guide the approach to clinical assessment, diagnosis, and intervention. Knowing the location, size, and characteristics of a discrete lesion—for example, from a traumatic brain injury (TBI) or stroke—maps well onto clinical neuropsychologists' knowledge of brain–behavior relationships and helps guide their clinical assessment and treatment approach. The specificity and clinical impact of information gleaned from neuroimaging is less clear for more diffuse findings such as general, widespread cortical atrophy, or scattered white matter changes often seen in normal aging.

Neuropsychologists and other healthcare providers can face some pitfalls when viewing brain images for clinical purposes. One pitfall is the risk of "premature closure" (a.k.a., anchoring heuristic) and the related risk of confirmatory bias (Blumenthal-Barby & Krieger, 2015; Graber et al., 2005). Knowledge of the size, location, and pathophysiological features of a discrete lesion seen on

Stephen Correia, Heidi Sarles-Whittlesey, and Lawrence H. Sweet, *Structural and Functional Imaging*. In: *Medical and Laboratory Examinations for Neuropsychologists*. Edited by: Thomas J. Farrer and R. Brock Frost, Oxford University Press. © Oxford University Press (2026). DOI: 10.1093/9780197619537.003.0003 brain imaging could bias the neuropsychologist to implement an assessment battery designed only to confirm the cognitive effects of the discrete lesion, while overly minimizing coverage of other cognitive abilities. For example, focusing clinical assessment on a new area of post-TBI frontal encephalomalacia while overlooking the possible cognitive sequelae of an old posterior subcortical lacunar infarction potentially risks diagnostic errors due do narrowed breadth of assessment. Solely focusing a neuropsychological assessment and diagnostic formulation on the likely cognitive effects of a new and clinically salient lesion also risks underappreciation of other factors such as the patient's psychological, psychosocial, socioeconomic, ethnoracial, and educational statuses and history that may be significantly contributing to test performance. These decisional biases could also influence interpretation, prognostication, and intervention planning. Additionally, clinical experience shows that neuroimaging findings do not always map perfectly onto neuropsychological findings. Many factors, such as brain and cognitive reserve, injury recovery, general health status, and others not captured in the images can influence the patient's neuropsychological presentation. When and why past brain imaging was performed might not always relate to the reason for neuropsychological referral. For example, a CT scan acquired at age 60 due to headaches might not be informative about the patient's current brain status 10 years later when sent for assessment of possible Alzheimer's disease. Other things to consider (Lebby, 2013b) include awareness that there are anatomical variations across individuals, many of which are normal and lack clinical relevance. A pathological finding may or may not have clinical relevance, and normal-appearing brain images do not rule out cognitive impairment. Also, a discrete lesion can impact network functioning and thereby result in broad cognitive effects. Neuropsychologists must also guard against interpretations that are overly constrained by the neuroimaging findings. Further, clinical neuropsychologists consuming neuroimaging findings are urged to be mindful of the limits of their scopes of practice and competencies regarding interpretation of the imaging report and answering questions that patients may pose during the assessment or feedback session.

Structural Imaging

The remainder of the chapter is divided into three broad sections covering CT, MRI, and PET (see Chapter 3 for a discussion of vascular imaging). Each section will provide a broad overview of how images are acquired in each modality. Then the chapter describes the clinical utility of those images in relation to the practice of clinical neuropsychology. Our approach to the discussion of the characteristics (e.g., gray-white contrast, colorization) of the various images is qualitative, as is typically how radiologists and neuroradiologists report their interpretation of the images. Computational software is available that may complement (not substitute for) neuroradiologists' interpretation by providing quantitative and normative values of brain region volumes, lesion volumes, and other metrics (e.g., NeuroQuant software—https://www.cortechs.ai/products/neuroquant/ and NeuroGage https://www.neurogage.com/) (Ross et al., 2022); however, their use, while growing, is not currently widespread, as not all centers have such software.

Artificial intelligence (AI) methods are being incorporated into radiology, as evidenced by the Radiological Society of North America's journal *Radiology: Artificial Intelligence* (https://pubs.rsna.org/journal/ai). Most neuropsychologists are end consumers of radiological interpretation of clinical images. However, in the authors' view, it is uncertain how the incorporation of AI in radiology will affect neuropsychology practice, and we recognize that the impact might differ depending on the clinical purview (e.g., characterization of cognitive status post-TBI versus surgical planning for epilepsy treatment).

Helpful Guidelines for Viewing Brain Scans

A few general guidelines are important when looking at clinical brain images. The first is clinical images are displayed in radiologic convention, in which the patient's right side is displayed on the left when viewing the image. Second, when viewing structural images, it is important to know whether you are looking at a CT scan or MRI, as the meaning of signal characteristics differs. Third, it is important to be mindful of what anatomical orientation you are looking at (i.e., sagittal, axial, coronal). Looking at a lesion in multiple planes (if available) may improve understanding of its anatomic extent and shape. Anatomy may appear asymmetric due to head tilt in the scanner rather than due to anatomic variation. Be mindful that there are various sources of image artifact that can cause tissue to appear abnormal to the viewer who is unfamiliar with those artifacts (Krupa & Bekiesinska-Figatowska, 2015; Lloyd-Jones, 2019). If unsure about an abnormal tissue, consultation with the reading radiologist may be helpful. Finally, when viewing images, it is helpful to have anatomic landmarks in mind—this helps one answer the questions "Where in the brain am I looking?" and "What structures am I seeing?" Salient landmarks might include the brain stem, cerebellum, cerebral aqueduct, superior and inferior colliculi, pons, cerebral peduncles, pituitary gland, mammillary bodies, basal ganglia, thalamus, corpus collosum (splenium, genu, body), ventricles (including their main body and anterior, posterior, and temporal horns), and hippocampus. Keeping such landmarks in mind will help with your anatomic orientation within the brain as you scroll through images slices in axial, sagittal, or coronal planes.

Computed Tomography (CT)

Computed tomography (CT), also called computed axial tomography (CAT) is an x-ray imaging technique that acquires several sequential axial cross-sectional brain images (Patel & De Jesus, 2022). These two-dimensional (2D) images can be computationally reformatted to a coronal or sagittal view (Wei et al., 2010) to aid diagnosis and determine the spatial extent and shape of lesions. Images are collected axially as an x-ray generator and opposing detectors rotate around the patient in a circular gantry. With each rotation, the patient is moved slightly to acquire the next slice. The slice thickness can vary from 1 mm to 10 mm. The 2D raw data is fed into the scanner's computer and reconstructed into 2D brain images in which pixel values are assigned grayscale intensities (Patel & De Jesus, 2022). Tissues that are denser than water absorb more x-rays and are rendered to appear relatively bright (hyperdense); less dense tissues appear relatively darker (hypodense). Bone and calcifications are highly radiodense and appear bright. Air is the darkest "tissue" on CT scan and is seen in the sinuses but also can be seen intraparenchymally in the uncommon and potentially catastrophic condition of an air embolism. Tissues with similar radiodensity are described as being "isodense" to one another. Acute extravascular intracranial blood is hyperdense, which makes CT very useful for identifying fresh extravasated blood from, for example, an acute hemorrhagic stroke, small vessel microbleed, TBI, or aneurysm rupture. Figure 2.1 shows the relative signal intensities of various intracranial tissues on an axial CT brain image. Figure 2.2 shows acute intraventricular blood due to head trauma caused by a fall.

CT images are collected either with or without intravenous administration of an iodinated or other contrast agent that facilitates visualization of blood vessels and vascularized lesions. The contrast agent is radiodense and causes enhancement (i.e., appearing brighter or hyperintense) of intracranial vessels and parenchymal capillary beds. Contrast enhancing lesions include, but are not limited to, brain abscesses, primary or metastatic tumors, arteriovascular malformations, perivascular infection, and other conditions (Garg & Sinha, 2010; Huynh et al., 2020; Kastrup et al., 2005; Piszczor et al., 1985; Smirniotopoulos et al., 2007).

The neuropsychological implications of a positive (i.e., clinically significant) brain CT finding depend on multiple factors such as lesion size, location, signal

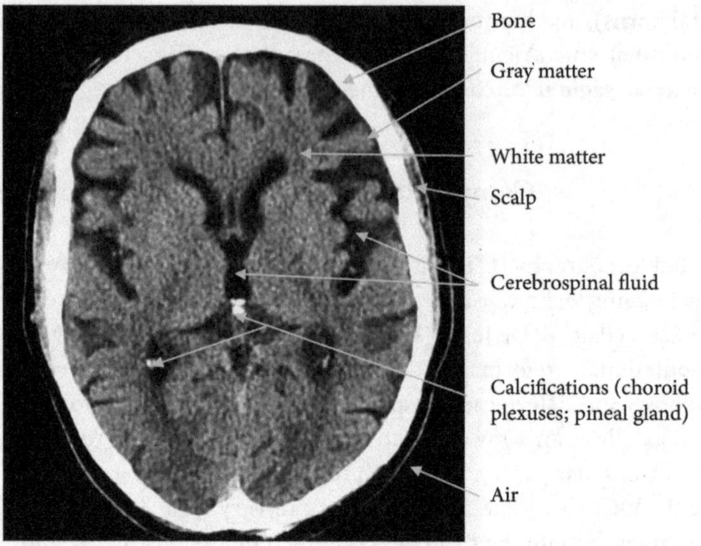

Figure 2.1 Appearance of various tissues on normal CT scan. This is an axial slice approximately through a region often referred to as the trigone, in which the anterior and posterior horns of the lateral ventricles are visualized; the third ventricle is also visualized in this slice (clinical image and labels provided by S. Correia).

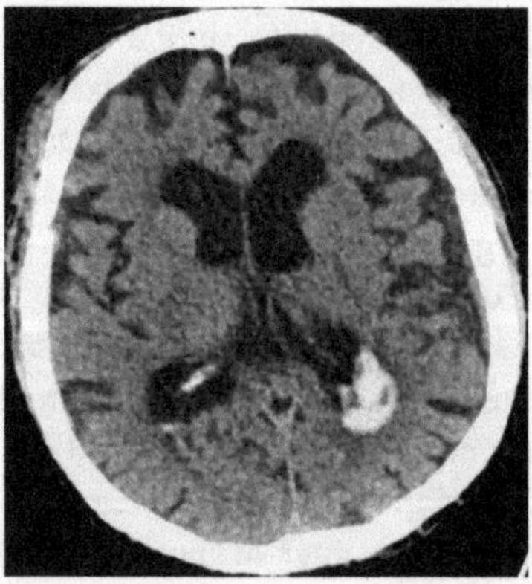

Figure 2.2 Acute blood in the posterior horn of the left lateral ventricle secondary to head trauma due to fall in an 84-year-old with dementia. The hyperdense signal in the opposite right posterior horn is a calcified choroid plexus. The choroid plexus in the left posterior horn is hard to visualize in the presence of hyperintense blood (clinical image and labels provided by S. Correia).

characteristics (e.g., hypodense, hyperdense, enhancing post-contrast), whether it is space-occupying, and whether its boundaries are discrete or diffuse. Cortical or subcortical atrophy that is clearly greater than expected for age may indicate a neurodegenerative disease. Enlarged ventricles that are greater than expected for the degree of cortical atrophy can, in an older adult, indicate normal pressure hydrocephalus. When serial images are available, the temporal course of the abnormal finding (e.g., stable, progressive, resolving) can have implications for assessment and treatment. Neuropsychologists must rely on their understanding of brain–behavior relationships as well as their understanding of neurological and medical conditions in determining the extent, if any, to which the finding warrants alteration of their typical approach to assessment and intervention. Alterations might include adding, removing, or substituting assessment measures to gain clinically relevant greater information about the cognitive and behavioral impacts of a finding, or to accommodate possible sensorimotor limitations caused by it.

Visual rating scales have been developed to aid the interpretation of CT findings in different clinical situations. Two such conditions, TBI and subcortical cerebrovascular changes, may be encountered by many neuropsychologists and are covered here briefly. The Marshall Classification System (MCS) (Charry et al., 2017; Marshall et al., 1992) was designed to predict patient outcomes and risk of intracranial pressure following severe TBI. The Rotterdam Scoring System (RSS) (Charry et al., 2017; Maas et al., 2005) was introduced in 2005 to provide a more fine-grained scale of severity and prognosis. Raj et al. (2014) reportedly improved the prediction of long-term outcomes using the Helsinki Computerized Tomography Scale, which adds bleeding type and size, intraventricular hemorrhage, and supersellar cisterns to the MCS and RSS.

Visual scales for rating severity of subcortical vascular changes on CT images, including white matter hypodensities and lacunar infarctions, have been developed (Geroldi et al., 2003; Wahlund et al., 2001; Wahlund et al., 2017), and the degree of subcortical vascular changes on CT have been shown to be related to reduced performance on tasks of processing speed, immediate and delayed memory, executive functions, visuoconstruction, language, and measures of global cognitive functioning (Gunning-Dixon & Raz, 2000; Hachinski et al., 2006; Longstreth et al., 2005; Schmidt et al., 2007; Tomimoto, 2015; van den Heuvel et al., 2006). Figure 2.3 shows an example of severe subcortical white matter hypodensities on CT.

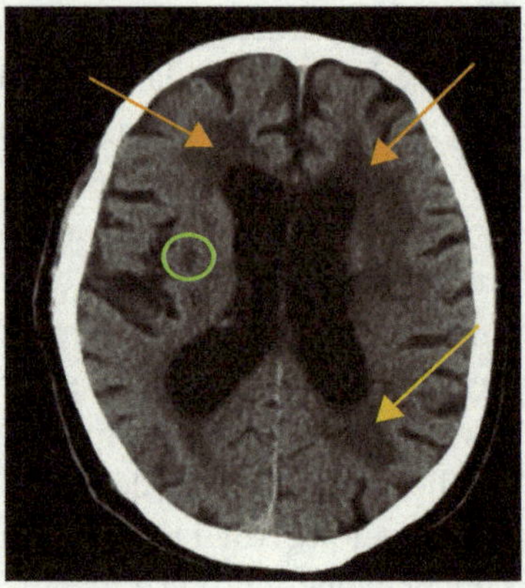

Figure 2.3 Head CT showing periventricular subcortical white matter hypodensity extending into deep white matter from the anterior and posterior horns (orange arrows) of the lateral ventricles bilaterally. The green circle shows a possible chronic lacunar infarction. There is ventriculomegaly in the lateral ventricles that appear due to parenchymal volume loss, as can be seen in widened sulci. Also note the asymmetric appearance of the brain such that the right, but not the left, Sylvian fissure is captured in this slice. Review of images of adjacent and more distal inferior anatomic slices (e.g., at the level of the eyes) revealed that this asymmetry is likely due to head tilt in the scanner and not due to anatomic variance (clinical image provided by S. Correia).

Case Example

A 63-year-old male was referred for neuropsychological assessment to characterize cognitive status in the context of guardianship. The patient was on the inpatient psychiatry service, and the referring provider suspected the patient met diagnostic criteria for dementia.

The patient's medical history was noteworthy for schizophrenia, alcohol abuse (reportedly in remission), cocaine abuse (also reportedly in remission), hepatitis C (treated), and multiple prior head injuries. The patient had multiple prior hospitalizations and underwent non-contrast CT scan of the head in 2004, 2012, and 2016 (current admission).

The panel of images below shows axial slices at roughly comparable anatomic levels across the three imaging time points. Slices across the *top row* are at the level of the Circle of Willis and show the anterior horns of the lateral ventricles, as well as the fourth ventricle. Note the horizontal streaking artifact in the cerebellum in the top left image (Figure 2.4). Slices in the *bottom row* are at the level of the anterior horns of the lateral ventricles and capture the third ventricle (see Figure 2.4).

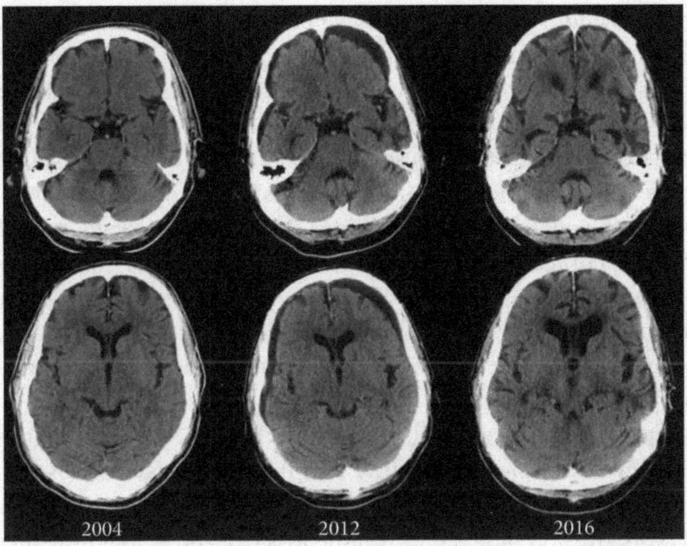

Figure 2.4 Serial CT images on patient described in case report (clinical images provided by S. Correia).

Observations of Clinical Interest

The patient was 51 years old at the time of the 2004 images. No acute abnormality is noted. There is slight anterior sulcal widening, but the anterior horns of the lateral ventricles are not overly dilated, indicating that atrophy is likely more localized than diffuse. Note the anteriorly split septum pellucidum. This is called a cavum septum pellucidum and is considered a normal anatomical variation. The temporal horns of the lateral ventricles appear slit-like and unremarkable. The 2012 images, taken at age 59, are noteworthy for subdural hematoma (SDH), which can be seen as a fluid collection (dark areas that appear isodense to the CSF in the ventricles) between the brain and the skull in the patient's left frontal region, running along the left hemisphere. Both SDHs are the result of head trauma (patient versus car) and can be seen on both 2012 slices. Although the anterior horns of the lateral ventricles and third ventricle look similar to the 2004 images, the temporal horns are now somewhat more prominent. Four years later in 2016, the areas of SDH have resolved—the extra-axial fluid collections are no longer visible. Note, however, the greater degree of sulcal widening compared to the 2004 image. The anterior and temporal horns of the lateral ventricles now have greater prominence, suggesting parenchymal volume loss. Additionally, the white matter anteriorly adjacent to the anterior horn of the patient's left lateral ventricle appears hypodense to the surrounding white matter and is suggestive of subcortical small vessel changes. The incidental finding of the split septum pellucidum (cavum septum) is now seen more easily.

The patient had undergone cognitive testing at each time point. Salient mild impairments were noted in memory encoding and retrieval, executive functions, and complex attention in 2004. The 2012 results were mostly stable apart from slight declines in some aspects of executive function. In 2016, there was generally mild impairment across most cognitive domains with relatively unchanged language functions. Memory test results now showed evidence of storage deficit. Additionally, recent records showed that the patient's functional status had declined noticeably to the point that guardianship was now being pursued. The neuropsychologists on this case concluded that the patient met the diagnostic criteria for dementia with multifactorial etiology.

This case provides an example of imaging changes over years aligning generally well with declines in cognitive and functional status. It also provides an opportunity to appreciate normal anatomical variants, the appearance of SDH, progression of cerebral volume loss over time, and how presumed subcortical small vessel disease can manifest in periventricular white matter.

Magnetic Resonance Imaging (MRI)

Full discussions of the physics of MRI and of how MRI scanners work to produce images are well beyond the scope of this chapter. Nonetheless, some familiarity with the workings of an MRI system may be of value in understanding the various types of images that can be acquired in MRI and in consuming neuroradiology reports and literature involving MRI. The following paragraphs provide a basic and simplified description of the key aspects of image acquisition (Dale, 2015; Lebby, 2013a; Mitchell, 2004).

Unlike x-ray-based CT imaging, MRI involves no radiation. MRI uses powerful magnetic fields and radiofrequencies to manipulate the normal energy state of the body's naturally occurring spinning hydrogen atoms (H^+; i.e., protons) to generate images of soft tissues. A traditional (i.e., non-open) MRI scanner consists of a horizontal cylindrical magnetic bore that produces a static magnetic field measured in Tesla units (T). Most clinical scanners have a static field strength, referred to as B_0, of 1.5 T or 3.0 T, although higher field strengths (e.g., 7.0 T) are available in some research centers (Nowogrodzki, 2018).

MRI scanners take advantage of two naturally occurring phenomena. The first is a property of electromagnetism: a spinning charged particle produces a magnetic dipole. In the case of H^+, this magnetic field is quite small. The second phenomenon is that while hydrogen protons spin on their axis, they also "precess" about their axis (like a wobbling spinning top) (Elster, 2021). Once the patient is situated in the scanner, the machine repeatedly applies a sequence of carefully selected and timed, brief radio frequency (RF) pulses and magnetic gradients. The RF pulses impart energy to the protons, causing them to flip their orientation (e.g., 90 degrees) away from the scanner's main magnetic field. The RF pulse also causes the proton precessions to phase align with one another (see Figure 2.5). The magnetic gradients are superimposed on the static magnetic field to produce a gradually increasing magnetic field in X, Y, or Z (or any) direction. Collectively, the application, timing, and duration of the RF and gradient pulses are commonly called the image acquisition "sequence." Different sequences produce different types of images (e.g., T1-weighted, T2-weighted, FLAIR—all described below) across which the appearance of different tissues (e.g., gray matter, white matter, CSF) may differ. For example, CSF appears bright (hyperintense) compared to gray and white matter on T2-weighted images, whereas it appears dark (hypointense) compared to those same tissues when seen on T1-weighted images. Different image types have different capacities to visualize different types of lesions (Dale, 2015; Lebby, 2013a; Mitchell, 2004).

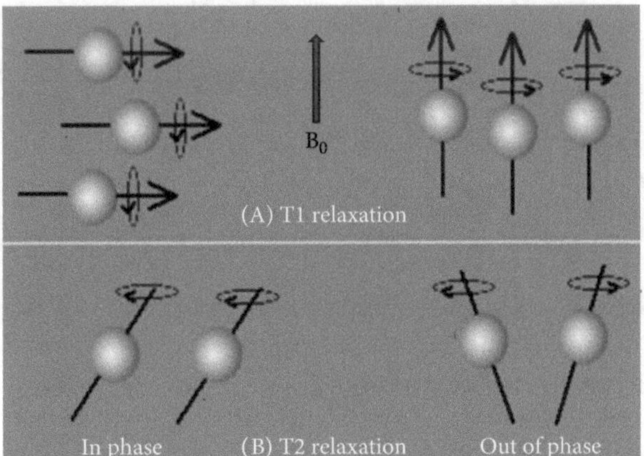

Figure 2.5 Schematic of T1 and T2 relaxation of protons. The direction of the scanner magnetic field is indicated as B_0 (adopted from Malloy et al., 2007). Reprinted by permission of the publisher (Taylor & Francis Ltd., http://www.tandfonline.com).

When the RF pulses and gradients are turned off, the protons begin to "relax," or dissipate the energy that they acquired from the pulse. The dissipated energy comes from two processes. The first is called T1 relaxation—the energy dissipated by the process of protons reverting from the RF-induced flipped state to their original alignment with the scanner's magnetic field. The second is called T2-relaxation—the energy dissipated by the dephasing of the precession frequencies (see Figure 2.5). The dephasing is caused by the molecular environment in the surrounding tissue or "lattice," but also due to inhomogeneities in the scanner's main magnetic field (i.e., "B_0"). T1 and T2 relaxation occur simultaneously, but T2 relaxation is much faster. However, the rate at which T1 and T2 relaxation occurs differs by tissue type (e.g., gray matter, white matter, cerebrospinal fluid). The RF energy dissipated during relaxation is detected by the scanner and, through a series of computational steps performed by the scanner's computer, is rendered as an image. Because T1 and T2 relaxation rates differ by tissue type, the timing of signal readout is critical for determining the tissue contrast in the image. A readout can be timed to be dominated by T1 or T2 relaxation. A T1-weighted image is one in which the signal readout was timed so that the T1 signal coming from different tissues is maximally different (Figure 2.6). In a T1-weighted image, CSF is dark, gray matter is gray, and white matter is lighter gray (Figure 2.7). A T2-weighted image is one in which the signal readout was timed so that the T2 signal coming from different tissues is maximally different (Figure 2.6). In a T2-weighted image, CSF is bright, white matter is dark gray,

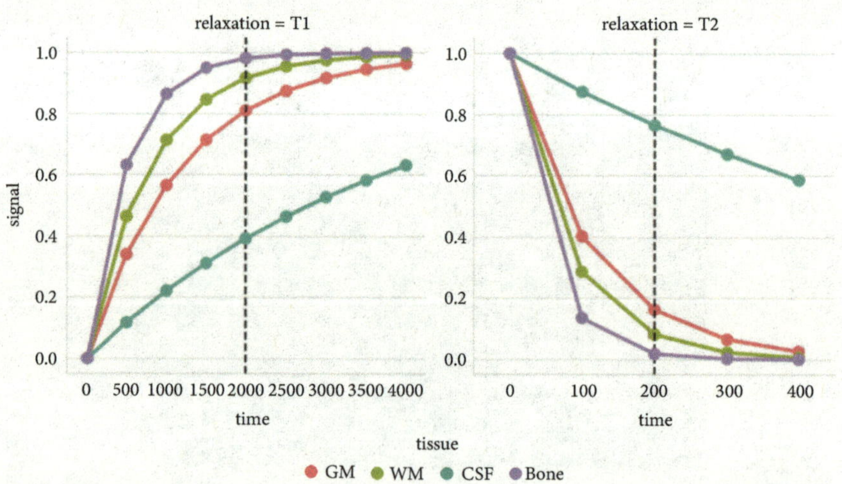

Figure 2.6 T1 and T2 relaxation times by tissue type. X-axis is in milliseconds. The vertical dotted line represents image acquisition time (Adapted image courtesy of https://carpentries-incubator.github.io/SDC-BIDS-sMRI/aio/index.html; Carpentries, 2023; (https://creativecommons.org/licenses/by-sa/4.0/).

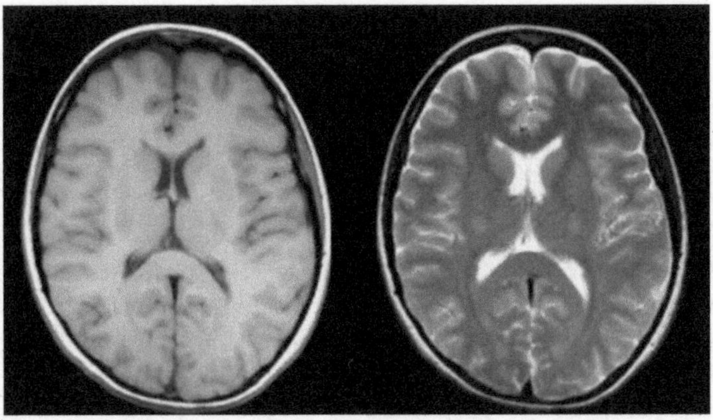

Figure 2.7 T1-weighted (left) and T2-weighted (right) axial brain images (image courtesy of and adapted from https://case.edu/med/neurology/NR/MRI%20Basics.htm).

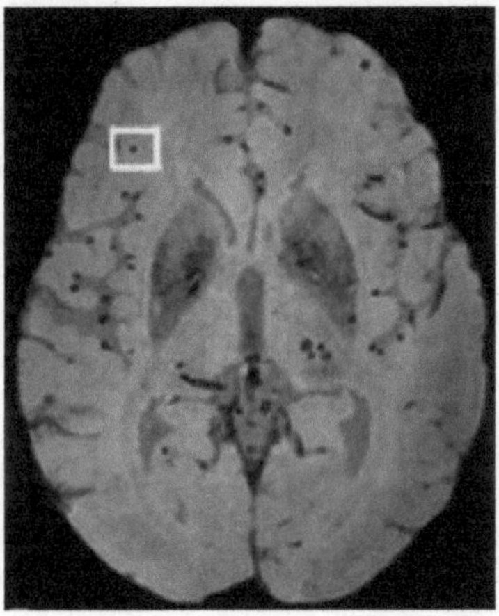

Figure 2.8 Axial T2* GRE image (skull-stripped) showing multiple cerebral microbleeds, which are black-appearing circular or ovular hypointensities in brain parenchyma. The white box shows one of these (Chesebro et al., 2021; https://creativecommons.org/licenses/by/4.0/).

and gray matter is lighter gray (Figure 2.7). Note that while bone and calcifications are very bright on CT scan, they are black on MRI scan due to bone's lack of hydrogen. A T2-weighted image differs from a T2* ("Tee-two-star") image in that the T2-weighted image is dominated by lattice-mediated dephasing without the effects of dephasing due to B_0 inhomogeneities (Figure 2.8). This correction for B_0 inhomogeneities is achieved through a RF/gradient sequence known as a "spin-echo" sequence.

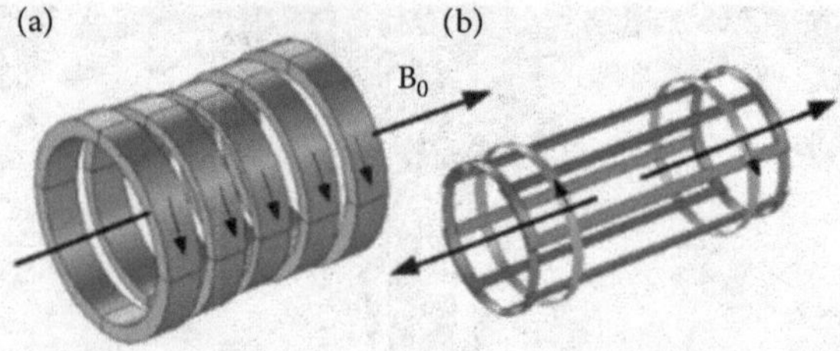

Figure 2.9 Schematic of (a) the main electromagnet of an MRI scanner (i.e., bore, B_0), and (b) schematic of gradient coils capable of producing magnetic field gradients in the x, y, or z direction that, though shown separately for illustration, would be superimposed on the bore (adapted from Razek, 2018; https://creativecommons.org/licenses/by/4.0).

The foregoing describes, in broad strokes, how image contrast between tissue types is achieved. Slice designation and spatial location within a slice is achieved through application of carefully timed magnetic gradients superimposed on the static B_0 field. Affixed to the exterior surface of the scanner bore are wire coils that, when electrical current is quickly passed through them, produce magnetic gradients in the x, y, or z direction (Figure 2.9). For example, a gradient can be applied such that the brain is exposed to a magnetic field that gradually increases above B_0 from left to right, or inferior to superior, or anterior to posterior. Application of these gradients at specific times during the image acquisition process permits slice selection (slice selection gradient) and resolves spatial location of tissue within a slice (phase encoding gradients).

Clinical Utility: Strengths of CT Versus MRI

CT scans are typically acquired more quickly, are less noisy, and less costly compared to MRI. For a CT scan, the patient places their head in a donut-shaped scanner rather than the bore shape of the MRI scanner (Figure 2.10). Therefore, patients may find the process of undergoing a CT scan more tolerable than undergoing MRI scanning. This may be particularly true for individuals with some degree of claustrophobia. CT also has advantages over MRI for acute detection of intracranial bleeding (e.g., TBI, suspicion of hemorrhagic stroke) and instances where the patient has or might have ferro-magnetic implants or shrapnel.

MRI has several strengths over CT. While modern CT scanners, especially flat-panel volume scanners, have superior spatial resolution to MRI, MRI has higher anatomic tissue contrast resolution with better visualization of gray

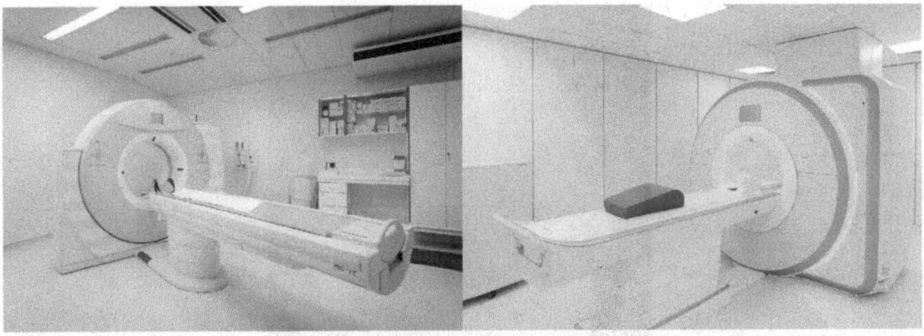

Figure 2.10 CT (left) and MRI scanner (right). Note the difference in bore depth, although the MRI scanner depicted has a relatively short bore (image sources: CT scanner: https://www.shutterstock.com/image-photo/ct-computerized-tomography-scan-device-hospital-1006650835; MRI scanner: https://www.shutterstock.com/image-photo/mri-magnetic-resonance-imaging-scan-device-1017608032).

versus white versus CSF tissues and their boundaries (Lin & Alessio, 2009). MRI scanners can acquire images at multiple slice angles such as axial, coronal, and sagittal planes. This allows more accurate visual appraisal of the location, size, and shape of an abnormality.

Another key strength of MRI is its ability to acquire images with different tissue contrast characteristics (e.g., T1-weighted, T2-weighted). Some types of lesions are best visualized by a particular type of image. For example, hemosiderin deposits due to small vessel hemorrhage from trauma or disease such as cerebral amyloid angiopathy (CAA), are best visualized on gradient-recalled echo (GRE) MRI scan (Figure 2.11) or susceptibility-weighted image. Deep white matter edema due to arteriosclerotic subcortical small vessel changes is easier to visualize on fluid attenuated inversion recovery (FLAIR) images than on T2-weighted images, although it can be seen on both (see Figure 2.12).

Types of MRI Images

T1-Weighted

A T1-weighted MRI image is one whose appearance is dominated by the signal given off by the T1 relaxation process. In general, the higher the water content of a tissue, the darker it appears on T1-weighted images. Accordingly, CSF appears dark while cortex and subcortical nuclei appear medium gray and relatively hypointense (darker) than white matter, which generally appears light gray (Figure 2.7 left). Among the various image types collected in a clinical brain MRI

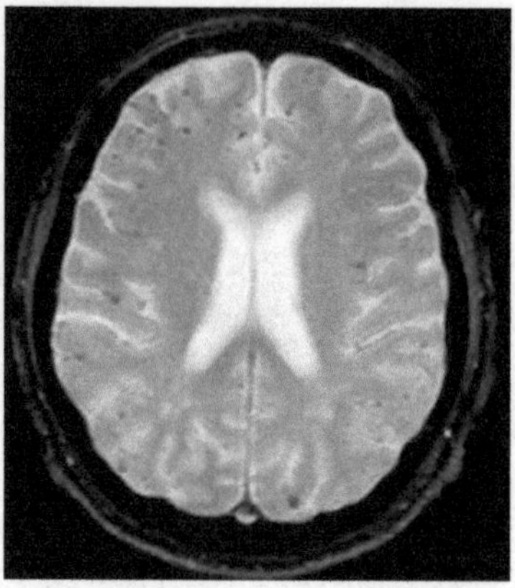

Figure 2.11 Axial GRE scan through the body of the lateral ventricles. Note multiple small hypointense (black) lesions in brain parenchyma due to old hemosiderin deposits from prior microhemorrhage in a case of cerebral amyloid angiopathy (image source: Wikimedia commons: https://en.m.wikipedia.org/wiki/File:Compare_SWI_and_GRE_CAA.png; https://creativecommons.org/licenses/by-sa/3.0/deed.en).

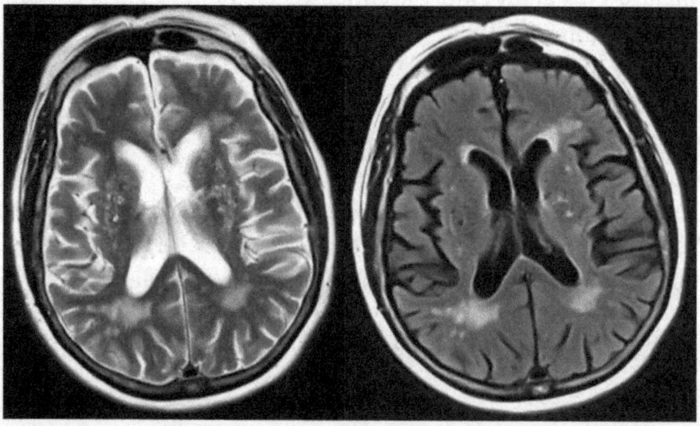

Figure 2.12 T2-weighted (left) and FLAIR (right) images of the same 82-year-old patient with a history of hypertension and other cardiovascular risks. Note that both images show hyperintense lesions in the periventricular deep white matter but CSF in the FLAIR image is dark, which makes it easier to ascertain the lesions compared to the T2-weighted image, in which both the lesions and CSF are bright. The lesions are thought to reflect subcortical ischemic vascular changes (clinical images provided by S. Correia).

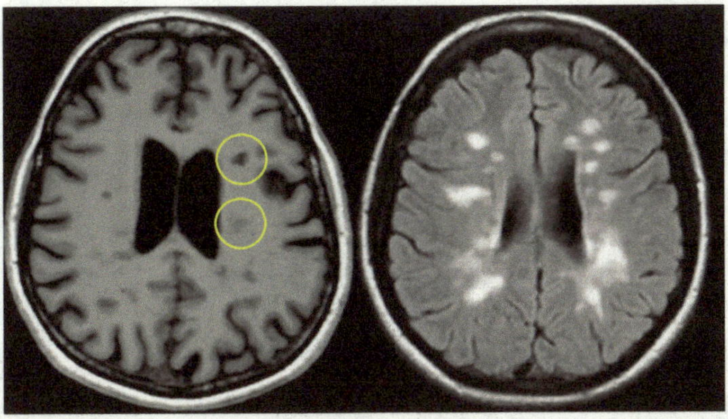

Figure 2.13 Multiple hypointense sclerosis lesions can be seen in deep white matter bilaterally on the T1-weighted image (left). Yellow circles indicate hypointense lesions in the right hemisphere, but others can be seen in the white matter bilaterally (Thaler et al., 2015). In contrast, multiple sclerosis lesions appear *hyper*intense on T2-weighted images (not shown) and on FLAIR (right). NOTE: Different patients. Note that in both right and left images, some lesions are aligned perpendicular to the long axis of the ventricles—this characteristic is commonly referred to as "Dawson's fingers" (FLAIR image courtesy of: https://case.edu/med/neurology/NR/MSAdvancedCourse5.htm).

scan, the T1-weighted image is generally considered to be the best for visualizing anatomy (Lebby, 2013a). Interstitial edema and other lesions with high water content, for example due to inflammatory edema in multiple sclerosis appear hypointense to their surroundings (Figure 2.13). Fat-containing tissue appears relatively hyperintense to its surroundings, as is the case for white matter in comparison to gray matter and CSF. Fatty lesions such as lipomas also appear hyperintense (Figure 2.14). Tissue loss due to encephalomalacia, for example, following TBI, appears isointense to CSF (Figure 2.15).

As is the case with CT images, MRI scans can be acquired after intravenous administration of gadolinium contrast to enhance the appearance of brain vasculature and vascularized lesions. The contrast agent is a safe chelated form of gadolinium (Gd), a rare-earth metal that, in its free non-chelated ionic form is toxic (Ibrahim et al., 2023). Gadolinium contrast, being paramagnetic that interacts weakly with an external magnetic field and shortens T1 relaxation rates. As a result, the cerebral vasculature and vascularized lesions such as tumors appear brighter (enhanced) on T1-weighted and other images (see Chapter 3 on vascular imaging). Gadolinium is eliminated through the kidneys (Ibrahim et al., 2023; Saade et al., 2018) and therefore might be contraindicated in patients with kidney disease. Figure 2.16 shows post-gadolinium enhancement

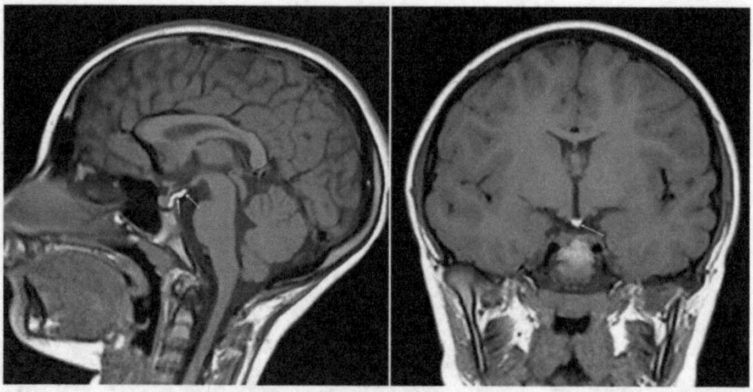

Figure 2.14 Hyperintense region (arrows) on sagittal (left) and coronal (right) images showing a lipoma of the tuber cinereum (Giacomozzi et al., 2021; https://creativecommons.org/licenses/by/4.0/).

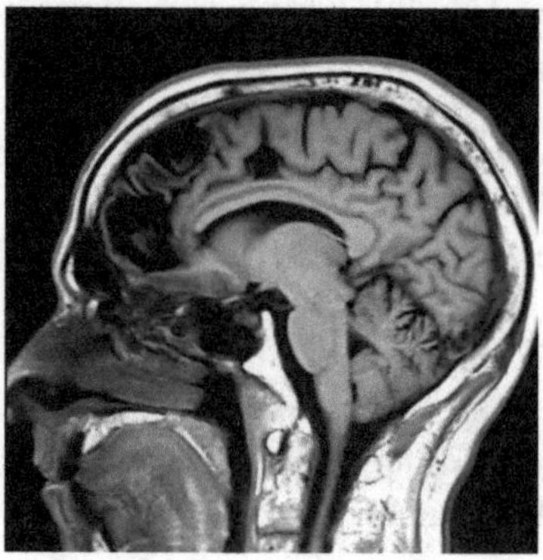

Figure 2.15 Sagittal T1 MRI at approximately the brain midline showing frontal encephalomalacia following TBI due to a fall of approximately 35 feet onto a hard surface (clinical image provided by S. Correia).

of metastatic melanoma tumors, which can appear hyperintense on non-contrast T1-weighted images (Escott, 2001).

T2-Weighted Images

A T2-weighted image is one whose appearance is dominated by the signal given off by the T2 relaxation process. Tissues with higher water content appear relatively brighter on T2 compared to tissues with low water content. Accordingly, CSF appears bright as does interstitial water from processes such as edema or inflammation. Gray matter appears light gray while white matter appears relatively darker with a charcoal gray appearance (Figure 2.7).

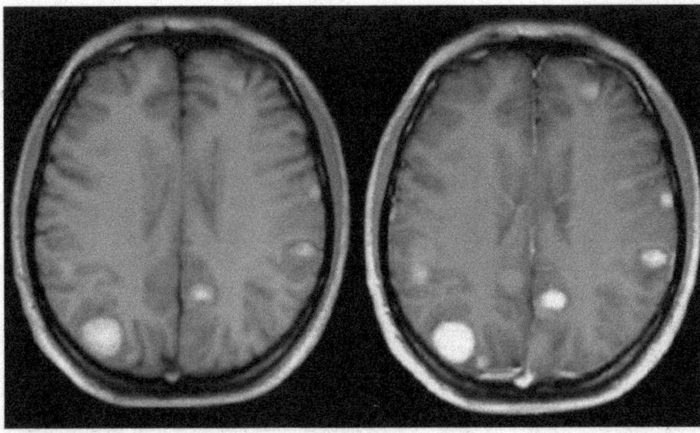

Figure 2.16 Metastatic malignant melanoma on non-contrast T1-weighted MRI (left). Tumor visualization is enhanced after administration of gadolinium contrast agent (right) (Atia et al., 2022; https://creativecommons.org/licenses/by/4.0/).

T2 MRI images have somewhat lower resolution of soft tissue anatomy than T1 images, but because of their sensitivity to water, both free flowing (e.g., CSF in the ventricles and sulci) and interstitial, are better for visualizing lesions, most of which have high water content (Lebby, 2013a). Examples of such lesions include cysts (Figure 2.17), abscesses, white matter lesions due to subcortical ischemic vascular disease (see Figure 2.12), and inflammatory processes such as multiple sclerosis (see Figure 2.18, left). Table 2.1 summarizes the appearance of various tissues and lesion types on T1- and T2-weighted images.

Fluid-Attenuated Inversion Recovery (FLAIR) Images

FLAIR images rely on an MRI scanning technique called *inversion recovery* (IR). Inversion recovery involves the application of a preparatory 180-degree RF pulse that "suppresses" hyperintense T2 signal due to water contained in large CSF-filled compartments such as the ventricles, sulci, and extra-axial spaces while retaining hyperintense T2 signal from excess interstitial water (i.e., edema). Essentially, the technique leverages the 180-degree RF pulse against the different T2 relaxation times of CSF versus white matter (Figure 2.6) and the timing of image readout such that the resulting images show CSF as dark, but excess interstitial water as bright. Table 2.2 below provides a useful summary of how different tissues appear on CT, T1-weighted, T2-weighted, and FLAIR images.

Compared to T2-weighted images, in which excess water is also bright, the contrast between dark CSF water and bright interstitial water makes it easier to visually perceive excess interstitial water due to edema from various causes,

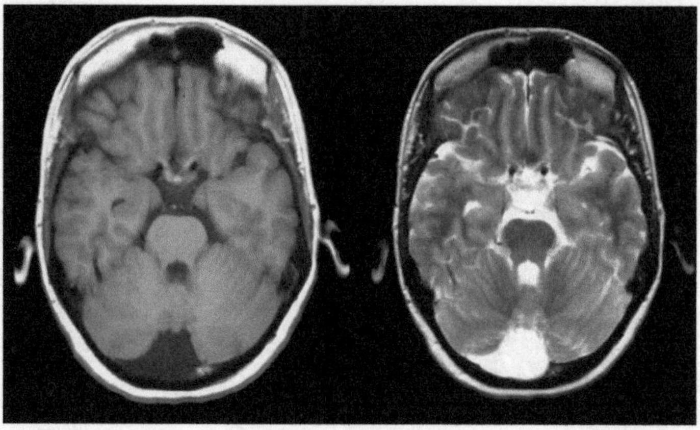

Figure 2.17 T1-weighted (left) and T2-weighted (right) MRI. Note the different signal characteristics of the posterior fossa arachnoid cyst. Its appearance is dark on T1 and bright on T2. In both images the signal characteristics of the cyst is isodense to the CSF owing to it being filled with fluid (image courtesy of https://case.edu/med/neurology/NR/PCyst.htm).

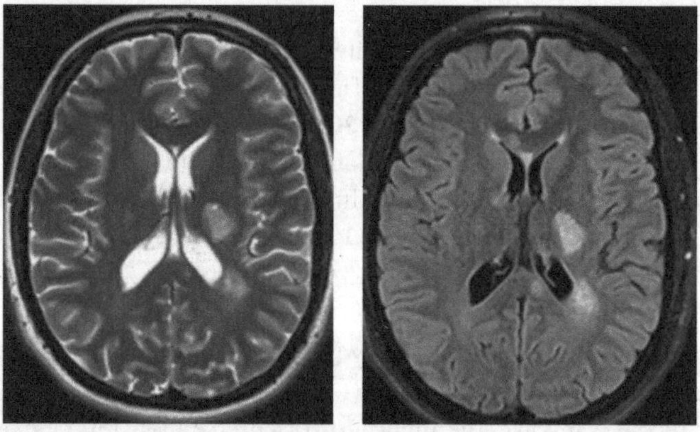

Figure 2.18 Multiple sclerosis. Left panel shows hyperintense lesions in deep white matter on T2-weighted image. Right panel shows the same slice and hyperintense lesions on a FLAIR image (adapted from Rovira et al., 2015; https://creativecommons.org/licenses/by/4.0/).

such as demyelination, gliosis, or ischemic vascular changes in the subcortical periventricular white matter (DeCarli, 2014). Figure 2.12 demonstrates the greater visual pop-out effect of the white matter hyperintensities on FLAIR versus T2-weighted because of FLAIR's ability to render ventricular CSF as dark. FLAIR images acquired with gadolinium contrast can assist in detection of inflammation and disruption of the blood-brain barrier (BBB) (Lebby, 2013a).

Table 2.1 Guidelines to appearance of tissues and select lesion types on T1 and T2 MRI images

	Appears Relatively Dark	Appears Relatively Light
T1-weighted	CSF Gray matter Inflammation Infection Edema Infarction Encephalomalacia Tumor Calcification	Fat (e.g., bone marrow, scalp) White matter Marrow Slow flowing blood Gadolinium contrast agent (see below) Necrosis
T2-weighted	White matter Fibrous tissue Protein rich fluid CSF movement (flow void)	Gray matter Inflammation Infection Encephalomalacia Edema Gliosis

Adopted from Lebby (2013b).

Table 2.2 Appearance of cranial tissues on computed tomography and magnetic resonance imaging

Tissue Type	Computed Tomography	MRI T1-weighted	MRI T2-weighted	MRI FLAIR
Bone	White	Black	Black	Black
Calcified Tissue	White	Variable, usually gray	Variable, usually gray	Variable, usually gray
Gray Matter	Light gray	Medium gray	Medium gray	Medium gray
White Matter	Medium gray	Light gray	Dark gray	Dark gray
Cerebrospinal Fluid	Nearly black	Black	White	Black
Air	Black	Black	Black	Black
Other pathology	Gray	Gray	White	White

Adapted from Hurley (2013).

Several MRI visual rating scales are available for categorizing the severity of hyperintense white matter changes due to factors such as subcortical ischemic vascular disease (SIVD; see CT section of this chapter for CT-based scales) and cortical atrophy related to age or neurodegenerative disorders. These scales may typically rely on FLAIR images although some use separate criteria for MRI and

CT. The scales typically label the degree of SIVD or atrophy as absent, mild, moderate, or severe (e.g., Pantoni et al., 2002; Scheltens et al., 1998), but some provide more fine-grained ratings.

Gradient Echo (GRE)

Iron infiltration into brain tissue due to hemorrhage becomes trapped in a lysosomal-ferritin molecular storage complex called hemosiderin. Gradient echo MRI images are sensitive to iron trapped in hemosiderin, but also to other paramagnetic iron-containing complexes of deoxyhemoglobin and ferritin. In the scanner, the iron in these complexes interacts with the local magnetic field, causing rapid loss of the T2 signal. This results in the affected tissue appearing black. Diamagnetic compounds such as calcium deposits also cause signal disruption and appear dark on GRE images (Janaway et al., 2014). Evidence of abnormal interstitial iron on GRE can be seen due to hemorrhage caused by several conditions including TBI, CAA, sickle cell anemia, leaking arteriovenous malformations, ruptured aneurysm, moyamoya disease, cavernous malformations, and others.

GRE lesions in brain tissue appear black and typically round or oblong (Figure 2.18). When viewing GRE images, bear in mind that small round black findings found in sulcal gaps or other non-parenchymal regions may be signal voids caused by fluid movement (flow voids) in healthy veins or small arteries (https://www.stroke-manual.com/cerebral-microbleeds-cmb/). Additionally, the choroid plexuses and the pineal gland may appear dark in older patients due to calcification. GRE lesions can also appear along the surface of cerebral gyri, a finding referred to as *superficial siderosis*. This can also be seen in Figure 2.19.

The appearance of hemorrhage and infarction can vary over time on CT and on T1-weighted, T2-weighted, FLAIR, and GRE images. This is due to temporal changes in the form of hemoglobin from oxyhemoglobin to deoxyhemoglobin to methemoglobin. These forms have different magnetic properties. Oxyhemoglobin is *diamagnetic*, or slightly repelled by the scanner magnetic field; whereas deoxyhemoglobin and methemoglobin are *paramagnetic*, or slightly attracted to the magnetic field. The evolution from oxyhemoglobin to methemoglobin, in turn, accounts for the different appearances of blood products over the hyperacute (<14 hours) to chronic (>14 days) of injury. These same tissue changes alter the radiodensity of the hemorrhage and cause an evolution of change on CT scan. Similarly, the image characteristics of an ischemic stroke change on CT and on T1 and T2 MRI scans due to evolution over time of the lesion core and penumbra. Table 2.3 provides a summary of these changes over time.

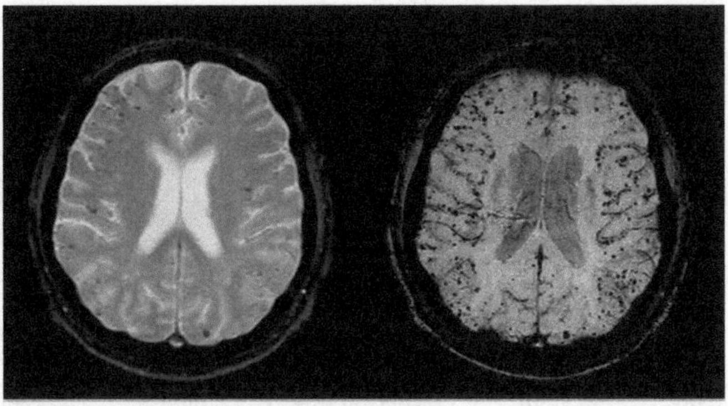

Figure 2.19 Case of cerebral amyloid angiopathy showing multiple cerebral microbleeds bilaterally on GRE (left image, green circles provide examples, although there are many) and superficial siderosis, which appears as hypointense signal along the margins of the sulci (yellow oval). The imaging on the right shows the increased sensitivity of SWI cerebral microbleeds and superficial siderosis. The SWI image is also more sensitive to deoxyhemoglobin in venous blood and, as a result, veins visualized on cross-section and longitudinally can be difficult to differentiate from parenchymal hemorrhage without careful inspection. Both images are in the axial plane through the body of the lateral ventricles.

Source: Haacke, E. M., DelProposto, Z. S., Chaturvedi, S., Sehgal, V., Tenzer, M., Neelavalli, J., & Kido, D. (2007). Imaging cerebral amyloid angiopathy with susceptibility-weighted imaging. AJNR. American journal of neuroradiology, 28(2), 316–317.

Susceptibility-Weighted Imaging

Susceptibility-weighted imaging (SWI) is a GRE-related (T2*) technique developed to produce greater contrast by taking advantage of the susceptibility of brain regions to T2* signal loss because of paramagnetic or diamagnetic compounds. Briefly, from a technical perspective, a key feature is that during image construction, a phase correction is applied that results in enhanced visualization of paramagnetic (e.g., iron-containing molecules such as deoxyhemoglobin, ferritin, and hemosiderin) and diamagnetic (e.g., calcium) intracranial compounds. The main clinical application of SWI has been the same as in T2* GRE—visualizing deoxyhemoglobin in veins and evidence of hemosiderin deposits from cerebral microbleeds including superficial siderosis, and calcifications such as is seen in older individuals' choroid plexuses and pineal gland and in abnormal calcium deposits due to conditions such as pseudohypoparathyroidism. As in GRE, veins, old microbleeds, and calcifications appear black (Haller et al., 2021). Figure 2.19 provides a useful visual comparison of T2* weighted GRE and SWI and demonstrates the level of detail seen in SWI including visualization of deoxyhemoglobin-containing venous vasculature.

Table 2.3 Appearance of blood on CT and MRI scans by post-infarction stage

Stages	CT	T1-weighted MRI	T2-weighted MRI	GRE MRI
Hyperacute (<12 h)	High density	Isointense or mildly hyperintense	Hyperintense	Hypointense rim
Acute (12 h to 2 days)	High density	Isointense of hypointense	Hypointense	Hypointense rim that gradually progresses to center
Early subacute (2–7 days)	High density	Hyperintense	Hypointense	Hypointense
Late subacute (8 days to 1 month)	Isodense	Hyperintense	Hyperintense with rim of low intensity	Hyperintense with rim of low intensity
Chronic (> 1 month)	bru	Hypointense	Hypointense	Slit-like hyperintense or isointense core that is surrounded by a hypointense rim

CT = computed tomography, MRI = magnetic resonance imaging, GRE = gradient-recalled echo.
Reproduced from Siddiqui et al. (2011)

Diffusion-Weighted Imaging

Diffusion-weighted imaging (DWI) images are commonly acquired in routine clinical imaging via MRI to detect acute ischemic stroke and tumors (Maier et al., 2010; van Everdingen et al., 1998). Contrast in these images is based on different directions and speeds of diffusion of water in tissue. The human body is composed of about 60%–70% water (Baliyan et al., 2016), and the brain is about 80% water (Oros-Peusqueηs et al., 2019). In a homogeneous environment, such as a glass of water, the movement of water molecules is random (isotropic diffusion) and speed of movement is driven by temperature—a phenomenon called Brownian motion. In the brain (or other soft tissues in the body), movement of water molecules occurs in intracellular and extracellular compartments. The different cellular architecture of these tissue compartments impacts the characteristics of diffusion, including the relative extents to which the movement of water molecules is restricted (anisotropic diffusion) by that architecture. Water diffusion in the intracellular compartment is restricted due to cell walls and various organelles that can present obstacles. In comparison, the direction of movement of an extracellular water molecule is relatively less restricted. One way to think of this is that because of the relatively crowded interior of a cell, in any given period, an intracellular water molecule is likely to bump into more things causing it to traverse less distance compared to one in the relatively less crowded extracellular space (Beaulieu, 2009). The relative amount of water between these compartments is influenced by pathological processes such as tumors, TBI, and infections (Nehring, 2023).

Ischemic stroke sets off a cascade of processes that lead to cerebral edema. Cytotoxic edema occurs in the early stages of ischemic stroke. Ionic edema follows due to the disruption of ionic gradients caused by cytotoxic edema. In turn, vasogenic edema develops due to BBB breakdown (Gu et al., 2022). The influx of water into the intracellular compartment during the early phases of acute stroke causes increased directional restriction in diffusion (i.e., anisotropy increases) due to the presence of cellular organelles. Additionally, edema-induced swelling of the intracellular compartment causes narrowing of the extracellular compartment. DWI is sensitive to the changes (Bodini, 2009).

DWI uses rapidly applied and then rapidly reversed 180° magnetic gradients (Figure 2.9) to assess water diffusion in tissue. These reversed gradients are applied in the three orthogonal directions, typically in the X, Y, and Z planes. Two types of images are produced: an apparent diffusion coefficient (ADC) map and a DWI map. Tissue with restricted diffusion appears hypointense (dark) on an ADC map. The DWI map shows the degree to which diffusion is restricted (i.e., anisotropic diffusion)—specifically, tissue with increased anisotropy appears bright. The ADC map is mathematically related to the DWI map and to the magnitude of water diffusion—tissue where the magnitude of diffusion is lower appears darker. These maps provide information about the magnitude (or speed) of diffusion and degree to which that diffusion is restricted, but does not provide directional information. Put differently, these are scalar value maps not vector maps (Beaulieu, 2009; Bodini, 2009; Higaki et al., 2018).

DWI has been shown to be more sensitive to early ischemic stroke than FLAIR or T2-weighted images. In humans, DWI has been shown to detect ischemic changes as soon as 2 to 6 hours after the onset of stroke symptoms (Dmytriw et al., 2017; van Everdingen et al., 1998). Figure 2.20 shows the evolution of an ischemic stroke from 19 hours to 27 days on both DWI and ADC maps. Note the signal differences in the right middle cerebral artery distribution, which is increasingly hyperintense over time, showing increasingly restricted diffusion; whereas the signal characteristics in the ADC map are hypointense at 19 hours and become hyperintense by day 27, showing increased magnitude of diffusion. The high signal on DWI maps seen in acute ischemic stroke is sometimes referred to as a “lightbulb sign.”

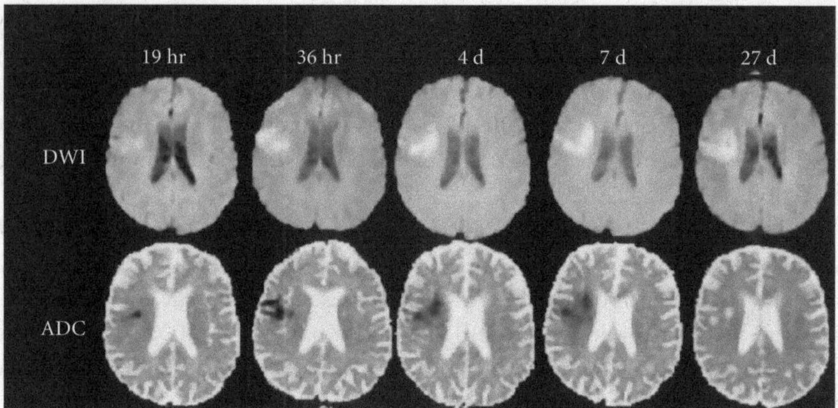

Figure 2.20 Evolution of left hemisphere ischemic stroke on DWI and ADC maps (adopted from Lansberg et al., 2001).

In chronic stroke (i.e., >3 weeks), DWI signal diminishes, indicating reduction in diffusion restriction as the injury evolves, whereas the ADC map shows continued high signal as the magnitude of diffusion increases (Bhuta, 2022).

DWI is an extremely useful MRI imaging method for detecting acute ischemic stroke, but it also can assist in ruling out stroke in conditions that mimic it. An anecdotal example is one of a patient with a history of right hemisphere stroke who, during the diagnostic interview portion of a neuropsychological evaluation, began showing mild facial droop, increasing dysarthria, and complaints of left-hand numbness. He was sent to the emergency room. DWI and other procedures failed to show evidence of a new ischemic stroke. Rather, the diagnosis was post-stroke recrudescence—a not-well-understood phenomenon in which stroke symptoms can reemerge up to 4 years after the stroke (Topcuoglu et al., 2017).

Diffusion Tensor Imaging (DTI)

Diffusion tensor imaging is a widely used MRI method for measuring and visualizing the structural integrity of brain white matter. The basic premise is that water diffuses more rapidly along the long axis of axons than across them. This is because the direction of water diffusion in white matter is constrained by the axon membrane fatty hydrophobic myelin sheathes. Of course, water diffuses in gray matter, but there is no coherent dominant direction. The direction of water diffusion in CSF spaces such as the lateral ventricles is largely random and constrained only by the ventricle walls or by gyri and membranes in the case of extra-axial CSF spaces (much like the glass of water example previously mentioned). Figure 2.21 provides a representation of directionally non-restricted or

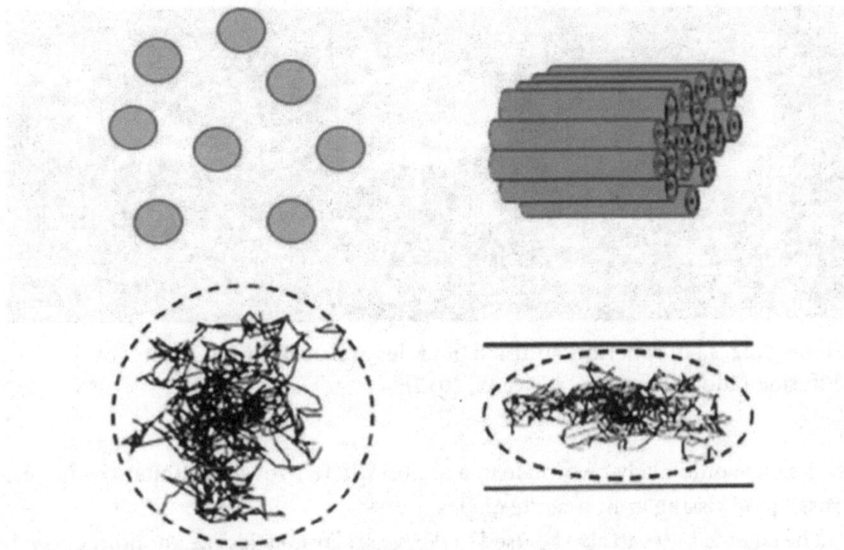

Figure 2.21 High angular resolution of diffusion imaging (HARDI) (adopted from Descoteaux, 2015).

isotropic diffusion in a large space such as the lateral ventricles (upper and bottom left) and directionally restricted or anisotropic diffusion along a myelinated white matter tract.

While a detailed discussion of the scanning parameters and settings involved in DTI and related methods of white matter imaging is beyond the scope of this chapter, in the most basic sense, DTI is an extension of clinical DWI. In DWI, diffusion-encoding magnetic gradients are applied orthogonally, typically in the X, Y, and Z planes. In DTI, at least six orthogonal directions are needed, although it is common to collect many more (e.g., 60 or more). A minimum of six directions is necessary to provide information about the direction of diffusion. The diffusion signal acquired from each direction is summarized mathematically as a matrix or tensor. The tensor can then be decomposed to produce three eigenvalues, which provide information about the magnitude of diffusion, and three eigenvectors, which provide information about the direction of diffusion. The relative magnitude of the eigenvalues can be graphically represented in X, Y, and Z coordinates to visually represent an idealized likely shape (ellipsoid) of diffusion in each voxel. Figure 2.22 shows, from left to right, spherical, prolate, and oblate diffusion shapes. The shape of diffusion in the lateral ventricle, being random and likely about equal in all directions would be spheroid, the shape in a myelinated white matter tract such as the cingulate bundle would be oblate, and

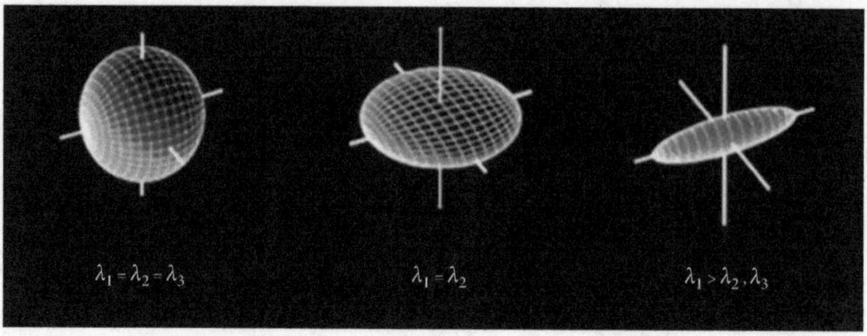

Figure 2.22 Left: spherical diffusion, middle: prolate diffusion, right: oblate diffusion (adopted from Descoteaux, 2015).

the shape would likely be prolate in a region where two white matter tracts were crossing or kissing in non-acute angles.

The eigenvalues can also be used to derive scalar metrics (i.e., nondirectional) that describe the degree of anisotropic diffusion in each voxel. The two most reported values are mean diffusivity (MD), which reflects the average amount of diffusion in each voxel (i.e., average of the three eigenvalues) and fractional anisotropy (FA), which reflects the degree to which diffusion is restricted. Other scalar metrics include radial and axial diffusivity (RD and AD, respectively) and trace, which is the magnitude of diffusion in the voxel (trace differs from MD, which is the average magnitude of diffusion). The magnitude of these scalar values ranges from 0 to 1. The values can be rendered as images. The scalar metric, FA, can also be combined with the eigenvector information to render images that use a widely accepted colorizing scheme that allows visualization of the brain's white matter architecture (red represents fibers running in the left-right direction, blue for superior-inferior, and green for anterior-posterior). Figure 2.23 shows FA, MD, RD, and AD maps and colorized FA maps in axial, coronal, and mid-sagittal views.

DTI data can also be displayed using a method called tractography, which provides a model of the shape, length, and direction of white matter pathways in the brain. There are several methods for rendering the data in this manner, including probabilistic data-driven methods and anatomically informed deterministic methods. A detailed discussion of these methods is beyond the scope of this chapter. An example of deterministic tractography rendering is seen in Figure 2.24.

Much of the clinical application of DTI and related images has been done in the context of research, including brain development, traumatic brain injury, demyelinating disorders, cerebrovascular conditions, infectious disorders, substance abuse, psychiatric conditions, neurodegenerative diseases, epilepsy, and

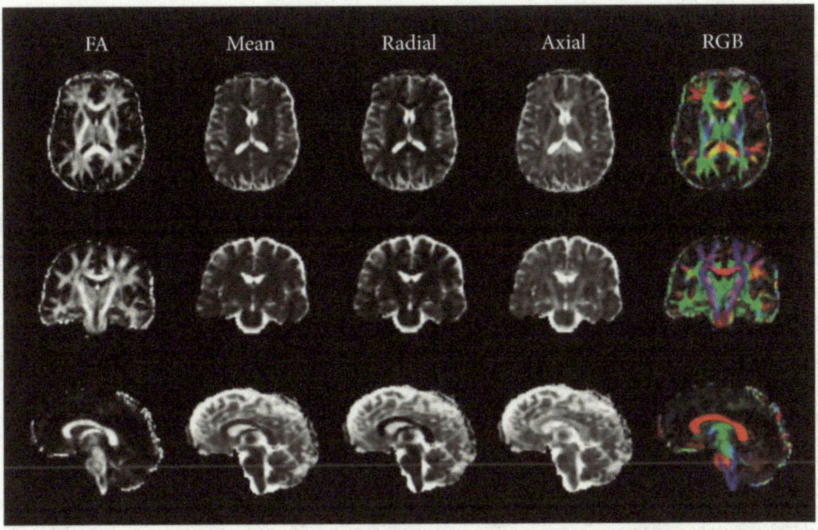

Figure 2.23 FA, MD, RD, and AD maps and colorized FA maps in axial, coronal, and mid-sagittal views (adopted from Descoteaux, 2015).

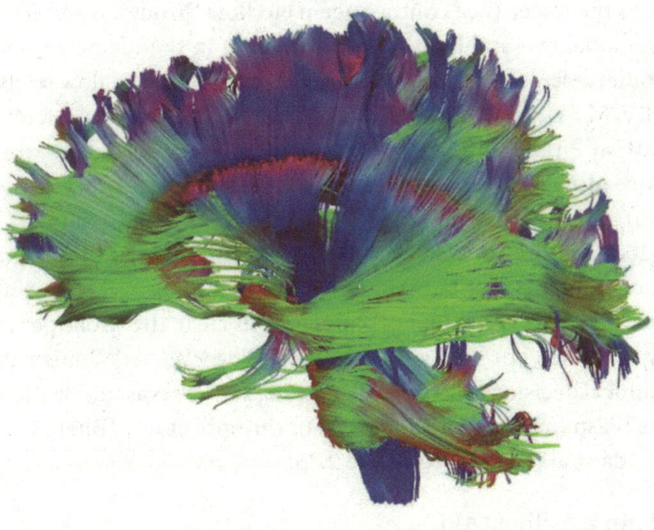

Figure 2.24 Whole brain white matter in a normal volunteer rendered in sagittal view (left is anterior, top is superior) as 3D tractography images (image source: https://wellcomecollection.org/works/ru5kef96/items; https://creativecommons.org/licenses/by/4.0/).

many others. The technique is gaining clinical adoption, in assessing factors such as scope of injury and trajectory of recovery following traumatic brain injury, surgical planning for brain tumors, gauging stroke rehabilitation outcome, and others (Tae et al., 2018).

Perfusion-Weighted Imaging (PWI)

MRI perfusion-weighted imaging provides information about the integrity of the vascular supply to the brain. There are two methods for measuring cerebral perfusion. The first is known as dynamic susceptibility contrast-enhanced (DSC) perfusion, which uses an exogenous contrast agent, typically gadolinium. Following injection of gadolinium, images are collected rapidly and repeatedly. The paramagnetic properties of the chelated gadolinium (see "Types of MRI Images/T1-Weighted," above) causes shortening of T1 relaxation and, to a lesser extent, T2 shortening as the gadolinium passes through the tissue. The MRI scanner's computer calculates the signal difference between the baseline images and the post-contrast images. The computational outputs are rendered images (maps) of various metrics derived from these calculations. These metrics and their corresponding images provide useful hemodynamic information about the brain's vascular integrity. Key among these are mean transit time (MTT)—the time it takes the tracer (i.e., contrast agent) to flow through the vascular area of interest; regional cerebral blood volume (rCBV) in the region-of-interest relative to another reference region; and regional cerebral blood flow (rCBF), which is the rCBV/MTT ratio in the region-of-interest relative to a reference region (Lebby, 2013a; Pirko, 2010).

PWI is used in the evaluation of many conditions such as vasospasm (Lad et al., 2006), TBI (Smith et al., 2019), vascular assessment of tumors (Rowley & Roberts, 2004), and more. PWI has been particularly useful in diagnosing the spatial extent and treatment options on ischemic stroke. In the hyperacute stage of ischemic stroke, PWI can identify an at-risk tissue in the stroke penumbra that is not fully infarcted nor fully visible on DWI. The PWI>DWI mismatch is used to guide clinical decisions about implementing stroke revascularization methods such tissue plasminogen activation (tPA) or thrombectomy (Bhattacharjee et al., 2021; Kakuda et al., 2008). See Figure 2.25.

Arterial Spin Labeling (ASL)

ASL is a noninvasive MRI method for measuring blood perfusion in tissue. In brain MRI, ASL allows the measurement of cerebral blood flow (CBF) or volume of blood per volume of brain tissue per minute. ASL enables the observation of both global and regional perfusion, typically in the context of ischemic or hypoxic events but also in the case of chronic and systemic changes to blood flow (e.g., from cardiac conditions like atrial fibrillation or heart failure, systemic

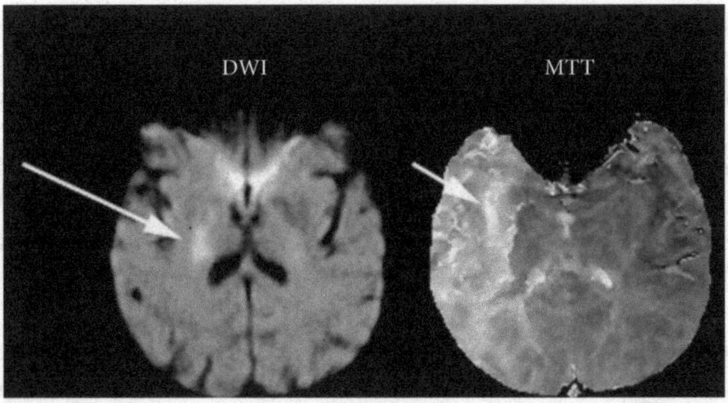

Figure 2.25 The diffusion-perfusion mismatch. A small lesion on the diffusion weighted image (DWI) in a 3 h stroke relative to the larger perfusion-weighted image abnormality on a relative mean transit time (MTT) image (Warach, 2001). Copyright ©2001 Warach; license BioMed Central Ltd. Verbatim copying and redistribution of this article are permitted in any medium for any purpose, provided this notice is preserved along with the article's original URL https://trialsjournal.biomedcentral.com/articles/10.1186/cvm-2-1-038.

cardiovascular conditions like hypo/hypertension, and chronic conditions like carotid arterial occlusion).

In ASL, radio frequencies (RFs) and gradients are applied to the free-flowing water molecules naturally found in the arterial blood (Grade et al., 2015; Williams et al., 1992). The RFs and gradients are selectively and briefly applied to the incoming arterial blood supply, specifically targeted at the internal carotid arteries at about the level of C4 vertebra. This RF/gradient "pulse" introduces energy to the tissue thereby inverting the spin of hydrogen atoms in the targeted region including those in blood that are flowing upstream to the brain through the internal carotids (Clement et al., 2022). This spin inversion in ASL is commonly called spin "tagging" or "labeling." Once the blood is tagged or labeled, time is allowed to elapse during which the labeled blood continues to flow upstream into the capillaries of the brain and the inverted spins relax—dissipating the energy induced by the RF/gradient pulse. The energy signal is detected and quantified by the scanner, which is then subtracted from the initial reference image to get the unitless perfusion-weighted image (PWI; Detre et al., 1992; Williams et al., 1992). A separate image measuring the magnetization of the brain tissue is used to determine the intensity of the signal and ultimately derive CBF.

CBF is typically determined by one of two ASL sequences: pulsed ASL (PASL; Wong et al., 1998) and pseudo-continuous ASL (PCASL; Alsop et al., 2015; Dai et al., 2008). The primary difference in the two main ASL sequences lies in the approach to labeling. Where PASL employs a relatively short, single pulse of RF and gradients on a thick slab of blood, PCASL uses a series of several short bursts of RF and gradient pulses over a longer interval on a smaller bolus of blood (Alsop et al., 2015; Dai et al., 2008; Grade et al., 2015; Wu et al., 2007). While both are accepted methods with high labeling efficiency, PCASL offers comparatively better signal-to-noise (SNR) ratio and higher image quality than PASL (Wu et al., 2007) which can only achieve 70% of the maximal SNR of PCASL (Buxton et al., 1998). Figure 2.26 provides a final CBF image after post-processing of raw PCASL image data is complete (Clement et al., 2022).

ASL offers distinct advantages over other methods of measuring perfusion such as MRI perfusion obtained using gadolinium-contrast agent and PET (Wolk & Detre, 2012). Most importantly, ASL MRI does not require the application of exogenous tracers, which can be radioactive or toxic and thus unsafe for pediatric or repeat imaging. ASL is also typically more accessible, less expensive, and can be safely used longitudinally, because it uses an endogenous tracer. It also has the added benefit of structural imaging being collected in tandem to support CBF mapping and region-specific perfusion. A disadvantage is that the technique has a low SNR and thus may require prolonged scanning time, which can increase susceptibility to patient motion in the scanner (Golay et al., 2004; Petersen et al., 2006).

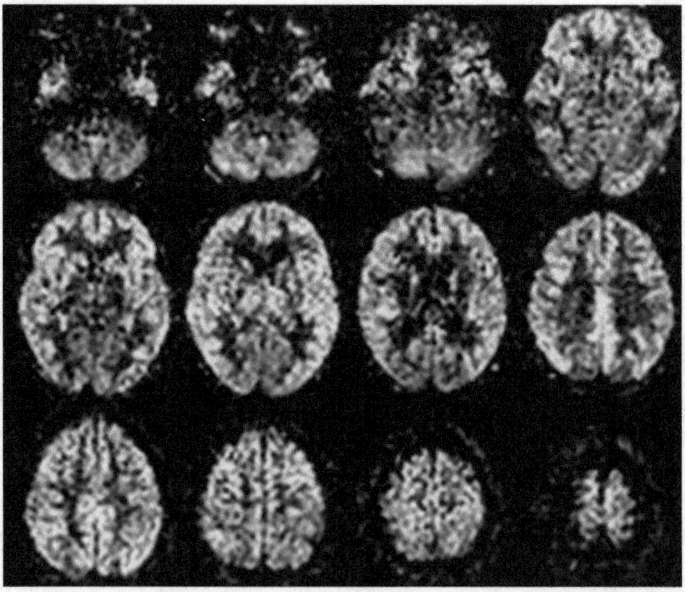

Figure 2.26 Final cerebral blood flow (CBF) image after post-processing of raw PCASL imaging data (Clement et al., 2022; https://creativecommons.org/licenses/by/4.0/).

Functional MRI

Functional magnetic resonance imaging (FMRI) is a noninvasive MRI technique to index changes in neural function based on capillary oxygen hemodynamics. Specifically, MRI relies on neurovascular coupling—the observation that blood supply to a brain region increases in response to neural activation—as an indirect marker of the level of neural activity in the whole brain or in a particular region of interest (Chen & Glover, 2015). This blood oxygen level dependent (BOLD) technique has shown great utility in clinical neuroscience research due to its capacity to capture localizable patterns of neural metabolic demand associated with discrete cognitive and emotional functions in the time frame during which they occur (i.e., from seconds to minutes). Figure 2.27 shows brain activation during a gustatory stimulation task. Although clinical applications are currently limited to presurgical functional brain mapping, emerging research suggests much greater utility, such as improved diagnosis and prognostics and monitoring treatment outcomes.

In the three decades following its first demonstrations in human visual and motor experiments, FMRI has been widely adopted as an invaluable research tool to help understand brain function. There are two broad methodological approaches to gleaning insights into brain function: task-based FMRI (tbFMRI) and resting-state FMRI (rsFMRI). Task-based FMRI approaches elicit

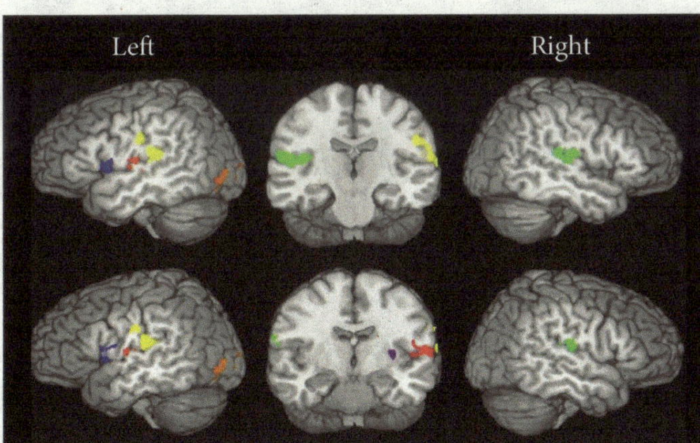

Figure 2.27 Brain activity associated with sustained food stimulation (lemon lollipop) among all participants (n = 49) in a study of three groups of individuals who met criteria for either obesity, normal weight, or successful weight-loss maintainers. Images are in left sagittal, coronal, and right sagittal views. Colored blobs show areas of activation: Red = left middle insula, Blue = left inferior frontal gyrus, Yellow = left posterior insula, Purple = left putamen, Orange = left occipital, Green = right insula (Sweet et al., 2012).

task-associated neural activation responses (Bandettini et al., 1993; Kwong et al., 1992; Ogawa et al., 1992). There are two general and widely used approaches to tbFMRI. In one approach, the mental task is presented in a blocked design in which brain activation information is acquired during alternating blocks of task performance followed by rest (or other activity). In the other approach, information about brain activation is derived by response to specific target events that occur relatively infrequently amid non-target events (e.g., oddball tasks).

Resting-state FMRI approaches emerged in the late 1990s (Biswal et al., 1995), when it was noted that some brain regions show BOLD deactivation responses during mental task performance or when the subject in the scanner was mentally at rest (Shulman et al., 1997). Studies utilizing the rsFMRI approach have revealed the existence of intrinsic networks that appeared to be active during resting states, without external cognitive demands, and appeared to exhibit relative suppression during engagement with tasks requiring external demands (Buckner et al., 2008). The identification of resting state networks relies on post-acquisition quantitative statistical modeling and produces images that show a variety of resting state networks (Adriaanse et al., 2014; Lee et al., 2013; Yeo et al., 2011). See Figure 2.28.

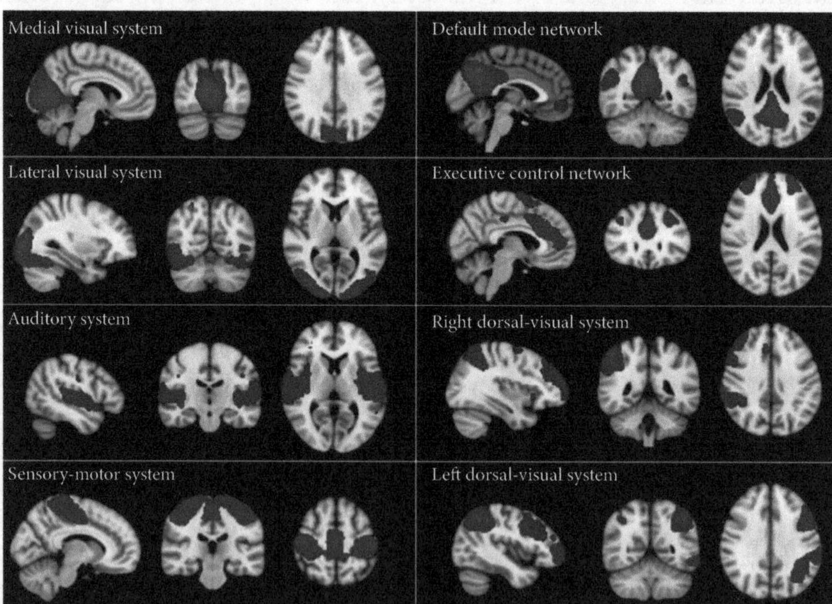

Figure 2.28 Blue regions are "masks" of eight resting state functional networks (images were produced by Adriaanse et al., 2014, based on image processing masks downloaded from http://www.fmrib.ox.ac.uk/analysis/royalsoc8/).

BOLD FMRI is most frequently accomplished using T2*-weighted echo-planar imaging (EPI) sequences, which allows the tracking of changes in the magnetic properties of hemoglobin as an endogenous contrast agent. T2* EPI applies the same principles as T2-weighted imaging, except that it also captures signal inhomogeneity due to B_0 inhomogeneity, which increases its sensitivity to deoxyhemoglobin. This method takes advantage of the fact that deoxyhemoglobin is paramagnetic while oxyhemoglobin is not. A smaller deoxyhemoglobin-to-oxyhemoglobin ratio in the capillaries surrounding active neurons yields a more homogeneous magnetic field and a greater T2* signal. Once these images are collected there are multiple computational steps to producing an actual image. This is done by interrogating patterns of task-based or resting-state activation or deactivation. Image rendering is typically done by superimposing spatially aligned functional images with a T1-weighted anatomical image (see Figure 2.29).

Functional neuroimaging has proven useful in localization of normal neural functions, as well as in increasing understanding of neural and psychiatric disorders. Functional neuroimaging has great potential for assessment of severity, prognosis, disease course in patients, the search for effective treatments, and a better understanding of neural mechanisms in general as they fail. The great utility of functional neuroimaging in clinical research is only beginning to extend into clinical practice. An excellent example of this is the use of

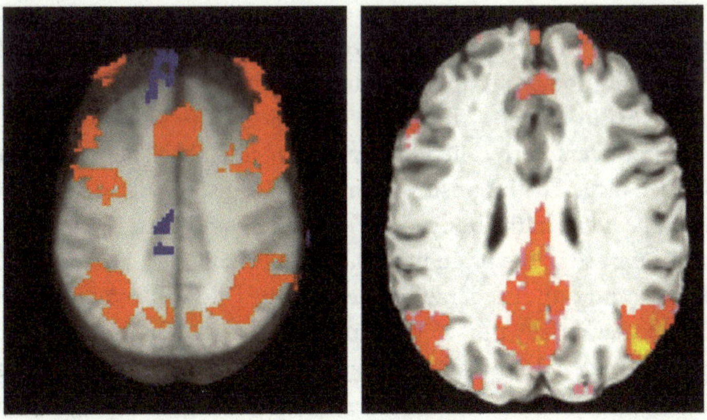

Figure 2.29 Functional MRI images. The left image shows a group-level result from 16 people performing a 2-back task (red indicates greater activity compared to a 0-back control task), overlaid on their average T1-weighted anatomy. The right image shows an individual's resting state functional connectivity data based on a posterior cingulate seed, overlaid on their T1 image (unpublished data provided by Dr. Sweet).

BOLD FMRI in presurgical mapping of brain function (Chakraborty & McEvoy, 2008). Language, memory, and motor systems are mapped with selective behavioral challenges and avoided during neurosurgery. Technical advances such as higher field-strength magnets and more powerful preprocessing and statistical analysis techniques will continue to produce significant advances in FMRI research. Another future direction relates to the fact that standardized tests are not available for FMRI. FMRI studies have employed paradigms derived from a wide range of different cognitive tasks, but for the most part have employed relatively small sample sizes. Most of these studies were not conducted with the goal of developing normative data on behavioral performance or on how the brain normally responds during various paradigms. Limited availability of valid and demographically stratified normative databases hinders interpretation of findings from a single patient. Statistically significant group differences in brain activity may reflect relatively subtle effects that are not readily apparent in the results from a single patient. Accordingly, before most techniques can become clinically useful, there is much standardization that needs to be accomplished for each of the functional brain imaging methods.

Molecular Imaging

Positron Emission Tomography (PET)

Positron emission tomography (PET) is a molecular imaging technique that provides information about physiological functions in the body, including the brain. It is widely used in oncology for tumor detection (Kapoor & Kasi, 2023). PET relies on radiotracers (also called radiopharmaceuticals). These are compounds that have been tagged with a positron emitting radioisotope and designed to bind with certain proteins in the body or to be taken up and used by cells. Radiotracers intended for use in the brain have the added challenge of being able to cross the BBB. A commonly used radiotracer is [18F] fluorodeoxyglucose (FDG), which cells preferentially uptake instead of glucose. Other commonly used isotopes include oxygen-15, nitrogen-13, and carbon-11 (Berger, 2003; Kapoor & Kasi, 2023; Pike, 2009). While considered safe, sites performing PET imaging or PET/CT imaging are expected to monitor radiation exposure and safety of patients and technicians.

The radiotracers utilized in PET imaging are administered intravenously, orally, or by inhalation depending on the body site being imaged. A waiting period of approximately 30 to 60 minutes occurs between administration and image acquisition to allow time for circulation and uptake. Factors such as tissue perfusion, changes in metabolism, and the presence of abnormal proteins (e.g., beta-amyloid-42 in Alzheimer's disease) affect uptake and can be seen in the PET image. The raw signal used to generate the PET image comes from atomic decay of the unstable radioisotope, which emits positrons that, when combined with neighboring electrons, give off high-energy photons in the form of gamma rays. The gamma rays are then detected by a set of detectors (sometimes referred to as a "camera") focused on the tissue of interest (e.g., brain). The PET scanner computer then transforms the signal data into a set of images (Berger, 2003; Kapoor & Kasi, 2023; Pike, 2009). See Figure 2.30 for a schematic of the basics of PET image acquisition.

Tissues that have taken up the radiotracer can then be visualized, for example, based on color or grayscale images. With CT and MRI, the images can be entered into analytic-statistical processing pipelines for quantitative measurement of the amount of radiotracer uptake that has occurred (Berger, 2003; Kapoor & Kasi, 2023; Pike, 2009).

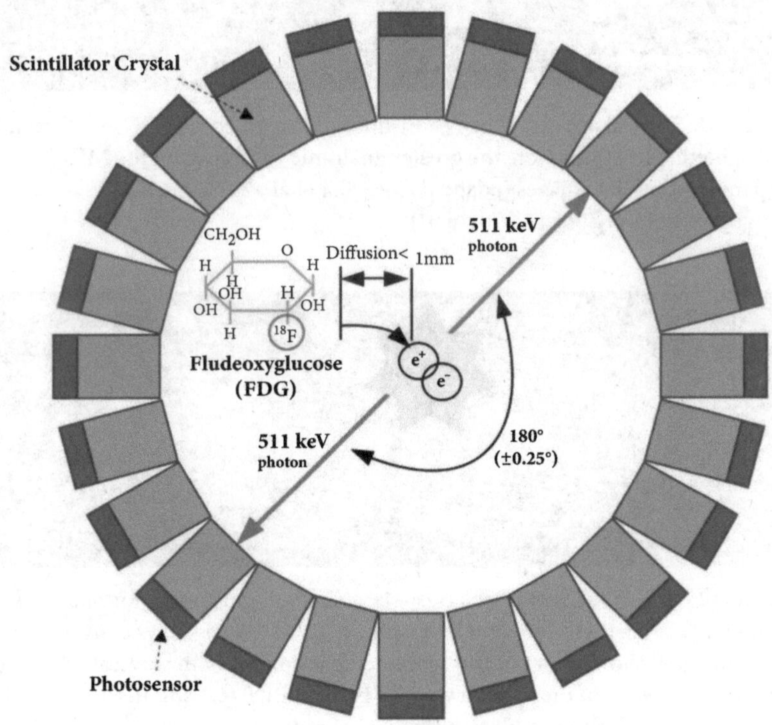

Figure 2.30 Schematic illustration of PET imaging; keV = kiloelectronvolt (image adapted from (Jiang et al., 2019; https://creativecommons.org/licenses/by/4.0/).

PET images have lower anatomic resolution than CT or MRI (Figure 2.31). However, techniques that combine PET and CT (PET-CT) or PET and MRI (PET-MRI) have improved on this limitation, as can be seen in Figure 2.32 below.

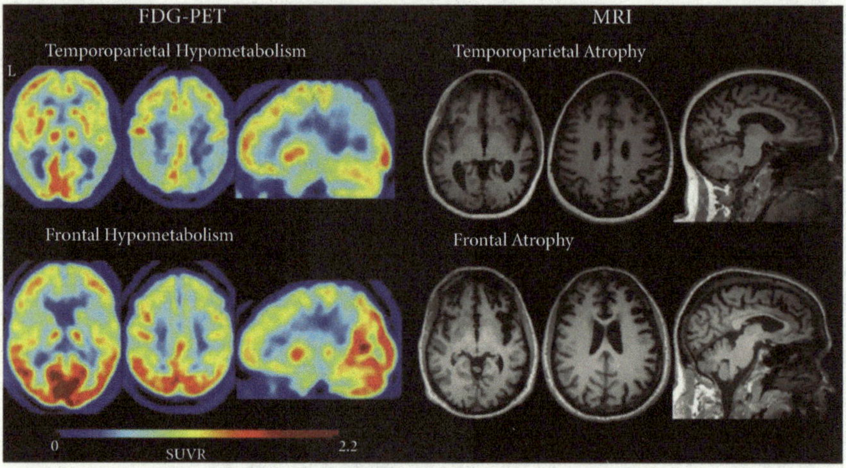

Figure 2.31 Typical appearance of [^{18}F]-fluorodeoxyglucose (FDG) PET versus T1-weighted MRI scans. Note the greater anatomic resolution of the MRI compared to the PET images (adapted from Sha et al., 2015; https://creativecommons.org/licenses/by/4.0/).

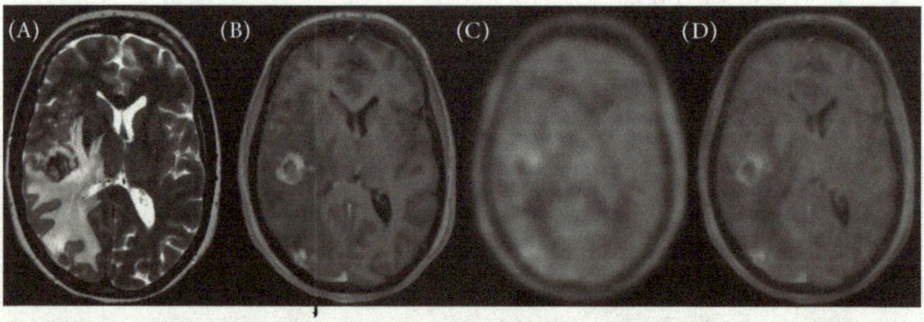

Figure 2.32 Axial images obtained post-stereotactic radiosurgery for malignant melanoma at two sites in the right hemisphere. T2 MRI (A) shows considerable edema. Image B shows new contrast enhancing lesions at both surgical sites. Slight tracer uptake is seen in those regions on [^{18}F]-FET PET (C) and fused PET/MRI. These imaging findings were deemed typical of radiation necrosis, which was subsequently confirmed by histopathology; FET = Fluoro-*O*-(2) fluoroethyl-l-tyrosine (adapted from Unterrainer et al., 2020; https://creativecommons.org/licenses/by/4.0/).

Neuropsychologists may encounter PET imaging in a variety of conditions including, but not limited to, the diagnosis and treatment/recovery outcome monitoring of brain tumors, metabolic changes following TBI (Byrnes et al., 2014), stroke (Czap & Sheth, 2021), multiple sclerosis (Niccolini et al., 2015), frontotemporal dementia (Ward et al., 2023), and Alzheimer's disease (Ruan & Sun, 2023). The utilization of PET scanning for identifying beta-amyloid-42 plaques in the brains of patients with clinical Alzheimer's disease is likely to increase as to determine patient eligibility for current and emerging amyloid-lowering monoclonal antibody treatments to slow disease and symptom progression.

Conclusion

This chapter provided broad coverage of commonly encountered structural, functional, and molecular imaging methods. Having working knowledge of various imaging modalities presented here will help neuropsychologists be knowledgeable consumers of clinical brain imaging results and images including key strengths and limitations of the most common neuroimaging modalities. Such knowledge can aid neuropsychologists in their approach to clinical assessment, interpretation of results and diagnostic formulation, and treatment planning and implementation. Neuroimaging has been and will continue to be a rapidly developing field. Recent and future innovations in statistical approaches to large datasets and application of artificial intelligence approaches will impact the future of diagnostic brain imaging and research. It is important for neuropsychologists to stay abreast of these advancements to best serve their patients.

Acknowledgment

The authors acknowledge the assistance of Ms. Sophie Elizabeth Slyman, BS, Psychology, University of Georgia, in proofreading, reference management, and obtaining copyright permissions for this chapter.

References

Adriaanse, S. M., Binnewijzend, M. A., Ossenkoppele, R., Tijms, B. M., van der Flier, W. M., Koene, T., Smits, L. L., Wink, A. M., Scheltens, P., van Berckel, B. N., & Barkhof, F. (2014). Widespread disruption of functional brain organization in early-onset alzheimer's disease. *PLoS One*, *9*(7), e102995. https://doi.org/10.1371/journal.pone.0102995

Alsop, D. C., Detre, J. A., Golay, X., Gunther, M., Hendrikse, J., Hernandez-Garcia, L., Lu, H., MacIntosh, B. J., Parkes, L. M., Smits, M., van Osch, M. J., Wang, D. J., Wong, E. C., & Zaharchuk, G. (2015). Recommended implementation of arterial spin-labeled perfusion MRI for clinical applications: A consensus of the ISMRM perfusion study group and the European consortium for ASL in dementia. *Magn Reson Med*, *73*(1), 102–116. https://doi.org/10.1002/mrm.25197

Atia, N., Benzaoui, A., Jacques, S., Hamiane, M., Kourd, K. E., Bouakaz, A., & Ouahabi, A. (2022). Particle swarm optimization and two-way fixed-effects analysis of variance for efficient brain tumor segmentation. *Cancers (Basel)*, *14*(18). https://doi.org/10.3390/cancers14184399

Baliyan, V., Das, C. J., Sharma, R., & Gupta, A. K. (2016). Diffusion weighted imaging: Technique and applications. *World J Radiol*, *8*(9), 785–798. https://doi.org/10.4329/wjr.v8.i9.785

Bandettini, P. A., Jesmanowicz, A., Wong, E. C., & Hyde, J. S. (1993). Processing strategies for time-course data sets in functional MRI of the human brain. *Magn Reson Med*, *30*(2), 161–173. https://doi.org/10.1002/mrm.1910300204

Beaulieu, C. (2009). The biological basis of diffusion anisotropy. In T. E. J. B. Heidi Johansen-Berg (Ed.), *Diffusion MRI: From Quantitative Measurement to In Vivo Neuroanatomy* (pp. 105–126). Academic Press.

Berger, A. (2003). How does it work? Positron emission tomography. *BMJ*, *326*(7404), 1449. https://doi.org/10.1136/bmj.326.7404.1449

Bhattacharjee, R., Gupta, R. K., Das, B., Dixit, V. K., Gupta, P., & Singh, A. (2021). Penumbra quantification from MR SWI-DWI mismatch and its comparison with MR ASL PWI-DWI mismatch in patients with acute ischemic stroke. *NMR Biomed*, *34*(7), e4526. https://doi.org/10.1002/nbm.4526

Bhuta, S., Knipe, H., Bell. D., et al. (2022). *Diffusion-weighted imaging in acute ischemic stroke.* Retrieved 11/02/2023 from https://radiopaedia.org/articles/13401

Biswal, B., Yetkin, F. Z., Haughton, V. M., & Hyde, J. S. (1995). Functional connectivity in the motor cortex of resting human brain using echo-planar MRI. *Magn Reson Med*, *34*(4), 537–541. https://doi.org/10.1002/mrm.1910340409

Blumenthal-Barby, J. S., & Krieger, H. (2015). Cognitive biases and heuristics in medical decision making: A critical review using a systematic search strategy. *Med Decis Making*, *35*(4), 539–557. https://doi.org/10.1177/0272989X14547740

Bodini, B. C., O. (2009). Diffusion MRI in neurological disorders. In T. E. J. B. Heidi Johansen-Berg (Ed.), *Diffusion MRI: From Quantitative Measurement to In Vivo Neuroanatomy* (pp. 176–203). Academic Press.

Buckner, R. L., Andrews-Hanna, J. R., & Schacter, D. L. (2008). The brain's default network: Anatomy, function, and relevance to disease. *Ann N Y Acad Sci*, *1124*, 1–38. https://doi.org/10.1196/annals.1440.011

Buxton, R. B., Frank, L. R., Wong, E. C., Siewert, B., Warach, S., & Edelman, R. R. (1998). A general kinetic model for quantitative perfusion imaging with arterial spin labeling. *Magn Reson Med*, *40*(3), 383–396. https://doi.org/10.1002/mrm.1910400308

Byrnes, K. R., Wilson, C. M., Brabazon, F., von Leden, R., Jurgens, J. S., Oakes, T. R., & Selwyn, R. G. (2014). FDG-PET imaging in mild traumatic brain injury: A critical review. *Front Neuroenergetics*, *5*, 13. https://doi.org/10.3389/fnene.2013.00013

Carpentries, T. (2023). *Structural MRI (Pre)processing and Neuroimaging Analysis: sMRI Acquisition and Modalities.* https://carpentries-incubator.github.io/SDC-BIDS-sMRI/aio/index.html

Chakraborty, A., & McEvoy, A. W. (2008). Presurgical functional mapping with functional MRI. *Curr Opin Neurol, 21*(4), 446–451. https://doi.org/10.1097/WCO.0b013e32830866e2

Charry, J. D., Falla, J. D., Ochoa, J. D., Pinzon, M. A., Tejada, J. H., Henriquez, M. J., Solano, J. P., & Calvache, C. (2017). External validation of the Rotterdam Computed Tomography Score in the prediction of mortality in severe traumatic brain injury. *J Neurosci Rural Pract, 8*(Suppl 1), S23–S26. https://doi.org/10.4103/jnrp.jnrp_434_16

Chen, J. E., & Glover, G. H. (2015). Functional magnetic resonance imaging methods. *Neuropsychol Rev, 25*(3), 289–313. https://doi.org/10.1007/s11065-015-9294-9

Chesebro, A. G., Amarante, E., Lao, P. J., Meier, I. B., Mayeux, R., & Brickman, A. M. (2021). Automated detection of cerebral microbleeds on T2*-weighted MRI. *Sci Rep, 11*(1), 4004. https://doi.org/10.1038/s41598-021-83607-0

Clement, P., Petr, J., Dijsselhof, M. B. J., Padrela, B., Pasternak, M., Dolui, S., Jarutyte, L., Pinter, N., Hernandez-Garcia, L., Jahn, A., Kuijer, J. P. A., Barkhof, F., Mutsaerts, H., & Keil, V. C. (2022). A beginner's guide to arterial spin labeling (ASL) image processing. *Front Radiol, 2*, 929533. https://doi.org/10.3389/fradi.2022.929533

Czap, A. L., & Sheth, S. A. (2021). Overview of Imaging Modalities in Stroke. *Neurology, 97*(20 Suppl 2), S42–S51. https://doi.org/10.1212/WNL.0000000000012794

Dai, W., Garcia, D., de Bazelaire, C., & Alsop, D. C. (2008). Continuous flow-driven inversion for arterial spin labeling using pulsed radio frequency and gradient fields. *Magn Reson Med, 60*(6), 1488–1497. https://doi.org/10.1002/mrm.21790

Dale, B. M., Brown, M. A., & Semelka, R. C. (2015). *MRI Basic Principles and Applications* (5th ed.). Wiley Blackwell.

DeCarli, C., & Lockhart, S. N. (2014). Cerebrovascular disease. In M. J. Aminoff & R. B. Daroff (Eds.), *Encyclopedia of Neurological Sciences* (2nd ed., pp. 722–726). Academic Press.https://doi.org/10.1016/B978-0-12-385157-4.00443-7

Descoteaux, M. (2015). High angular resolution of diffusion imaging (HARDI). In J. Webster (Ed.), *Wiley Encyclopedia of Electrical and Electronics Engineering* (pp. 1–25). Wiley.

Detre, J. A., Leigh, J. S., Williams, D. S., & Koretsky, A. P. (1992). Perfusion imaging. *Magn Reson Med, 23*(1), 37–45. https://doi.org/10.1002/mrm.1910230106

Dmytriw, A. A., Sawlani, V., & Shankar, J. (2017). Diffusion-weighted imaging of the brain: Beyond stroke. *Can Assoc Radiol J, 68*(2), 131–146. https://doi.org/10.1016/j.carj.2016.10.001

Elster, A. D. (2021). *MRIQuestions.com*. ELSTER LLC. Retrieved June 2025 from https://mriquestions.com/index.html

Escott, E. J. (2001). A variety of appearances of malignant melanoma in the head: A review. *Radiographics, 21*(3), 625–639. https://doi.org/10.1148/radiographics.21.3.g01ma19625

Garg, R. K., & Sinha, M. K. (2010). Multiple ring-enhancing lesions of the brain. *J Postgrad Med, 56*(4), 307–316. https://doi.org/10.4103/0022-3859.70939

Geroldi, C., Galluzzi, S., Testa, C., Zanetti, O., & Frisoni, G. B. (2003). Validation study of a CT-based weighted rating scale for subcortical ischemic vascular disease in patients with mild cognitive deterioration. *Eur Neurol, 49*(4), 193–209. https://doi.org/10.1159/000070183

Giacomozzi, C., Nicoli, L., Sozzi, C., Piovan, E., & Maghnie, M. (2021). Case report: Lipoma of the tuber cinereum mimicking a pituitary gland abnormality in a girl with central precocious puberty. *Front Endocrinol (Lausanne), 12*, 766253. https://doi.org/10.3389/fendo.2021.766253

Golay, X., Hendrikse, J., & Lim, T. C. (2004). Perfusion imaging using arterial spin labeling. *Top Magn Reson Imaging, 15*(1), 10–27. https://doi.org/10.1097/00002142-200402000-00003

Graber, M. L., Franklin, N., & Gordon, R. (2005). Diagnostic error in internal medicine. *Arch Intern Med, 165*(13), 1493–1499. https://doi.org/10.1001/archinte.165.13.1493

Grade, M., Hernandez Tamames, J. A., Pizzini, F. B., Achten, E., Golay, X., & Smits, M. (2015). A neuroradiologist's guide to arterial spin labeling MRI in clinical practice. *Neuroradiology, 57*(12), 1181–1202. https://doi.org/10.1007/s00234-015-1571-z

Gu, Y., Zhou, C., Piao, Z., Yuan, H., Jiang, H., Wei, H., Zhou, Y., Nan, G., & Ji, X. (2022). Cerebral edema after ischemic stroke: Pathophysiology and underlying mechanisms. *Front Neurosci, 16*, 988283. https://doi.org/10.3389/fnins.2022.988283

Gunning-Dixon, F. M., & Raz, N. (2000). The cognitive correlates of white matter abnormalities in normal aging: A quantitative review. *Neuropsychology, 14*(2), 224–232. https://doi.org/10.1037//0894-4105.14.2.224

Hachinski, V., Iadecola, C., Petersen, R. C., Breteler, M. M., Nyenhuis, D. L., Black, S. E., Powers, W. J., DeCarli, C., Merino, J. G., Kalaria, R. N., Vinters, H. V., Holtzman, D. M., Rosenberg, G. A., Wallin, A., Dichgans, M., Marler, J. R., & Leblanc, G. G. (2006). National Institute of Neurological Disorders and Stroke–Canadian Stroke Network vascular cognitive impairment harmonization standards. *Stroke, 37*(9), 2220–2241. https://doi.org/10.1161/01.STR.0000237236.88823.47

Haller, S., Haacke, E. M., Thurnher, M. M., & Barkhof, F. (2021). Susceptibility-weighted imaging: Technical essentials and clinical neurologic applications. *Radiology, 299*(1), 3–26. https://doi.org/10.1148/radiol.2021203071

Higaki, T., Nakamura, Y., Tatsugami, F., Kaichi, Y., Akagi, M., Akiyama, Y., Baba, Y., Iida, M., & Awai, K. (2018). Introduction to the technical aspects of computed diffusion-weighted imaging for radiologists. *Radiographics, 38*(4), 1131–1144. https://doi.org/10.1148/rg.2018170115

Hurley, R. A., Lucas, D. M., & Taber, K. H. (2013). Structural imaging. In D. B. Arciniegas, C. A. Anderson, & C. M. Filley (Ed.), *Behavioral Neurology and Neuropsychiatry*. Cambridge University Press. https://doi.org/10.1017/CBO9781139016919

Huynh, K., Baghdanian, A. H., Baghdanian, A. A., Sun, D. S., Kolli, K. P., & Zagoria, R. J. (2020). Updated guidelines for intravenous contrast use for CT and MRI. *Emerg Radiol, 27*(2), 115–126. https://doi.org/10.1007/s10140-020-01751-y

Ibrahim, M. A., Hazhirkarzar, B., & Dublin, A. B. (2023). Gadolinium magnetic resonance imaging. *StatPearls*. https://www.ncbi.nlm.nih.gov/pubmed/29494094

Janaway, B. M., Simpson, J. E., Hoggard, N., Highley, J. R., Forster, G., Drew, D., Gebril, O. H., Matthews, F. E., Brayne, C., Wharton, S. B., Ince, P. G., & MRC Cognitive Function and Ageing Neuropathology Study. (2014). Brain haemosiderin in older people: Pathological evidence for an ischaemic origin of magnetic resonance imaging (MRI) microbleeds. *Neuropathol Appl Neurobiol, 40*(3), 258–269. https://doi.org/10.1111/nan.12062

Jiang, W., Chalich, Y., & Deen, M. J. (2019). Sensors for positron emission tomography applications. *Sensors (Basel), 19*(22). https://doi.org/10.3390/s19225019

Kakuda, W., Lansberg, M. G., Thijs, V. N., Kemp, S. M., Bammer, R., Wechsler, L. R., Moseley, M. E., Marks, M. P., Albers, G. W., & Investigators, D. (2008). Optimal definition for PWI/DWI mismatch in acute ischemic stroke patients. *J Cereb Blood Flow Metab, 28*(5), 887–891. https://doi.org/10.1038/sj.jcbfm.9600604

Kapoor, M., & Kasi, A. (2023). PET scanning. *StatPearls*. https://www.ncbi.nlm.nih.gov/pubmed/32644515

Kastrup, O., Wanke, I., & Maschke, M. (2005). Neuroimaging of infections. *NeuroRx*, *2*(2), 324–332. https://doi.org/10.1602/neurorx.2.2.324

Krupa, K., & Bekiesinska-Figatowska, M. (2015). Artifacts in magnetic resonance imaging. *Pol J Radiol*, *80*, 93–106. https://doi.org/10.12659/PJR.892628

Kwong, K. K., Belliveau, J. W., Chesler, D. A., Goldberg, I. E., Weisskoff, R. M., Poncelet, B. P., Kennedy, D. N., Hoppel, B. E., Cohen, M. S., Turner, R., et al. (1992). Dynamic magnetic resonance imaging of human brain activity during primary sensory stimulation. *Proc Natl Acad Sci U S A*, *89*(12), 5675–5679. https://doi.org/10.1073/pnas.89.12.5675

Lad, S. P., Guzman, R., Kelly, M. E., Li, G., Lim, M., Lovbald, K., & Steinberg, G. K. (2006). Cerebral perfusion imaging in vasospasm. *Neurosurg Focus*, *21*(3), E7. https://doi.org/10.3171/foc.2006.21.3.7

Lansberg, M. G., Thijs, V. N., O'Brien, M. W., Ali, J. O., de Crespigny, A. J., Tong, D. C., Moseley, M. E., & Albers, G. W. (2001). Evolution of apparent diffusion coefficient, diffusion-weighted, and T2-weighted signal intensity of acute stroke. *AJNR Am J Neuroradiol*, *22*(4), 637–644. https://www.ncbi.nlm.nih.gov/pubmed/11290470

Lebby, P. C. (2013a). Brain imaging techniques. In *Brain Imaging: A Guide for Clinicians* (pp. 155–235). Oxford.

Lebby, P. C. (2013b). Viewing and interpreting brain scans. In *Brain Imaging: A Guide for Clinicians* (pp. 29–70). Oxford.

Lee, M. H., Smyser, C. D., & Shimony, J. S. (2013). Resting-state fMRI: A review of methods and clinical applications. *AJNR Am J Neuroradiol*, *34*(10), 1866–1872. https://doi.org/10.3174/ajnr.A3263

Lin, E., & Alessio, A. (2009). What are the basic concepts of temporal, contrast, and spatial resolution in cardiac CT. *J Cardiovasc Comput Tomogr*, *3*(6), 403–408. https://doi.org/10.1016/j.jcct.2009.07.003

Lloyd-Jones, G. (2019). *Acute CT: Brain CT quality and artifacts*. Retrieved 11/09/2023 from https://www.radiologymasterclass.co.uk/tutorials/ct/ct_acute_brain/ct_brain_quality

Longstreth, W. T., Jr., Arnold, A. M., Beauchamp, N. J., Jr., Manolio, T. A., Lefkowitz, D., Jungreis, C., Hirsch, C. H., O'Leary, D. H., & Furberg, C. D. (2005). Incidence, manifestations, and predictors of worsening white matter on serial cranial magnetic resonance imaging in the elderly: The Cardiovascular Health Study. *Stroke*, *36*(1), 56–61. https://doi.org/10.1161/01.STR.0000149625.99732.69

Maas, A. I., Hukkelhoven, C. W., Marshall, L. F., & Steyerberg, E. W. (2005). Prediction of outcome in traumatic brain injury with computed tomographic characteristics: A comparison between the computed tomographic classification and combinations of computed tomographic predictors. *Neurosurgery*, *57*(6), 1173–1182; discussion 1173–1182. https://doi.org/10.1227/01.neu.0000186013.63046.6b

Maier, S. E., Sun, Y., & Mulkern, R. V. (2010). Diffusion imaging of brain tumors. *NMR Biomed*, *23*(7), 849–864. https://doi.org/10.1002/nbm.1544

Malloy, P., Correia, S., Stebbins, G., & Laidlaw, D. H. (2007). Neuroimaging of white matter in aging and dementia. *Clin Neuropsychol*, *21*(1), 73–109. https://doi.org/10.1080/13854040500263583

Marshall, L. F., Marshall, S. B., Klauber, M. R., Van Berkum Clark, M., Eisenberg, H., Jane, J. A., Luerssen, T. G., Marmarou, A., & Foulkes, M. A. (1992). The diagnosis of head injury requires a classification based on computed axial tomography. *J Neurotrauma, 9*(Suppl 1), S287–292. https://www.ncbi.nlm.nih.gov/pubmed/1588618

Mitchell, D. G., & Cohen, M. S. (2004). *MRI Principles* (2nd ed.). Saunders.

Nehring S. M., T. P., Tenney, S. (2023, July 3). Cerebral edema. *StatPearls*. https://www.ncbi.nlm.nih.gov/books/NBK537272/

Niccolini, F., Su, P., & Politis, M. (2015). PET in multiple sclerosis. *Clin Nucl Med, 40*(1), e46–52. https://doi.org/10.1097/RLU.0000000000000359

Nowogrodzki, A. (2018). The world's strongest MRI machines are pushing human imaging to new limits. *Nature, 563*(7729), 24–26. https://doi.org/10.1038/d41586-018-07182-7

Ogawa, S., Tank, D. W., Menon, R., Ellermann, J. M., Kim, S. G., Merkle, H., & Ugurbil, K. (1992). Intrinsic signal changes accompanying sensory stimulation: functional brain mapping with magnetic resonance imaging. *Proc Natl Acad Sci U S A, 89*(13), 5951–5955. https://doi.org/10.1073/pnas.89.13.5951

Oros-Peusquens, A. M., Loucao, R., Abbas, Z., Gras, V., Zimmermann, M., & Shah, N. J. (2019). A single-scan, rapid whole-brain protocol for quantitative water content mapping with neurobiological implications. *Front Neurol, 10*, 1333. https://doi.org/10.3389/fneur.2019.01333

Pantoni, L., Simoni, M., Pracucci, G., Schmidt, R., Barkhof, F., & Inzitari, D. (2002). Visual rating scales for age-related white matter changes (leukoaraiosis): Can the heterogeneity be reduced? *Stroke, 33*(12), 2827–2833. https://doi.org/10.1161/01.str.0000038424.70926.5e

Patel, P. R., & De Jesus, O. (2022). CT scan. *StatPearls*. https://www.ncbi.nlm.nih.gov/pubmed/33620865

Petersen, E. T., Zimine, I., Ho, Y. C., & Golay, X. (2006). Non-invasive measurement of perfusion: A critical review of arterial spin labelling techniques. *Br J Radiol, 79*(944), 688–701. https://doi.org/10.1259/bjr/67705974

Pike, V. W. (2009). PET radiotracers: Crossing the blood-brain barrier and surviving metabolism. *Trends Pharmacol Sci, 30*(8), 431–440. https://doi.org/10.1016/j.tips.2009.05.005

Pirko, I. (2010). Advances in multiple sclerosis imaging. In C. F. Lucchinetti & R. Hohlfeld (Ed.), *Multiple Sclerosis 3* (Vol. 35, pp. 83–119). Elsevier.

Piszczor, M., Thornton, G., & Bia, F. J. (1985). The evaluation of contrast-enhancing brain lesions: pitfalls in current practice. *Yale J Biol Med, 58*(1), 19–27. https://www.ncbi.nlm.nih.gov/pubmed/4013370

Raj, R., Siironen, J., Skrifvars, M. B., Hernesniemi, J., & Kivisaari, R. (2014). Predicting outcome in traumatic brain injury: Development of a novel computerized tomography classification system (Helsinki computerized tomography score). *Neurosurgery, 75*(6), 632–646; discussion 646-637. https://doi.org/10.1227/NEU.0000000000000533

Razek, A. (2018). Towards an image-guided restricted drug release in friendly implanted therapeutics. *European Physical Journal-Applied Physics, 82*(3). https://doi.org/10.1051/epjap/2018180201

Ross, D. E., Seabaugh, J., Seabaugh, J. M., Barcelona, J., Seabaugh, D., Wright, K., Norwind, L., King, Z., Graham, T. J., Baker, J., & Lewis, T. (2022). Updated review of the evidence supporting the medical and legal use of NeuroQuant(R) and NeuroGage(R) in patients with traumatic brain injury. *Front Hum Neurosci, 16*, 715807. https://doi.org/10.3389/fnhum.2022.715807

Rovira, A., Wattjes, M. P., Tintore, M., Tur, C., Yousry, T. A., Sormani, M. P., De Stefano, N., Filippi, M., Auger, C., Rocca, M. A., Barkhof, F., Fazekas, F., Kappos, L., Polman, C., Miller, D., Montalban, X., & MAGNIMS study group. (2015). MAGNIMS consensus guidelines on the use of MRI in multiple sclerosis-clinical implementation in the diagnostic process. *Nat Rev Neurol, 11*(8), 471–482. https://doi.org/10.1038/nrneurol.2015.106

Rowley, H. A., & Roberts, T. P. (2004). Clinical perspectives in perfusion: Neuroradiologic applications. *Top Magn Reson Imaging, 15*(1), 28–40. https://doi.org/10.1097/00002142-200402000-00004

Ruan, D., & Sun, L. (2023). Amyloid-beta PET in Alzheimer's disease: A systematic review and Bayesian meta-analysis. *Brain Behav, 13*(1), e2850. https://doi.org/10.1002/brb3.2850

Saade, C., Bou-Fakhredin, R., Yousem, D. M., Asmar, K., Naffaa, L., & El-Merhi, F. (2018). Gadolinium and multiple sclerosis: Vessels, barriers of the brain, and glymphatics. *AJNR Am J Neuroradiol, 39*(12), 2168–2176. https://doi.org/10.3174/ajnr.A5773

Scheltens, P., Erkinjunti, T., Leys, D., Wahlund, L. O., Inzitari, D., Del Ser, T., Pasquier, F., Barkhof, F., Mantyla, R., Bowler, J., Wallin, A., Ghika, J., Fazekas, F., & Pantoni, L. (1998). White matter changes on CT and MRI: an overview of visual rating scales. European Task Force on Age-Related White Matter Changes. *Eur Neurol, 39*(2), 80–89. https://doi.org/10.1159/000007921

Schmidt, R., Petrovic, K., Ropele, S., Enzinger, C., & Fazekas, F. (2007). Progression of leukoaraiosis and cognition. *Stroke, 38*(9), 2619–2625. https://doi.org/10.1161/STROKEAHA.107.489112

Sha, S. J., Ghosh, P. M., Lee, S. E., Corbetta-Rastelli, C., Jagust, W. J., Kornak, J., Rankin, K. P., Grinberg, L. T., Vinters, H. V., Mendez, M. F., Dickson, D. W., Seeley, W. W., Gorno-Tempini, M., Kramer, J., Miller, B. L., Boxer, A. L., & Rabinovici, G. D. (2015). Predicting amyloid status in corticobasal syndrome using modified clinical criteria, magnetic resonance imaging and fluorodeoxyglucose positron emission tomography. *Alzheimers Res Ther, 7*(1), 8. https://doi.org/10.1186/s13195-014-0093-y

Shulman, G. L., Fiez, J. A., Corbetta, M., Buckner, R. L., Miezin, F. M., Raichle, M. E., & Petersen, S. E. (1997). Common blood flow changes across visual tasks: II. Decreases in cerebral cortex. *J Cogn Neurosci, 9*(5), 648–663. https://doi.org/10.1162/jocn.1997.9.5.648

Siddiqui, F. M., Bekker, S. V., & Qureshi, A. I. (2011). Neuroimaging of hemorrhage and vascular defects. *Neurotherapeutics, 8*(1), 28–38. https://doi.org/10.1007/s13311-010-0009-x

Smirniotopoulos, J. G., Murphy, F. M., Rushing, E. J., Rees, J. H., & Schroeder, J. W. (2007). Patterns of contrast enhancement in the brain and meninges. *Radiographics, 27*(2), 525–551. https://doi.org/10.1148/rg.272065155

Smith, L. G. F., Milliron, E., Ho, M. L., Hu, H. H., Rusin, J., Leonard, J., & Sribnick, E. A. (2019). Advanced neuroimaging in traumatic brain injury: An overview. *Neurosurg Focus, 47*(6), E17. https://doi.org/10.3171/2019.9.FOCUS19652

Sweet, L. H., Hassenstab, J. J., McCaffery, J. M., Raynor, H. A., Bond, D. S., Demos, K. E., Haley, A. P., Cohen, R. A., Del Parigi, A., & Wing, R. R. (2012). Brain response to food stimulation in obese, normal weight, and successful weight loss maintainers. *Obesity (Silver Spring), 20*(11), 2220–2225. https://doi.org/10.1038/oby.2012.125

Tae, W. S., Ham, B. J., Pyun, S. B., Kang, S. H., & Kim, B. J. (2018). Current clinical applications of diffusion-tensor imaging in neurological disorders. *J Clin Neurol*, *14*(2), 129–140. https://doi.org/10.3988/jcn.2018.14.2.129

Thaler, C., Faizy, T., Sedlacik, J., Holst, B., Stellmann, J. P., Young, K. L., Heesen, C., Fiehler, J., & Siemonsen, S. (2015). T1- thresholds in black holes increase clinical-radiological correlation in multiple sclerosis patients. *PLoS One*, *10*(12), e0144693. https://doi.org/10.1371/journal.pone.0144693

Tomimoto, H. (2015). White matter integrity and cognitive dysfunction: Radiological and neuropsychological correlations. *Geriatr Gerontol Int*, *15 Suppl 1*, 3–9. https://doi.org/10.1111/ggi.12661

Topcuoglu, M. A., Saka, E., Silverman, S. B., Schwamm, L. H., & Singhal, A. B. (2017). Recrudescence of deficits after stroke: Clinical and imaging phenotype, triggers, and risk factors. *JAMA Neurol*, *74*(9), 1048–1055. https://doi.org/10.1001/jamaneurol.2017.1668

Unterrainer, M., Eze, C., Ilhan, H., Marschner, S., Roengvoraphoj, O., Schmidt-Hegemann, N. S., Walter, F., Kunz, W. G., Rosenschold, P. M. A., Jeraj, R., Albert, N. L., Grosu, A. L., Niyazi, M., Bartenstein, P., & Belka, C. (2020). Recent advances of PET imaging in clinical radiation oncology. *Radiat Oncol*, *15*(1), 88. https://doi.org/10.1186/s13014-020-01519-1

van den Heuvel, D. M., ten Dam, V. H., de Craen, A. J., Admiraal-Behloul, F., Olofsen, H., Bollen, E. L., Jolles, J., Murray, H. M., Blauw, G. J., Westendorp, R. G., & van Buchem, M. A. (2006). Increase in periventricular white matter hyperintensities parallels decline in mental processing speed in a non-demented elderly population. *J Neurol Neurosurg Psychiatry*, *77*(2), 149–153. https://doi.org/10.1136/jnnp.2005.070193

van Everdingen, K. J., van der Grond, J., Kappelle, L. J., Ramos, L. M., & Mali, W. P. (1998). Diffusion-weighted magnetic resonance imaging in acute stroke. *Stroke*, *29*(9), 1783–1790. https://doi.org/10.1161/01.str.29.9.1783

Wahlund, L. O., Barkhof, F., Fazekas, F., Bronge, L., Augustin, M., Sjogren, M., Wallin, A., Ader, H., Leys, D., Pantoni, L., Pasquier, F., Erkinjuntti, T., Scheltens, P. (2001). A new rating scale for age-related white matter changes applicable to MRI and CT. *Stroke*, *32*(6), 1318–1322. https://doi.org/10.1161/01.str.32.6.1318

Wahlund, L. O., Westman, E., van Westen, D., Wallin, A., Shams, S., Cavallin, L., Larsson, E. M., & from the Imaging Cognitive Impairment Network (ICINET). (2017). Imaging biomarkers of dementia: Recommended visual rating scales with teaching cases. *Insights Imaging*, *8*(1), 79–90. https://doi.org/10.1007/s13244-016-0521-6

Warach, S. (2001). Use of diffusion and perfusion magnetic resonance imaging as a tool in acute stroke clinical trials. *Curr Control Trials Cardiovasc Med*, *2*(1), 38–44. https://doi.org/10.1186/cvm-2-1-038

Ward, J., Ly, M., & Raji, C. A. (2023). Brain PET imaging: Frontotemporal dementia. *PET Clin*, *18*(1), 123–133. https://doi.org/10.1016/j.cpet.2022.09.010

Wei, S. C., Ulmer, S., Lev, M. H., Pomerantz, S. R., Gonzalez, R. G., & Henson, J. W. (2010). Value of coronal reformations in the CT evaluation of acute head trauma. *AJNR Am J Neuroradiol*, *31*(2), 334–339. https://doi.org/10.3174/ajnr.A1824

Williams, D. S., Detre, J. A., Leigh, J. S., & Koretsky, A. P. (1992). Magnetic resonance imaging of perfusion using spin inversion of arterial water. *Proc Natl Acad Sci U S A*, *89*(1), 212–216. https://doi.org/10.1073/pnas.89.1.212

Wolk, D. A., & Detre, J. A. (2012). Arterial spin labeling MRI: An emerging biomarker for Alzheimer's disease and other neurodegenerative conditions. *Curr Opin Neurol, 25*(4), 421–428. https://doi.org/10.1097/WCO.0b013e328354ff0a

Wong, E. C., Buxton, R. B., & Frank, L. R. (1998). A theoretical and experimental comparison of continuous and pulsed arterial spin labeling techniques for quantitative perfusion imaging. *Magn Reson Med, 40*(3), 348–355. https://doi.org/10.1002/mrm.1910400303

Wu, W. C., Fernandez-Seara, M., Detre, J. A., Wehrli, F. W., & Wang, J. (2007). A theoretical and experimental investigation of the tagging efficiency of pseudocontinuous arterial spin labeling. *Magn Reson Med, 58*(5), 1020–1027. https://doi.org/10.1002/mrm.21403

Yeo, B. T., Krienen, F. M., Sepulcre, J., Sabuncu, M. R., Lashkari, D., Hollinshead, M., Roffman, J. L., Smoller, J. W., Zollei, L., Polimeni, J. R., Fischl, B., Liu, H., & Buckner, R. L. (2011). The organization of the human cerebral cortex estimated by intrinsic functional connectivity. *J Neurophysiol, 106*(3), 1125–1165. https://doi.org/10.1152/jn.00338.2011

3
Vascular Imaging

Nada El Husseini and Thomas J. Farrer

Introduction

Vascular health is paramount to neurological and cognitive function. While the brain is only about 3 pounds, or 2% of body mass, it is highly vascularized and uses approximately 20% of a person's daily metabolic demand (Kalaria, 2010).

Vascular imaging techniques are crucial for assessing and monitoring cerebrovascular conditions such as stroke, aneurysms, vessel stenosis, vasculopathies, and arteriovenous malformations. In general, vascular imaging is in the purview of vascular neurology and radiology. However, clinical neuropsychologists typically evaluate patients who have imaging studies. Neuropsychologists should understand these techniques and how abnormal findings may affect neuropsychological function. As the brain is supplied by an elaborate vascular network that originates extracranially and reaches deep into the brain, the neurovascular function and regulation has major implications for brain health and disease and for brain imaging (Schaeffer & Iadecola, 2021). The integration of structural and functional neuroimaging findings with neuropsychology is important to continue to advance research discoveries into brain function and treatment modalities as well as to enhance patient-centered medical decision-making.

This chapter reviews common vascular imaging techniques, including magnetic resonance angiography (MRA), computed tomography angiography (CTA), digital subtraction angiography (DSA), and neurovascular ultrasound. Other structural and functional modalities are discussed in Chapter 2 of this text but are also briefly mentioned here given their role in assessing regional blood flow (e.g., positron emission tomography [PET] and single-photon emission computed tomography [SPECT]).

A comprehensive review of the cerebrovascular anatomy and the clinical syndromes associated with stroke is beyond the scope of this chapter. However, main branches of the cerebral vasculature are reviewed here. For simplicity, cerebral vasculature will be grouped into the anterior and posterior circulation. In addition, when describing the main arteries supplying the brain including

Nada El Husseini and Thomas J. Farrer, *Vascular Imaging*. In: *Medical and Laboratory Examinations for Neuropsychologists*. Edited by: Thomas J. Farrer and R. Brock Frost, Oxford University Press.
 DOI: 10.1093/9780197619537.003.0004

Brief Review of Cerebrovascular Anatomy

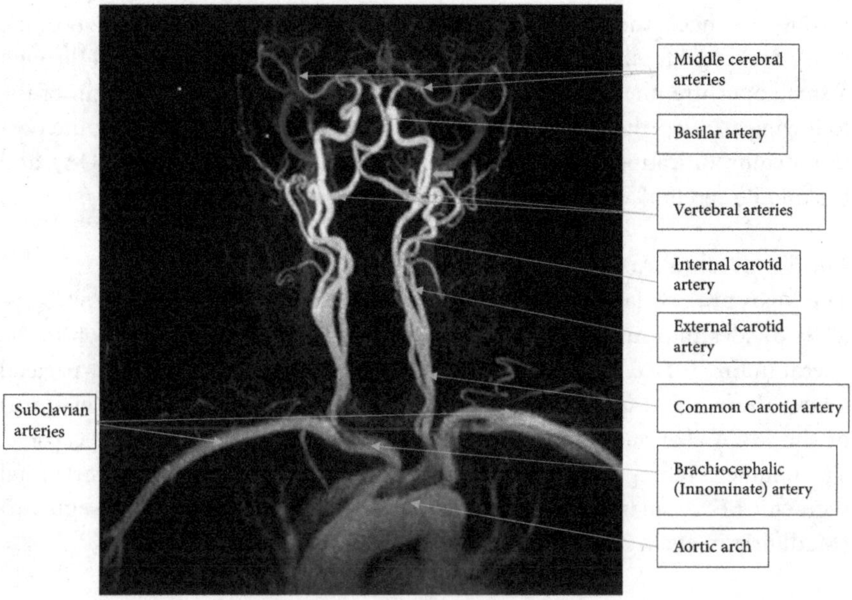

Figure 3.1 MRA head and neck: A chronic left internal carotid artery dissection is noted (thick arrow).

the carotids, middle cerebral arteries, vertebral arteries, and posterior cerebral arteries, the vessels are often divided into anatomical segments that makes the reference to them in medical communication more uniform (Madhuripan et al., 2017). These will be briefly mentioned below.

Three main arteries originate from the aortic arch: (1) the right brachiocephalic (innominate) artery, (2) the left common carotid artery, and (3) the left subclavian artery (Figure 3.1).

Anterior Circulation

The anterior circulation originates from the bilateral internal carotid arteries (ICAs) and supplies blood to more than half of the cerebral hemispheres, including the frontal lobes, parietal lobes, lateral temporal lobes, and the anterior part of the deep cerebral hemispheres.

Stemming off the right brachiocephalic artery is the right common carotid artery, which will bifurcate to become the right internal and right external carotid arteries. The left common carotid artery, which usually stems directly from the aortic arch, also bifurcates to become the left internal and left external carotid arteries. As the left and right internal and external carotid arteries ascend through the neck, the external portion will supply blood to several areas of the head, neck, and face. The ICA on each side enters the base of the skull through a small aperture, the carotid canal. Intracranially, the ICA is the origin of the following arteries: the ophthalmic artery, the anterior choroidal artery, the posterior communicating artery (PCOM), the anterior cerebral artery (ACA), and the middle cerebral artery (MCA), which are part of the circle of Willis.

Middle Cerebral Artery

The MCA projects laterally through the lateral sulcus (Sylvian fissure) and typically divides into superior and inferior divisions, providing circulation to the lateral brain surface, including portions of the frontal, temporal, and parietal lobes. There are also smaller deep penetrating branches (i.e., deep territory) of the MCA that supply portions of the basal ganglia and internal capsule. The middle cerebral artery is often divided into four segments (M1, horizontal segment; M2, insular segment; M3, opercular segment; M4, cortical segment) (Madhuripan et al., 2017).

Anterior Cerebral Artery

From the circle of Willis, the ACA projects anteriorly to supply the medial portion of the basal forebrain, medial portions of the prefrontal cortex, portions of the medial parietal lobes, and the anterior portion of the corpus callosum. It also supplies the head of the caudate nucleus via a branch known as the recurrent artery of Heubner (Feekes & Cassell, 2006). Clinically significant anatomical variations exist: for example, an azygos ACA is a rare variant (0.3%–2% of the population) characterized by the absence of the anterior communicating artery and the union of two proximal segments of the anterior cerebral arteries forming a single trunk (Figure 3.2) (de Sousa et al., 2017). An occlusion of the proximal azygos ACA typically leads to bilateral bifrontal infarctions with clinical manifestations that may include bilateral leg weakness, urinary incontinence, akinesia, rigidity, hypobulia/apathy, and hypophonia (Kobayashi et al., 2011). The ACA is usually divided into 3 segments (A1, horizontal segment; A2, vertical segment; A3, callosal segment) (Madhuripan et al., 2017).

Posterior Circulation

The posterior circulation of the brain originates from the vertebral arteries, which arise from the left and right subclavian arteries discussed earlier. As the vertebral arteries come off the subclavian arteries, they ascend through the transverse foramina of the cervical vertebrae entering the foramina usually at the level

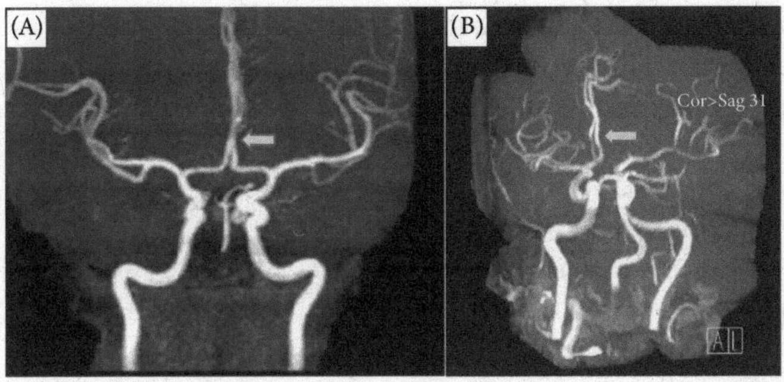

Figure 3.2 MRA showing bilateral ACAs originating from the respective carotid arteries (thick arrow) (A). MRA with an azygos ACA showing a common trunk originating from the right carotid artery (thick arrow) (B).

of C6, eventually entering the skull via the foramen magnum, running along the ventral side of the medulla. They then merge into a single basilar artery on the ventral portion of the pons, bifurcating again to become the posterior cerebral arteries (PCAs). Each vertebral artery is typically divided into four segments (V1, pre-foraminal segment; V2, foraminal segment; V3, atlantic, extradural, or extraspinal segment; V4, intradural or intracranial segment).

Additionally, there are multiple smaller arteries stemming from these branches that serve portions of the brainstem and cerebellum. These include the anterior spinal artery (supplying the anterior portion of the spinal cord and sometimes medial portions of the medulla; not discussed in detail here) and the posterior inferior cerebellar artery (PICA), which originate from the vertebral arteries, and the anterior inferior cerebellar artery (AICA) and superior cerebellar artery (SCA), which originate from the basilar artery.

Posterior Cerebral Artery

The PCA curves posteriorly from the circle of Willis and provides oxygenated blood to portions of the inferior temporal lobes, all portions of the occipital lobe, and medial aspects of the parietal lobes. Deep territories of the PCA also provide vascularization to thalamic regions and posterior portions of the corpus callosum, including the splenium. A fetal PCA, whereby the PCA originates from the ICA with no or small atretic connection to the basilar artery, is a common variant of the cerebral circulation occurring in about 30% of the population (Shaban et al., 2013). When a fetal PCA is present, carotid disease, which is typically not associated with posterior circulation strokes, may cause a stroke in the PCA distribution. The PCA is usually divided into four segments (P1, pre-communicating segment; P2, post-communicating segment; P3, quadrigeminal segment; P4, cortical segment).

Posterior Inferior Cerebellar Artery

The PICA arises from the vertebral arteries at the level of the mid medulla. It projects posteriorly and supplies the lateral portions of the medulla and inferior cerebellum.

Anterior Inferior Cerebellar Artery

Stemming from the basilar artery, adjacent to the abducens nerve, the AICA supplies portions of the pons and anterior portions of the cerebellum.

Superior Cerebellar Artery

Near the top of the basilar artery, resting above the pons and adjacent to crus cerebri, the SCA supplies the superior portions of the cerebellum and portions of the pons.

Summary of Vascular Anatomy

The above review of vascular anatomy is of significant clinical importance in clinical settings and greatly aids in differential diagnosis and lesion localization. The keen neurologist can determine where a hemorrhage or occlusion is located based on signs and symptoms and make critical decisions to patient care. Similarly, to the neuropsychologist, understanding these vascular territories and the underlying behavioral or cognitive functional anatomy can make all the difference in assessment techniques and interpretation of cognitive data from an assessment. Next, the imaging modalities discussed below provide the clinician with powerful tools to visualize and confirm suspected lesions. These techniques enhance diagnostic accuracy, guide treatment plans, and facilitate a clinician's interventions based on neuroanatomical findings.

Imaging Techniques

MRI and Magnetic Resonance Angiography and Venography

A detailed discussion of brain MRI is included in Chapter 2. Magnetic resonance angiography (MRA) utilizes MR procedures to visualize the vasculature of the head, neck, and brain.

MRI includes multiple imaging sequences designed to detect different aspects of neuropathology (Bigler, 2019). Each sequence depends on differences in the excitation pulses of strong magnetic fields separated by a repetition time (Bigler, 2019). MRI is contraindicated in patients with fragments of metal, for example, within the eyes and in patients with certain electrically or magnetically active implanted devices that could interact with the magnetic field (e.g., pacemakers). Additionally, claustrophobic patients may have difficulty tolerating the studies.

MRA is the MR technique that is used to image the large and medium vessels supplying the brain. It can be done with or without contrast and thus can still be performed in the case of contrast allergy or advanced kidney failure. MRA may be used to image intracranial and extracranial vessels and can help with diagnosing vessel stenosis, occlusion, aneurysm, or vascular malformation (Figure 3.3). Intracranially, phase contrast (PC) MRA may be performed in less than 2 minutes whereas time-of-flight (TOF) MRA provides more detailed anatomical images but takes more than 4 minutes for data acquisition. This may result in degraded motions in the event of movement or when scanning restless patients. In the neck, contrasted MRA of the carotid arteries usually leads to better imaging than non-contrast enhanced MRA. Provided the patient is able

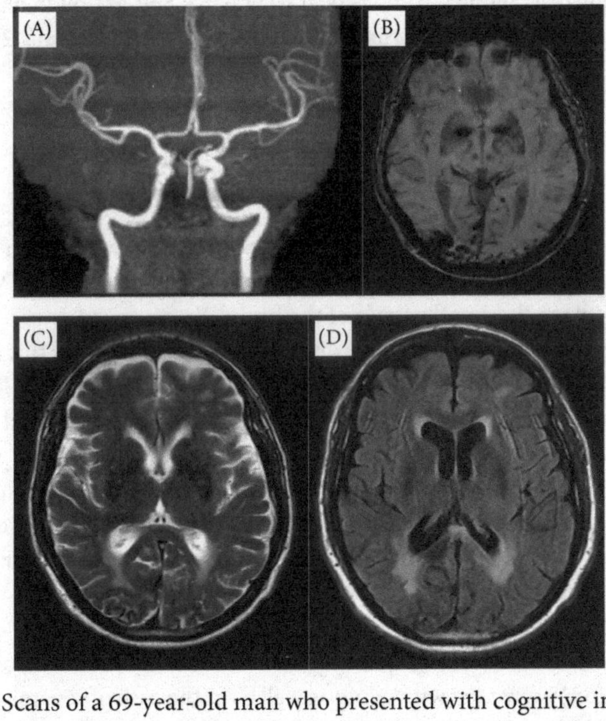

Figure 3.3 Scans of a 69-year-old man who presented with cognitive impairment and suspected amyloid angiopathy. MRA head without significant abnormality (A), SWI sequence showing microbleeds predominantly in a posterior and cortical distribution concerning for amyloid angiopathy (B), T2 imaging shows white matter disease assumed in this case to be from small vessel disease (C); FLAIR imaging shows white matter disease (D).

to cooperate, the addition of MRA to MRI brain makes it possible to image the major arteries of supply from the aortic arch to the circle of Willis in the same exam evaluating the brain parenchyma (Tress, 2003).

Magnetic resonance venography (MRV) is similar to MRA with arterial flow suppressed (Tress, 2003). It is often used to examine the intracranial venous system, particularly in the evaluation of dural sinus thrombosis (Ayanzen et al., 2000). However, transverse sinus flow gaps can be observed in as many as 31% of patients with normal MR imaging findings and should not be mistaken for dural sinus thrombosis (Ayanzen et al., 2000). MRV may be performed with or without contrast but phase contrast MRV is usually more reliable (Tress, 2003). Venous sinus thrombosis may lead to headaches, increased intracranial pressure, venous infarcts that don't necessarily follow arterial boundaries, and an increased risk of intracranial hemorrhage and seizures. Thrombosis of the deep venous system may lead to ischemia and edema involving both thalami and the upper brainstem, which may present acutely as altered level of consciousness, weakness, sensory changes, and vertical gaze palsy and are often accompanied by significant cognitive impairment and apathy (Kumral et al., 2001; Tress, 2003).

MR Imaging of Small Vessel Disease

Small vessel disease is another common cause of both ischemic and hemorrhagic strokes as well as neurocognitive impairment and dementia. The MRA is not adequate to image the small vessels of the brain.

While current MRI sequences commonly used in clinical practice do not provide direct imaging of the small vessels, several sequences of the MRI brain are commonly used as surrogates instead. The T2 sequence and fluid-attenuated inversion recovery (FLAIR) sequence are sensitive in detecting white matter pathology and dilated perivascular spaces. The FLAIR technique produces T2-weighted images while at the same time subtracting the normally bright cerebrospinal fluid (CSF), thus showing white matter changes more clearly (Figure 3.3) (Tress, 2003).

The gradient-recalled echo (GRE) and susceptibility-weighted imaging (SWI) sequences detect, localize, and quantify hemosiderin (a blood byproduct due to iron deposition) increased with aging as well as in the case of several small vessel disorders including in hypertensive small vessel disease and amyloid angiopathy (Figure 3.3) (Sullivan & Bigler, 2015). In addition, GRE and SWI imaging combined with other MRI sequences are also useful for imaging of small vascular malformations such cavernomas (Figure 3.4). See Chapter 2 for additional information on structural brain imaging modalities.

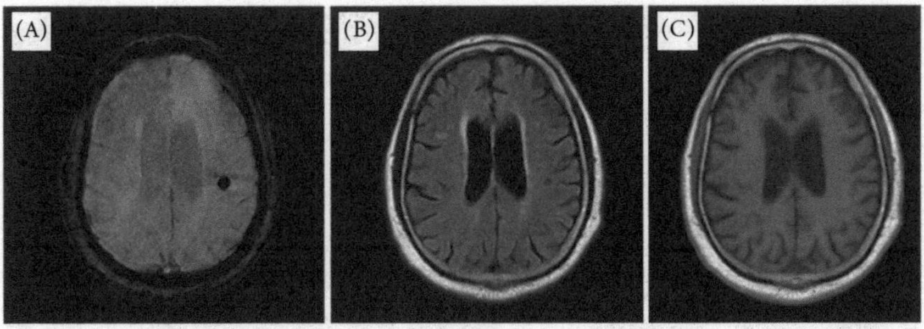

Figure 3.4 SWI MRI sequence (axial cuts) shows left postcentral gyrus cavernoma (A), FLAIR sequence (B) and T1-weighted sequence (C) showing the same cavernoma. Cavernomas may be asymptomatic or may be associated with headaches, focal neurological deficits, intracranial hemorrhage, and seizures.

Computed Tomography and Computed Tomography Angiography and Venography

The tomographic image obtained by CT is derived by detecting the passage of x-ray beams through tissue at different orientations, leading to image creation by computer reconstruction of a two-dimensional section of the tissue (Bigler, 2019; Wilde et al., 2012). The CT brain image reflects tissue density where the bone is bright white, gray matter is light gray, white matter is darker gray, and CSF is dark (Figures 3.5 and 3.6). CT brain is the most commonly used brain-imaging modality in emergency departments because it can usually be easily and rapidly obtained and has good overall sensitivity to detecting intracranial hemorrhage. Acute intracranial hemorrhage is hyperdense on CT brain without contrast (Figure 3.5) and becomes isodense and then hypodense with time. See Chapter 2 for additional information on structural brain imaging.

Computed tomographic angiography (CTA) is a rapid and reliable method for imaging the intracranial and extracranial vasculature. Data is acquired following the administration of an intravenous contrast agent. Reconstruction of this data produces images that can be displayed in three dimensions. CTA is able to delineate the site of vessel occlusion or stenosis as well as vessel tortuosity and relation of the vessels to other structures. It can be performed in patients who are critically ill or claustrophobic or who have metallic foreign bodies that preclude an MRI (e.g., pacemakers or spinal cord stimulators) (Barber & Davis, 2003). CTA is often obtained during acute stroke workup and is typically the modality of choice to diagnose acute and chronic vessel occlusions (Figure 3.6). CT venography is a CT technique to assess venous anatomy and is often used in the

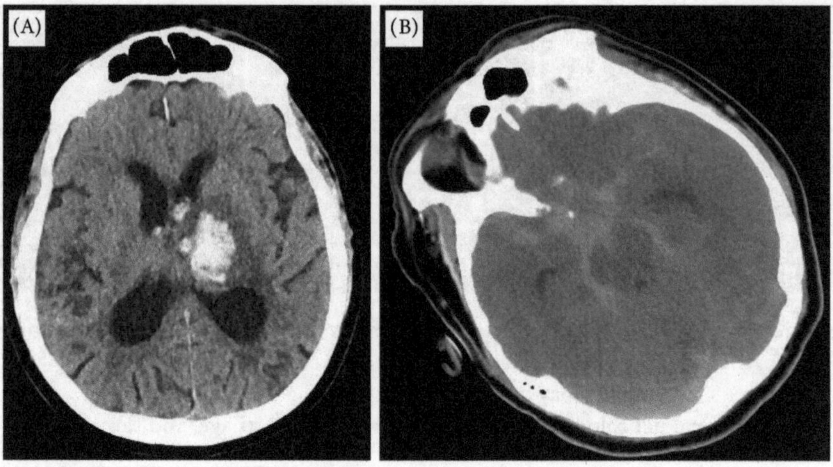

Figure 3.5 Left thalamic hemorrhage with intraventricular component (A). Subarachnoid hemorrhage associated with a ruptured right MCA aneurysm (B).

evaluation of venous sinus thrombosis. It has short acquisition time and may be done in combination with arterial phase CTA if needed.

CT Perfusion

Perfusion imaging uses an intravascular tracer and serial imaging to quantify blood flow through the brain parenchyma by following the change in Hounsfield number that occurs with the passage of an intravenous contrast agent through the cerebral vascular bed (Demeestere et al., 2020). The change in density is proportional to the concentration of the contrast agent and can therefore be used to calculate maps or tracer transient time and cerebral blood volume (Barber & Davis, 2003).

Acquiring high-quality perfusion maps requires a scanning protocol that is optimized for high contrast sensitivity and low image noise and also ensures that bolus passage is captured in full with adequate temporal resolution. For CTP, these requirements must be balanced against the need to minimize radiation dose (Demeestere et al., 2020). The fundamental hemodynamic properties of the tissue are cerebral blood volume, cerebral blood flow (CBF; the volume of blood flow per minute per 100 mL of tissue), and mean transit time (the average transit time for a tracer particle to traverse the capillary bed). In addition, the delay of the bolus from the proximal vasculature to the tissue (time-to-maximum [Tmax]) is a commonly used metric (Demeestere et al., 2020). For clinical use,

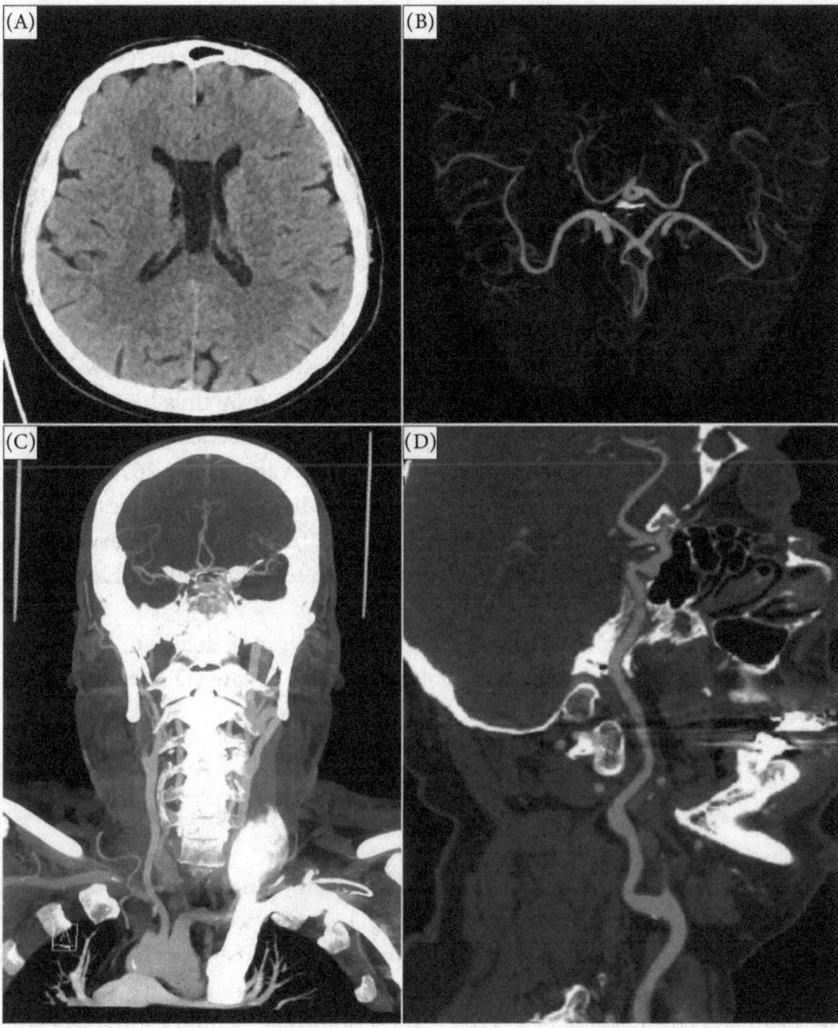

Figure 3.6 CT brain without contrast (axial cut) showing no acute pathology. Cavum vergae which is an anatomical variant of the septum pellucidum is noted (A), CTA reconstruction of the intracranial vessels shows no acute pathology in the intracranial vessels (B), carotid imaging in the coronal plane shows right carotid occlusion of unknown chronicity (C), follow-up CTA head and neck 6 months later shows complete recanalization of the right carotid occlusion (D) which was ultimately diagnosed as carotid dissection with complete recanalization. Patient had presented initially with neck pain and transient vision loss in the right eye.

the goal is often to quantify: (1) tissue that experiences significant hypoperfusion and is likely to infarct in the absence of reperfusion (often referred to as area at risk or penumbra) and (2) tissue that is likely irreversibly infarcted, often referred to as the ischemic core (Demeestere et al., 2020).

Multiple software products that produce perfusion maps and estimate volumes of ischemic core and tissue at risk are available (Figure 3.7). Occlusion of a cerebral blood vessel causes a variable decrease of blood flow in the downstream parenchyma proportionate to the degree of collateral circulation. In brain regions with poor collaterals, lack of oxygen and glucose may ultimately result in the failure of the cell's energy metabolism, leading to tissue infarction (Demeestere et al., 2020). In the absence of reperfusion, the penumbra will eventually grow into the ischemic core. The rate of infarct growth is highly variable between individuals and strongly dependents on the presence and structure of the collateral circulation (Demeestere et al., 2020; Vagal et al., 2018). Genetic and environmental factors likely explain interindividual differences between native collateral circulation (Demeestere et al., 2020). Furthermore, several stimuli (e.g., chronic hypoperfusion) may induce collateral formation (Liebeskind, 2003).

Lacunar or small subcortical infarcts are usually not detected using CTP thresholds, and sensitivity for infratentorial lesions is low (Demeestere et al., 2020; Rudilosso et al., 2015).

Digital Subtraction Angiography (DSA)

DSA permits a distinct visualization of the vasculature in a skeletal or dense soft tissue environment by providing real-time 2D subtraction imaging. The target tissue is initially exposed to x-ray or MRI, to obtain the first set of images; then a contrast agent is administered into the vasculature and additional x-ray or MRI are obtained. The mask, which is the first set of images, is then subtracted from the latter or contrast-enhanced images, allowing the visualization of the vessels free of the surrounding tissue (Figure 3.8). DSA is often considered to be the gold standard to visualize intracranial or extracranial vessels, vascular abnormalities such as arteriovenous malformations or fistulas, and aneurysm as well as grading Moyamoya disease and collateral flow (Carmody & Seeger, 1984; Hung et al., 2014; Laviña, 2016; Srinivasan et al., 2016; Wen et al., 2016; Wong et al., 2012). However, DSA is more invasive than other vascular studies and is typically performed by experienced operators including usually a neurointerventional trained physician. DSA of the cerebral vasculature carries a small risk of stroke in addition to other potential complications such as pseudoaneurysm or hematoma (Nam et al., 2022).

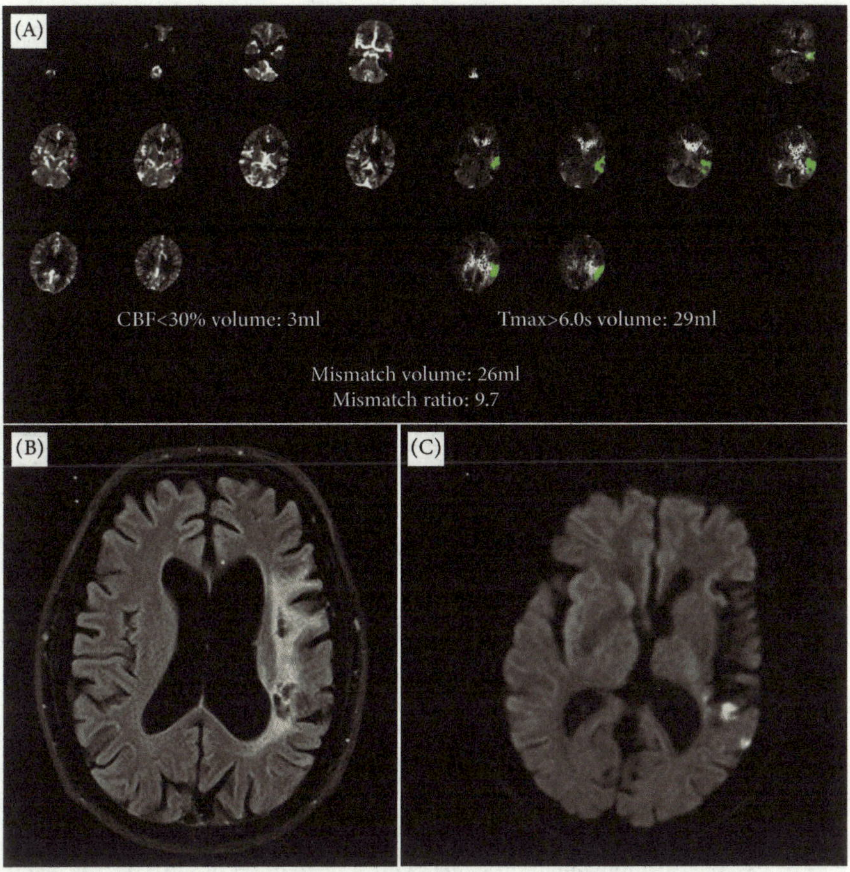

Figure 3.7 Acute left MCA stroke in a patient with previous history of left MCA stroke: Patient had a previous ischemic stroke with mild residual aphasia and sustained a recurrent ischemic stroke in the same vascular distribution. CT perfusion acutely showed a small area of irreversible tissue injury (CBF <30%) and a larger area of tissue at risk (difference between Tmax >6.0s volume and CBF <30% volume, which is the mismatch volume) (A). MRI brain shows new stroke on DWI (C) and old infarct with encephalomalacia in the left parietal-temporal area on FLAIR imaging (B). Patient subsequently had mild depressive and anxiety symptoms and cognitive impairment with difficulty talking and with speech processing. His Montreal Cognitive Assessment (MoCA) was 20/30 and mostly driven by language areas (language and verbal cues). He was referred to work with a speech therapist.

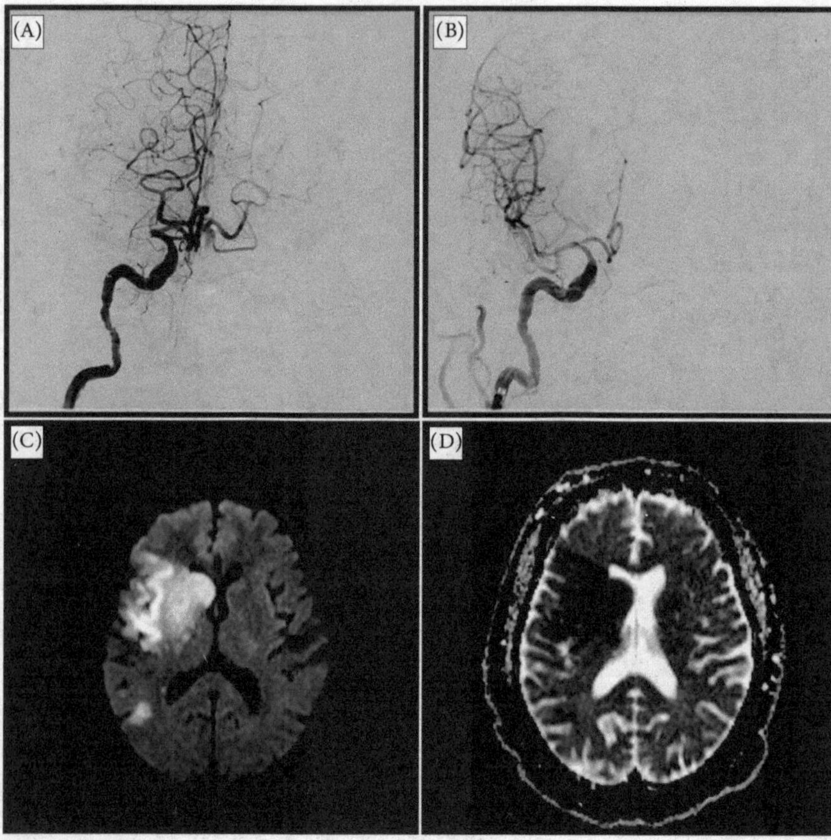

Figure 3.8 62-year-old woman presented with sudden onset of slurred speech and plegia of the left face, arm, and leg. On examination, her eyes were deviated to the right and she was also found to have decreased sensation on the left and left heminelgect. Based on her clinical exam, an ischemic stroke in the right MCA distribution was suspected. Thrombolysis with alteplase was administered within 4.5 hours of clinical presentation. A CTA angiogram acutely showed a right M1 occlusion and she was taken for thrombectomy. A DSA confirmed the right M1 occlusion (A). Endovascular thrombectomy was performed with a TICI 2B (i.e., complete filling of all of the expected vascular territory is visualized but the filling is slower than normal) reperfusion (B). Subsequent MRI brain DWI (C) with corresponding ADC sequences (D) confirmed ischemic changes in the right MCA distribution. Her left-sided weakness improved during her hospitalization and she was discharged initially to inpatient rehabilitation and subsequently went home with additional outpatient therapy.

Ultrasound-Based Techniques: Carotid and Transcranial Doppler Ultrasound (TCD)

Ultrasound testing is a noninvasive technique which uses ultrasound waves to evaluate the carotid arteries (carotid ultrasound) and the intracranial arteries (transcranial Doppler, TCD). Carotid ultrasound evaluates the presence of carotid stenosis and carotid plaques and helps in determining the hemodynamic effect of the stenosis (Figure 3.9). It may also be used to sequentially follow up on carotid stenosis.

TCD ultrasound uses a low-frequency (≤2 MHz) transducer probe to direct ultrasound waves (i.e., insonate) through specific thin regions of the skull, allowing for assessment of cerebral blood flow. TCD enables users to acquire images of some of the major intracranial vessels through the intact skull and monitor cerebral blood flow (CBF) velocity and vessel pulsatility over extended time periods (Laviña, 2016; Naqvi et al., 2013; Purkayastha & Sorond, 2012). TCD has multiple different applications and may be used to evaluate for vasospasm, hemodynamic effect of vessel stenosis, screen for sickle cell vasculopathy in children, determine the risk of embolization by evaluating for microembolic signals, evaluate for paradoxical embolism in stroke workup, evaluate for cerebrovascular reserve, and evaluate for changes in CBF with provoking head maneuvers (Piepgras et al., 1990; Pindzola et al., 2001; Purkayastha & Sorond, 2012).

TCD is often used to evaluate for vasospasm of the arteries following subarachnoid hemorrhage (SAH). Delayed cerebral ischemia which may be partially mediated by vasospasm has been associated with an increased risk of cognitive impairment after SAH (El Husseini et al., 2023; Stienen et al., 2014). TCD may also be used to evaluate the hemodynamic effect of intracranial stenosis or the hemodynamic effect of more proximal stenosis. In a study of 42 participants with 80% or more unilateral carotid stenosis, cognitive impairment has been found to correlate linearly with decreased middle cerebral artery mean flow velocity below a definable threshold by TCD (Marshall et al., 2020). Other uses for TCD include screening for sickle cell vasculopathy. The Stroke Prevention Trial in Sickle Cell Anemia (STOP) showed that exchange transfusion decreased the risk of initial stroke in children with sickle-cell anemia determined to be high risk as determined by TCD (Lee et al., 2006).

Positron Emission Tomography (PET) and Single-Photon Emission Computed Tomography (SPECT)

PET and SPECT are molecular imaging modalities and two nuclear medicine techniques that use specific radioactive tracers to allow visualization and measurement of physiological processes in intact living brains

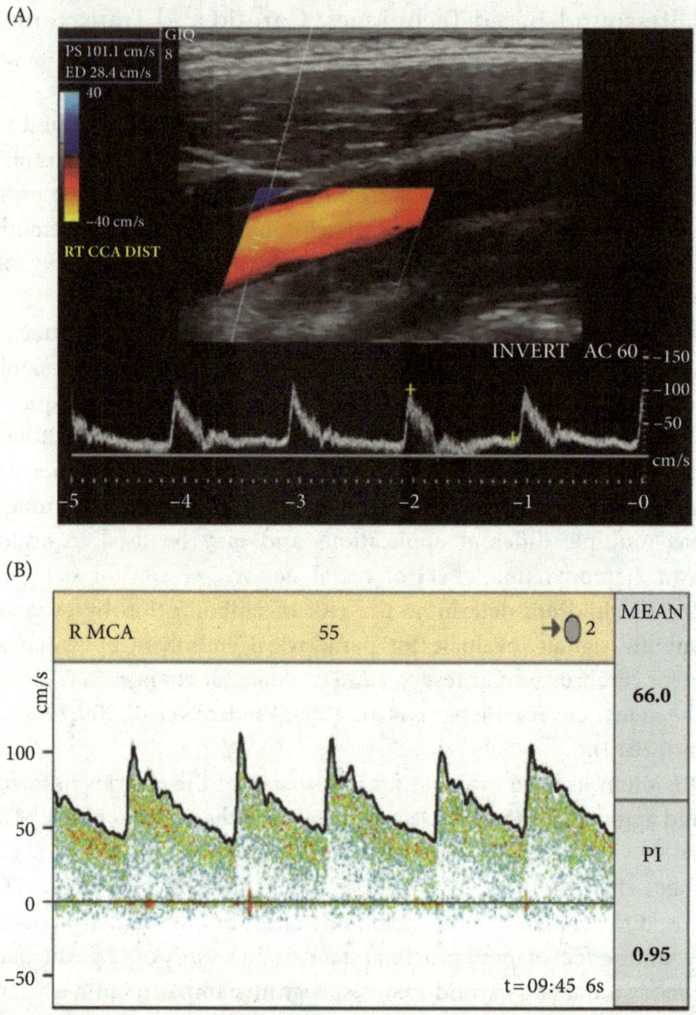

Figure 3.9 Output from carotid ultrasound that includes B-mode imaging, power Doppler, and spectral Doppler from the common carotid artery (A). Spectral Doppler output from transcranial Doppler of the right MCA showing the mean velocity and pulsatility index (PI) (B).

(Goffin & van Laere, 2016; Laviña, 2016). The main difference between SPECT and PET scans is the type of radiotracers used. The routine clinical use of PET studies in cerebrovascular disorders remains limited.

PET involves injecting rapidly decaying radioactive substances into the bloodstream. PET functional imaging is most useful in combination with anatomical imaging and hybrid imaging systems such as PET/MRI and PET/CT (Laviña, 2016). PET imaging in cerebrovascular disease has multiple different applications and has had a great impact on cerebrovascular research (Heiss, 2014). For example, it can be used to assess cerebrovascular reserve as well as the effect of stroke on the surrounding and contralateral primarily unaffected tissue in order to understand the symptoms caused by disturbances in functional networks (Heiss, 2014). It may also be used to detect metabolic disturbances in the setting of cognitive impairment associated with cerebrovascular disease (Heiss, 2014).

SPECT imaging in Moyamoya disease may more accurately identify affected areas of the brain than CT (Mountz et al., 1988). In addition, acetazolamide-stressed brain perfusion SPECT scan may help in determining vascular reserve in patients with vasculopathy such as Moyamoya disease.

Vascular Imaging and Neuropsychology

Both structural (see Chapter 2) and vascular imaging are of interest to the clinical neuropsychologist. This is because cognitive deficits are common post-stroke and in chronic cerebrovascular disease. Post-stroke cognitive impairment occurs in up to 60% of stroke survivors in the first year after stroke and ranges from mild to severe (El Husseini et al., 2023). Even among people with "good functional recovery" (modified Rankin Scale of 0–1), 71% continue to have impaired cognition in at least one cognitive domain up to 15 months after stroke, typically in the domains of memory or visual-spatial function (Jokinen et al., 2015). Interestingly, in this study, the mini-mental status exam (MMSE) was often normal (Jokinen et al., 2015). Only 6% had scores <24 and 31% had scores <27. This implies that many people with normal functional recovery and normal MMSE scores continue to have impaired cognitive function. Similarly, Planton et al. (2012) showed that after recovery of motor, sensory-perceptual, linguistic, and general functions, first-episode ischemic stroke patients scored lower than controls in all cognitive domains and had higher rates of apathy (109 days post-injury) (Planton et al., 2012). Like the findings from Jokinen et al. (2015), another study examined 136 acute stroke patients an average of 11 days post-stroke and showed that 30 (22%) had normal scores on the Montreal Cognitive Assessment (MoCA; normal defined as ≥ 25). However, even among those with "normal" MoCA scores, 77% have cognitive impairment in at least one domain. Among those with perfect MoCA scores (30/30), 59% had impaired attention in more formal testing and 35% had memory deficits (Chan et al., 2014). Taken together, neuropsychological evaluation is a complementary assessment in the care of stroke patients and can reveal ongoing deficits not routinely observed in other assessments.

Additionally, neuropsychiatric symptoms are also common, with 24%–60% of patients experiencing depression, 29% with anxiety, and 36% with apathy (Ferro et al., 2016). Neuropsychological assessment helps predict outcomes post-stroke. Revet et al. (2021) demonstrated that neuropsychological assessment is helpful for rehabilitation goal-setting, treatment strategies, discharge planning, patient education, and reduced healthcare costs (Revet et al., 2021). Aphasia and hemineglect also predict functional outcomes. Additionally, neuropsychologists usually provide extensive feedback on the test results, the nature of the condition, and best course of action to improve or maintain cognitive health. Studies show that neuropsychological assessment feedback can improve or increase all of the following outcomes: perceived everyday cognitive functioning; self-efficacy or sense of control; involvement in patient's own care and coping strategies; stress and depression about cognition; patient understanding of treatment goals; engagement in rehabilitation; and greater initiation of safety behaviors (Gruters et al., 2022; Pegg et al., 2005; Longley et al., 2023; Vandermorris et al., 2021).

Vascular Health and Cognition

There is a significant association between vascular health and cognitive function. Several studies showed that decreased cerebral perfusion impacts cognitive function. Wang and colleagues (2017) used a multimodal approach to understand the impact of decreased vascular perfusion among individuals with asymptomatic carotid artery stenosis (Wang et al., 2017). Asymptomatic stenosis implies decreased flow but without stroke or TIA. They showed that patients (vs. controls) had worse global cognition, memory, and processing speed. Patients also displayed decreased cerebral blood flow to left inferior frontal regions, and magnetic resonance spectroscopy (MRS) demonstrated the difference in concentrations of N-acetyl-aspartate and creatine ratio in the left hippocampus. Patients also had decreased functional connectivity to regions belonging to the default mode network (DMN) (Wang et al., 2017). Additionally, a 2020 meta-analysis assessed outcomes after carotid artery endarterectomy across 60 studies and over 4,800 patients (Aceto et al., 2020). This meta-analysis demonstrated that approximately 20% have delayed neurocognitive recovery but most demonstrate improved cognitive function after 3 months (Aceto et al., 2020). Additional studies have also demonstrated improved function after either endarterectomy or carotid stent placement. Cheng et al. (2012) studied asymptomatic groups with no difference in the medial temporal lobe volume (Cheng et al., 2012; Velickaite et al., 2020) and demonstrated that participants with a carotid stenosis of ≥70% stenosis had decreased fractional anisotropy (FA) on diffusion tensor imaging (DTI) and decreased attention and memory (Cheng et al., 2012). The sample demonstrated a small but significant improvement in cognition and FA with revascularization (Cheng et al., 2012). Additionally, Porcu et al. (2022) showed that carotid endarterectomy resulted in improved cognition and resting-state fMRI regional neural activation in multiple areas (Porcu et al., 2022).

Whooley et al. (2020) examined a sample of older adults with a mean age of 71 (n =75; 63% male) versus controls (no cognitive difference at baseline) (Whooley et al., 2020). The sample included those with asymptomatic carotid artery stenosis, with ≥80% stenosis on CTA or MRA. The sample underwent revascularization via carotid endarterectomy or carotid stent and were followed for 12 months. The control group included patients with stenosis as well, but who received standard medical management rather than surgery. At follow-up, those with revascularization had substantially greater internal carotid artery flow (68% of surgical sample versus 9% of control). There was also greater MCA flow on the ipsilateral side (26%). Those who underwent revascularization also displayed improved executive function, and greater improvement in blood flow as correlated with greater gains on executive tasks relative to those who underwent conservative medical treatment (Whooley et al., 2020).

In summary, it is well known that the integrity of the cerebrovascular system is central to brain health and that attenuated vascular flow, microvascular disease, stroke, or other perfusion deficits will have a deleterious impact on neuropsychological, psychiatric, motor, and sensory outcomes. Vascular brain imaging can aid in differential diagnosis and treatment planning.

Case 1 Presentation

A 35-year-old woman presented to the emergency room following increased eye pain and intermittent double vision (diplopia). She reported 2 years of intermittent headaches, with pain behind her left eye. Six months before her presentation, she began to have drooping of her left eyelid and dilation of her left pupil. She reported that all symptoms gradually worsened over time. Due to her symptoms, she discontinued work as an office manager because she had difficulty using her computer screen and double vision increased her headaches and made it difficult to drive. She has a long-standing history of depression which she reports is unchanged, but she has had increased anxiety over the last 6 months. Gait was normal. General motor and sensory examinations were unremarkable.

Assessment of cranial nerves

- CN I: Deferred
- CN II: WNL
- CN III, IV, & VI: Right pupil was responsive to light and with normal accommodation. Left pupil appeared to have reduced response to light and no accommodation. Right eye movements were Within Normal Limits. Left was reduced for upward and downward gaze and adduction.
- CN V, & VII–XII: Unremarkable.

Imaging:

- MRI revealed a 1.4 cm mass near the posterior communicating artery.
- CTA confirmed this as an aneurysm adjacent to the third cranial nerve.

Course: Neurosurgery was consulted, and the aneurysm was clipped and the blood was evacuated from the aneurysm. Six months post-surgery, the patient reported that her headaches and double vision had abated. The follow-up cranial nerve examination was essentially unremarkable. However, she entered cognitive behavior therapy for ongoing mood symptoms, and she resumed employment at half-time hours.

Case 2 Presentation

A 69-year-old man with 12 years of education was referred for follow-up neuropsychological assessment in the context of a prior diagnosis of mild cognitive impairment (MCI) and family report of continued declines in cognitive function and daily functioning. His premorbid history is significant for peripheral artery disease, atrial fibrillation, hypertension, and mild sleep apnea. He had a prior smoking history of 25 years, which was discontinued 5 years earlier. He presented to his primary care 6 months ago because his spouse reported the patient had ongoing declines in attention and concentration. He also seemed to have difficulty with planning and organizing, difficulty managing his medications, and difficulties with quick reaction times when driving. He was referred to neurology and neuropsychology. A prior neuropsychological assessment 26 months earlier documented difficulties with processing speed and multitasking, as well as reduced verbal fluency and difficulty with free recall on memory testing, with normal recognition. He was diagnosed with MCI at the prior assessment because he was still able to complete all activities of daily living (ADLs). Three months before the follow-up neuropsychological assessment, he had acute onset of aphasia and confusion, lasting approximately 30 minutes. He was taken to the emergency room and symptoms abated by the time he was evaluated. An MRI was performed during his ER visit, which revealed advanced parenchymal atrophy for age without any specific lobar predominance, as well as severe chronic microvascular white matter disease. The radiologist assigned a Fazekas score of 3, which suggested severe confluent white matter burden. There were no apparent acute findings to suggest an acute stroke. He was diagnosed with a transient ischemic attack (TIA). During a subsequent outpatient neurological exam, his performance on the Montreal Cognitive Assessment (MoCA) was 18/30. MRA was ordered, which revealed >80% bilateral ICA stenosis. Relative to the baseline neuropsychological assessment, he demonstrated a reduction in timed tests, general measures of executive functioning, and verbal fluency. He also demonstrated mild declines in memory performance, but he was not amnestic, being able to recognize information when given cues. Due to the decline in functioning and cognition, he was diagnosed with dementia, which appeared to be associated with his vascular health.

Conclusion

This chapter highlights the critical role of vascular imaging in enhancing the neuropsychological assessment and care of individuals with cerebrovascular disease. By understanding the intricate vascular anatomy and utilizing advanced imaging modalities, clinicians can better localize lesions and recognize patterns of cognitive and psychiatric impairment associated with cerebrovascular health. Vascular imaging techniques such as MRA, CTA, DSA, and TCD, among others, are invaluable in refining differential diagnoses, guiding treatment strategies, and monitoring patient progress over time. As demonstrated in the cases presented, neuropsychologists benefit from integrating vascular imaging findings into cognitive evaluations, enabling more precise assessments, improving diagnostic accuracy, and ultimately enhancing patient-centered care. This integration supports a multidisciplinary approach that is essential for comprehensive management of cerebrovascular conditions and their neurocognitive outcomes.

References

Aceto, P., Lai, C., De Crescenzo, F., Crea, M. A., Di Franco, V., Pellicano, G. R., Perilli, V., Lai, S., Papanice, D., & Sollazzi, L. (2020). Cognitive decline after carotid endarterectomy: systematic review and meta-analysis. *European Journal of Anaesthesiology*, *37*(11), 1066–1074. https://doi.org/10.1097/EJA.0000000000001130

Ayanzen, R. H., Bird, C. R., Keller, P. J., McCully, F. J., Theobald, M. R., & Heiserman, J. E. (2000). Cerebral MR venography: Normal anatomy and potential diagnostic pitfalls. *American Journal of Neuroradiology, 21*(1), 74–78.

Barber, P. A., & Davis, S. M. (2003). Limitations of current brain imaging modalities in stroke. In S. Davis, M. Fisher, & S. Warach (Eds.), *Magnetic Resonance Imaging in Stroke* (pp. 15–30). Cambridge University Press.

Bigler, E. D. (2019). Neuroimaging and neuropsychology. In K. M. Sanders (Ed.), *Physician's Field Guide to Neuropsychology: Collaboration Through Case Example* (pp. 421–434). Springer New York. https://doi.org/10.1007/978-1-4939-8722-1_25

Carmody, R. F., & Seeger, J. F. (1984). Intracranial applications of digital subtraction angiography. *Critical Reviews in Diagnostic Imaging, 23*(1), 1–40.

Chan, E., Khan, S., Oliver, R., Gill, S. K., Werring, D. J., & Cipolotti, L. (2014). Underestimation of cognitive impairments by the Montreal Cognitive Assessment (MoCA) in an acute stroke unit population. *Journal of the Neurological Sciences, 343*(1–2), 176–179. https://doi.org/10.1016/j.jns.2014.05.005

Cheng, H.-L., Lin, C.-J., Soong, B.-W., Wang, P.-N., Chang, F.-C., Wu, Y.-T., Chou, K.-H., Lin, C.-P., Tu, P.-C., & Lee, I.-H. (2012). Impairments in cognitive function and brain connectivity in severe asymptomatic carotid stenosis. *Stroke, 43*(10), 2567–2573. https://doi.org/10.1161/STROKEAHA.111.645614

Demeestere, J., Wouters, A., Christensen, S., Lemmens, R., & Lansberg, M. G. (2020). Review of perfusion imaging in acute ischemic stroke: From time to tissue. *Stroke, 51*(3), 1017–1024. https://doi.org/10.1161/STROKEAHA.119.028337

de Sousa, C. S. M., de Miranda, C. L. V. M., Avelino, M. C., & Bastos, B. B. (2017). Thrombosis of the azygos anterior cerebral artery. *Case Reports in Radiology*, 2017, 5409430. https://doi.org/10.1155/2017/5409430

El Husseini, N., Katzan, I. L., Rost, N. S., Blake, M. L., Byun, E., Pendlebury, S. T., Aparicio, H. J., Marquine, M. J., Gottesman, R. F., Smith, E. E., & American Heart Association Stroke Council; Council on Cardiovascular and Stroke Nursing; Council on Cardiovascular Radiology and Intervention; Council on Hypertension; and Council on Lifestyle and Cardiometabolic Health. (2023). Cognitive impairment after ischemic and hemorrhagic stroke: A scientific statement from the American Heart Association/American Stroke Association. *Stroke, 54*(6), e272–e291. https://doi.org/10.1161/STR.0000000000000430

Feekes, J. A., & Cassell, M. D. (2006). The vascular supply of the functional compartments of the human striatum. *Brain: A Journal of Neurology, 129*(Pt 8), 2189–2201. https://doi.org/10.1093/brain/awl158

Ferro, J. M., Caeiro, L., & Figueira, M. L. (2016). Neuropsychiatric sequelae of stroke. *Nature Reviews. Neurology, 12*(5), 269–280. https://doi.org/10.1038/nrneurol.2016.46

Goffin, K., & van Laere, K. (2016). Single-photon emission tomography. *Handbook of Clinical Neurology, 135*, 241–250. https://doi.org/10.1016/B978-0-444-53485-9.00013-1

Gruters, A. A., Ramakers, I. H., Verhey, F. R., Kessels, R. P., & de Vugt, M. E. (2022). A scoping review of communicating neuropsychological test results to patients and family members. *Neuropsychology Review, 32*(2), 294–315. https://doi.org/10.1007/s11065-021-09507-2

Heiss, W.-D. (2014). PET imaging in ischemic cerebrovascular disease: Current status and future directions. *Neuroscience Bulletin, 30*(5), 713–732. https://doi.org/10.1007/s12264-014-1463-y

Hung, S.-C., Liang, M.-L., Lin, C.-F., Lin, C.-J., Guo, W.-Y., Chang, F.-C., Wong, T.-T., & Chang, C.-Y. (2014). New grading of Moyamoya disease using color-coded parametric quantitative digital subtraction angiography. *Journal of the Chinese Medical Association, 77*(8), 437–442. https://doi.org/10.1016/j.jcma.2014.05.007

Jokinen, H., Melkas, S., Ylikoski, R., Pohjasvaara, T., Kaste, M., Erkinjuntti, T., & Hietanen, M. (2015). Post-stroke cognitive impairment is common even after successful clinical recovery. *European Journal of Neurology, 22*(9), 1288–1294. https://doi.org/10.1111/ene.12743

Kalaria, R. N. (2010). Vascular basis for brain degeneration: faltering controls and risk factors for dementia. *Nutrition Reviews, 68*(Suppl 2), S74–87. https://doi.org/10.1111/j.1753-4887.2010.00352.x

Kobayashi, S., Maki, T., & Kunimoto, M. (2011). Clinical symptoms of bilateral anterior cerebral artery territory infarction. *Journal of Clinical Neuroscience, 18*(2), 218–222. https://doi.org/10.1016/j.jocn.2010.05.013

Kumral, E., Evyapan, D., Balkir, K., & Kutluhan, S. (2001). Bilateral thalamic infarction: Clinical, etiological and MRI correlates. *Acta Neurologica Scandinavica, 103*(1), 35–42. https://doi.org/10.1034/j.1600-0404.2001.00141.x

Laviña, B. (2016). Brain vascular imaging techniques. *International Journal of Molecular Sciences, 18*(1). https://doi.org/10.3390/ijms18010070

Lee, M. T., Piomelli, S., Granger, S., Miller, S. T., Harkness, S., Brambilla, D. J., Adams, R. J., & STOP Study Investigators. (2006). Stroke Prevention Trial in Sickle Cell Anemia (STOP): Extended follow-up and final results. *Blood, 108*(3), 847–852. https://doi.org/10.1182/blood-2005-10-009506

Liebeskind, D. S. (2003). Collateral circulation. *Stroke, 34*(9), 2279–2284. https://doi.org/10.1161/01.STR.0000086465.41263.06

Longley, W. A., Tate, R. L., & Brown, R. F. (2023). The psychological benefits of neuropsychological assessment feedback as a psycho-educational therapeutic intervention: A randomized-controlled trial with cross-over in multiple sclerosis. *Neuropsychological Rehabilitation, 33*(5), 764–793. https://doi.org/10.1080/09602011.2022.2047734

Madhuripan, N., Atar, O. D., Zheng, R., & Tenenbaum, M. (2017). Computed tomography angiography in head and neck emergencies. *Seminars in Ultrasound, CT, and MR, 38*(4), 345–356. https://doi.org/10.1053/j.sult.2017.02.003

Marshall, R. S., Pavol, M. A., Cheung, Y. K., Asllani, I., & Lazar, R. M. (2020). Cognitive impairment correlates linearly with mean flow velocity by transcranial Doppler below a definable threshold. *Cerebrovascular Diseases Extra, 10*(1), 21–27. https://doi.org/10.1159/000506924

Mountz, J. M., Foster, N. L., Ackermann, R. J., Bluemlein, L., Petry, N. A., & Kuhl, D. E. (1988). SPECT imaging of moyamoya disease using 99mTc-HM-PAO: Comparison with computed tomography findings. *Journal of Computed Tomography, 12*(4), 247–250. https://doi.org/10.1016/0149-936x(88)90077-x

Nam, H. H., Jang, D. K., & Cho, B. R. (2022). Complications and risk factors after digital subtraction angiography: 1-year single-center study. *Journal of Cerebrovascular and Endovascular Neurosurgery, 24*(4), 335–340. https://doi.org/10.7461/jcen.2022.E2022.05.001

Naqvi, J., Yap, K. H., Ahmad, G., & Ghosh, J. (2013). Transcranial Doppler ultrasound: A review of the physical principles and major applications in critical care. *International Journal of Vascular Medicine, 2013*, 629378. https://doi.org/10.1155/2013/629378

Pegg, P. O., Jr., Auerbach, S. M., Seel, R. T., Buenaver, L. F., Kiesler, D. J., & Plybon, L. E. (2005). The impact of patient-centered information on patients' treatment satisfaction and outcomes in traumatic brain injury rehabilitation. *Rehabilitation Psychology, 50*(4), 366–374. https://doi.org/10.1037/0090-5550.50.4.366

Piepgras, A., Schmiedek, P., Leinsinger, G., Haberl, R. L., Kirsch, C. M., & Einhäupl, K. M. (1990). A simple test to assess cerebrovascular reserve capacity using transcranial Doppler sonography and acetazolamide. *Stroke, 21*(9), 1306–1311. https://doi.org/10.1161/01.str.21.9.1306

Pindzola, R. R., Balzer, J. R., Nemoto, E. M., Goldstein, S., & Yonas, H. (2001). Cerebrovascular reserve in patients with carotid occlusive disease assessed by stable xenon-enhanced CT cerebral blood flow and transcranial Doppler. *Stroke, 32*(8), 1811–1817. https://doi.org/10.1161/01.str.32.8.1811

Planton, M., Peiffer, S., Albucher, J. F., Barbeau, E. J., Tardy, J., Pastor, J., Januel, A. C., Bezy, C., Lemesle, B., Puel, M., Demonet, J. F., Chollet, F., & Pariente, J. (2012). Neuropsychological outcome after a first symptomatic ischaemic stroke with "good recovery." *European Journal of Neurology, 19*(2), 212–219. https://doi.org/10.1111/j.1468-1331.2011.03450.x

Porcu, M., Cocco, L., Cau, R., Suri, J. S., Mannelli, L., Yang, Q., Defazio, G., Wintermark, M., & Saba, L. (2022). The mid-term effects of carotid endarterectomy on cognition and regional neural activity analyzed with the amplitude of low frequency fluctuations technique. *Neuroradiology, 64*(3), 531–541. https://doi.org/10.1007/s00234-021-02815-7

Purkayastha, S., & Sorond, F. (2012). Transcranial Doppler ultrasound: Technique and application. *Seminars in Neurology, 32*(4), 411–420. https://doi.org/10.1055/s-0032-1331812

Revet, M., Immerzeel, J., Voogt, L., & Paulis, W. (2021). Patients with neuropsychological disorders short after stroke have worse functional outcome: A systematic review and meta-analysis. *Disability and Rehabilitation, 43*(16), 2233–2252. https://doi.org/10.1080/09638288.2019.1693642

Rudilosso, S., Urra, X., San Román, L., Laredo, C., López-Rueda, A., Amaro, S., Oleaga, L., & Chamorro, Á. (2015). Perfusion deficits and mismatch in patients with acute lacunar infarcts studied with whole-brain CT perfusion. *American Journal of Neuroradiology, 36*(8), 1407–1412. https://doi.org/10.3174/ajnr.A4303

Schaeffer, S., & Iadecola, C. (2021). Revisiting the neurovascular unit. *Nature Neuroscience, 24*(9), 1198–1209. https://doi.org/10.1038/s41593-021-00904-7

Shaban, A., Albright, K. C., Boehme, A. K., & Martin-Schild, S. (2013). Circle of Willis variants: Fetal PCA. *Stroke Research and Treatment, 2013*, 105937. https://doi.org/10.1155/2013/105937

Srinivasan, V. M., Chintalapani, G., Duckworth, E. A. M., & Kan, P. (2016). Application of 4-dimensional digital subtraction angiography for dural arteriovenous fistulas. *World Neurosurgery, 96*, 24–30. https://doi.org/10.1016/j.wneu.2016.05.021

Stienen, M. N., Smoll, N. R., Weisshaupt, R., Fandino, J., Hildebrandt, G., Studerus-Germann, A., & Schatlo, B. (2014). Delayed cerebral ischemia predicts neurocognitive impairment following aneurysmal subarachnoid hemorrhage. *World Neurosurgery, 82*(5), e599–605. https://doi.org/10.1016/j.wneu.2014.05.011

Sullivan, E. V., & Bigler, E. D. (2015). Neuroimaging's role in neuropsychology: Introduction to the special issue of neuropsychology review on neuroimaging in neuropsychology. *Neuropsychology Review, 25*(3), 221–223. https://doi.org/10.1007/s11065-015-9296-7

Tress, B. M. (2003). Clinical use of standard MRI. In S. Davis, M. Fisher, & S. Warach (Eds.), *Magnetic Resonance Imaging in Stroke* (pp. 69–84). Cambridge University Press. https://doi.org/10.1017/CBO9780511544927.007

Vagal, A., Aviv, R., Sucharew, H., Reddy, M., Hou, Q., Michel, P., Jovin, T., Tomsick, T., Wintermark, M., & Khatri, P. (2018). Collateral clock is more important than time clock for tissue fate. *Stroke, 49*(9), 2102–2107. https://doi.org/10.1161/STROKEAHA.118.021484

Vandermorris, S., Cochrane, M., & Troyer, A. K. (2021). Understanding the client and family experience of neuropsychological evaluation. *Professional Psychology: Research and Practice, 52*(3), 226–233. https://doi.org/10.1037/pro0000385

Velickaite, V., Ferreira, D., Lind, L., Ahlström, H., Kilander, L., Westman, E., & Larsson, E.-M. (2020). Visual rating versus volumetry of regional brain atrophy and longitudinal changes over a 5-year period in an elderly population. *Brain and Behavior, 10*(7), e01662. https://doi.org/10.1002/brb3.1662

Wang, T., Xiao, F., Wu, G., Fang, J., Sun, Z., Feng, H., Zhang, J., & Xu, H. (2017). Impairments in brain perfusion, metabolites, functional connectivity, and cognition in severe asymptomatic carotid stenosis patients: An integrated MRI study. *Neural Plasticity, 2017*, 8738714. https://doi.org/10.1155/2017/8738714

Wen, W., Fang, Y., Yang, P., Zhang, Y., Wu, Y., Shen, H., Ge, J., Xu, Y., Hong, B., Huang, Q., & Liu, J. (2016). Parametric digital subtraction angiography imaging for the objective grading of collateral flow in acute middle cerebral artery occlusion. *World Neurosurgery, 88*, 119–125. https://doi.org/10.1016/j.wneu.2015.12.084

Whooley, J. L., David, B. C., Woo, H. H., Hoh, B. L., Raftery, K. B., Hussain Siddiqui, A., Westerveld, M., Amin-Hanjani, S., & Ghogawala, Z. (2020). Carotid revascularization and its effect on cognitive function: A prospective nonrandomized multicenter clinical study. *Journal of Stroke and Cerebrovascular Diseases, 29*(5), 104702. https://doi.org/10.1016/j.jstrokecerebrovasdis.2020.104702

Wilde, E. A., Hunter, J. V., & Bigler, E. D. (2012). A primer of neuroimaging analysis in neurorehabilitation outcome research. *NeuroRehabilitation, 31*(3), 227–242. https://doi.org/10.3233/NRE-2012-0793

Wong, S. C., Nawawi, O., Ramli, N., & Abd Kadir, K. A. (2012). Benefits of 3D rotational DSA compared with 2D DSA in the evaluation of intracranial aneurysm. *Academic Radiology, 19*(6), 701–707. https://doi.org/10.1016/j.acra.2012.02.012

4
The Role of Systemic Imaging in Neuropsychology

Thomas A. Nelson, Jorg Dietrich, and Michael Parsons

Introduction

Prior to evaluation by a neuropsychologist, patients with neurocognitive or neurobehavioral concerns often have undergone extensive diagnostic testing including head imaging. In some circumstances, a variety of laboratory evaluations, electroencephalography, and systemic imaging also may have been obtained. As the role of the neuropsychologist in the diagnostic process varies by clinical setting, it is worthwhile to understand the utility of various imaging modalities as they relate to neuropsychology practice. Furthermore, insight into the indications and applications of these studies will help with interpretation of their results, which subsequently can inform the recommendations made both to the patient and consulting providers.

This chapter focuses on the role of whole-body imaging in the diagnostic workflow of neuropsychologic disorders. There is discussion regarding workup of a single or multiple brain lesions; the use of systemic imaging in paraneoplastic syndromes; and the role of systemic positron emission tomography (PET) within neuropsychology.

Whole-Body Imaging in Evaluation of Malignancy

A challenging, albeit common, clinical scenario within neurology, psychiatry, and neuropsychology is the role of additional, often systemic, imaging in the workup of a new brain lesion. This is especially challenging with solitary lesions for which various etiologies can be considered. The quandary involves both the indication for further testing and what type of imaging to obtain—such as computed tomography (CT) of the chest, abdomen, and pelvis with or without a PET study—particularly when the lesion does not correspond to noted physical exam findings, i.e., it is found incidentally. These decisions are often guided

Thomas A. Nelson, Jorg Dietrich, and Michael Parsons, *The Role of Systemic Imaging in Neuropsychology*.
In: *Medical and Laboratory Examinations for Neuropsychologists*. Edited by: Thomas J. Farrer and R. Brock Frost, Oxford University Press. © Oxford University Press (2026). DOI: 10.1093/9780197619537.003.0005

by patient demographics, imaging characteristics, and the initial reason for obtaining head CT or magnetic resonance imaging (MRI). In some circumstances, no additional testing is immediately indicated, but repeat head imaging would be indicated at an interval concordant with clinical suspicion of the suspected etiology.

As discussed in Chapters 2 and 3, head imaging may be obtained for a variety of indications, such as acute or chronic neurocognitive or behavioral changes; trauma with or without an associated neurologic alteration; focal neurologic deficits such as weakness, numbness, or language difficulties; and craniofacial or headache pain syndromes. Frequently, the first study performed is a CT scan, which is obtained by directing a rotating beam of x-rays toward a region of interest to build cross-sectional images of the body. Air, tissue, blood, and bone demonstrate variable attenuation of the x-ray beams, and the resultant grayscale image reflects these densities; brighter regions are seen with higher density substances. A quantitative assessment of density on CT imaging is possible with Hounsfield units (HU), which express how easily the study x-rays penetrate a tissue. This scale runs from −1000 to +3000, with air at −1000 HU, CSF at +5 to +15 HU, brain parenchyma at +25 to +40 HU, blood around +40 to +60 HU, and bone from +1000 to +3000 HU (1). Most image review software will display the HU at any region of interest to assist in interpretation. Computed tomography is less sensitive for parenchymal and soft tissue abnormalities but is sufficient for identification of blood products, bony and structural changes, edema, and lesions larger than 3 mm using current techniques (2). This imaging modality is often obtained in an emergency setting, where clinical decisions—including the administration of thrombolytic therapies for suspected thromboembolic phenomena—must be made quickly. Benefits of CT include the short duration of the study, low cost relative to other modalities, widespread availability, and the ability to identify many life-threatening pathologies rapidly. If CT is unrevealing and symptoms persist or if CT demonstrates a concerning or uncertain finding, further information is often obtained using MRI.

The principle of MRI is that externally applied radiofrequency pulses can produce predictable excitation of hydrogen atoms, abundant in living tissues, that upon relaxation exhibit characteristics that can be converted to a high-resolution grayscale image. Numerous MRI sequences are typically obtained, which when interpreted in concert allow for sophisticated, albeit imperfect, analysis of the body. Evaluation of gross anatomy, aging of blood products and areas of infarction, and detection of regions of edema and hypercellularity are among the most common uses of MRI. In patients for whom presenting symptoms are of a subacute or chronic nature, or if there is high suspicion that CT would be of inadequate resolution, MRI may be obtained upfront.

In many institutions, CT of the chest, abdomen, and pelvis will be obtained routinely when a space-occupying lesion, which may raise concern for an underlying malignant tumor or metastatic process, is noted in the brain. However, in resource-constrained or -stewarding environments or in facilities where subspecialty services are available, additional consideration often is given prior to ordering further testing. To assess whether additional imaging would contribute meaningfully to the patient's diagnostics, providers evaluate two primary factors: patient risk for malignancy and characteristics of the intracranial lesion(s) in obtained imaging.

Assessing Patient Risk for Malignancy

In a patient without known systemic cancer, we first consider patient demographics and risk factors for malignancy. To this end, a focused history of the present illness as well as medical, surgical, social, and family histories are obtained. Ideally, history is gathered from the patient directly, but this may not be feasible in individuals with decreased alertness, encephalopathy, or intellectual disability, as well as in those requiring immediate medical intervention. Collateral information in those circumstances relies on insight from family, friends, or even coworkers, preferentially gathered from individuals highly knowledgeable about the patient and their medical history irrespective of relation.

Through the discussion, the patient or individual sharing on their behalf may highlight inherited or acquired risk factors for cancer; some examples are found in Table 4.1.

A family history of malignancy may not correspond to an elevated risk for cancer in the patient being assessed. However, it is worth noting if multiple family

Table 4.1 Elements of history that may indicate an increased risk for malignancy

Potential Risk Factors for Malignancy
Tobacco and alcohol use disorders
History of or ongoing intravenous drug use
Environmental and occupation exposures to known carcinogens such as asbestos
Prior ionizing radiation or chemotherapy earlier in life, particularly in childhood
Immunocompromised state including HIV or long-term immunosuppressive medication use
Advanced age
Family history of malignancy or heritable cancer syndromes (see Table 4.2)

Table 4.2 Hereditary cancer syndromes, including their implicated genes and associated neoplasms[a]

Syndrome	Gene(s)	Associated Neoplasm(s)
Hereditary nonpolyposis colorectal	MSH2, MLH1, MSH6, PMS2	Colorectal, endometrial, gastric, small intestine, urothelial
Familial breast and ovarian	BRCA1, BRCA2	Breast, ovarian, prostate
Neurofibromatosis type1	NF1	Neurofibroma, optic nerve glioma, neurofibrosarcoma
Familial retinoblastoma	RB1	Retinoblastoma in childhood and secondary tumors later in life
Multiple endocrine neoplasia type2	RET	Medullary thyroid cancer, pheochromocytoma, hyperparathyroidism
Familial adenomatous polyposis	APC	Multiple colonic adenomas, upper gastrointestinal tumors, desmoids
Von Hippel-Lindau	VHL	Clear cell renal cell carcinoma, hemangioblastomas
Li-Fraumeni	TP53	Sarcomas, breast cancer, primary brain tumors, leukemia
MUTYH-associated polyposis	MUTYH	Colorectal cancer, colonic adenomas
Ataxia telangiectasis	ATM	Non-Hodgkin lymphoma, leukemia

[a]Adapted from Rahner and Steinke, 2008 (2).

members are reported to have developed the same type of cancer, particularly in first-degree relatives such as parents or siblings. Certain heritable genetic alterations also may predispose individuals to the development of neoplasms (3); Table 4.2 describes some relevant syndromes with their associated gene(s) and neoplasm(s).

Other elements in the history to elucidate are adherence to age-appropriate cancer screenings such as mammography and colonoscopy as well as recent symptoms including weight loss, fevers, night sweats, and chills.

Assessing Brain Imaging Characteristics

In a patient presenting with new neurologic symptoms, consideration of brain imaging results is a critical starting point for generating a differential diagnosis. Understanding of the characteristic features of masses discovered on brain imaging drives decision-making regarding the relevance or necessity of systemic imaging. As there can be significant overlap in patient characteristics and risk factors between metastatic malignancy and primary brain tumors, it is useful to consider the imaging characteristics of the lesion(s) in question for clues to the etiology.

Brain metastases represent the most common form of intracranial neoplasm, estimated at 70,000 to upward of 200,000 new cases per year in the United States alone. Metastases in the brain tend to be well-circumscribed masses arising at the junction of gray and white matter, often causing displacement of surrounding tissue due to mass effect. There can be a wide range of surrounding edema, best seen as T2/fluid-attenuated inversion recovery (FLAIR) hyperintensity on MRI or hypoattenuation—a darker area as compared to surrounding brain tissue—on CT. When contrast is administered, metastatic lesions may have heterogeneous or rim enhancement, and blood products can occasionally be seen in the mass. On CT scan, blood products appear hyperdense—brighter than normal brain parenchyma—whereas MRI sequences, including gradient echo (GRE) and susceptibility-weighted imaging (SWI), can also help identify hematoma or hemorrhage with varying imaging qualities based on the age of the blood product. Malignancies most often associated with hemorrhagic metastases are melanoma, renal cell carcinoma, choriocarcinoma, and thyroid carcinoma. It is rare for metastases to involve the periventricular white matter or the corpus callosum, but metastases occasionally arise from the dura mater; breast, prostate, and lung primary neoplasms do this most commonly (4). Dural metastases are, however, a separate entity from leptomeningeal disease, which is metastatic involvement of the cerebrospinal fluid. An infrequent finding is spinal cord metastatic disease, which is most often associated with lung cancer (5).

Although metastatic disease can arise as a single lesion—typically with breast, renal, and colorectal cancers—an estimated 60% to 75% of patients have multiple masses at the time of identification (6). Accordingly, when multiple lesions are demonstrated on imaging, clinical suspicion for metastatic disease increases in the correct clinical context. For instance, a patient with a history of malignancy, particularly cancers with a tendency to metastasize to the brain such as lung, breast, melanoma, renal, and gastrointestinal neoplasms, would be strongly considered to have metastatic disease unless a compelling alternative diagnosis is established. This differs from primary brain tumors, which often present as a single lesion. Notably, the imaging characteristics of primary brain tumors vary widely across the over 100 entities recognized in the 2021 World Health Organization Classification of Tumors of the Central Nervous System (7).

An important additional consideration is CNS lymphoma, which can arise as isolated intracranial—called primary CNS lymphoma—or systemic/metastatic (secondary CNS lymphoma) disease. Lymphoma can be radiographically variable, and there may be a single or multiple CNS lesions. However, the presence of one or multiple subcortical lesions, often near or abutting the lateral ventricles, is suspicious for intracranial involvement of a lymphoproliferative disorder. On MRI, the presence of homogeneous diffusion restriction and contrast enhancement would further support the possibility of CNS lymphoma that would merit CT chest/abdomen/pelvis, testicular ultrasound when indicated by associated signs and symptoms, ophthalmologic evaluation, and—in many cases—CSF studies to evaluate extent of disease for treatment planning.

An Approach to Systemic Imaging

If the gathered history and/or brain imaging characteristics raise concern for metastatic disease, systemic imaging typically starts with CT of the chest, abdomen, and pelvis to look for a primary malignancy. As noted previously, CT is widely available and balances test duration, cost, and adequate visualization of most tumor types. If the patient can tolerate intravenous and oral contrast, it is recommended to obtain contrast-enhanced sequences. In the setting of acute or chronic kidney injury, nil per os ("NPO" or "nothing by mouth") status, or prior severe allergic reactions to iodinated contrast, CT without contrast can be obtained but may not demonstrate subtle findings as clearly. Suspicious changes seen on CT may subsequently require MRI. Alternatively, MRI of the chest, abdomen, and pelvis can be considered up front, but due to scan duration and cost is not often the initial modality of choice. Exceptions to this paradigm are children and pregnant patients, in whom avoiding radiation and contrast exposure are prioritized; these patients preferentially undergo MRI unless imaging cannot be delayed and CT imaging is obtained.

When CT or MRI findings are indeterminate or equivocal, fluorodeoxyglucose- (FDG-) PET can be considered to evaluate for regions of increased FDG uptake concerning for hypermetabolism suggestive of malignancy. PET scans work by detecting short-lived radioactive tracers—biologic compounds bound to a positron-emitting isotope such as ^{11}C, ^{18}F, ^{68}Ga, ^{13}N, or ^{15}O—to localize areas of relatively high or low concentrations of the organic ligand (8). The biologic compound often serves as a substrate for an established metabolic process, such as 2-deoxy-2-glucose—as part of FDG—in carbohydrate metabolism, and its detected concentration ideally serves as a surrogate for a physiologic process (9). As the radioisotopes decay, they emit positrons that collide with electrons and produce two gamma photons, traveling opposite of each other and with a characteristic energy of 511 keV (8). The photons then strike detectors, which allow for localization of the collision within 3–10 mm (10,11). The lower limit for lesion size detection using current techniques is approximately 4 mm but varies by machine and processing algorithm (12). Image reconstruction occurs by applying corrective and interpretative algorithms to the raw data, producing a human-interpretable color gradient superimposed on the structure(s) of interest. The resultant image, often highlighting areas of increased metabolic activity as can be seen in malignancy, serves as a complement to structural diagnostics like CT and MRI by affording insight into the biochemical processes within tissues. Therefore, PET can increase or decrease suspicion for pathological behavior in an area of interest identified on CT or MRI, which can prompt further workup, such as dedicated imaging or, in most cases, biopsy for histologic diagnosis.

If systemic imaging is unrevealing, it is crucial to reflect on the differential diagnosis for the brain mass, which may also include primary CNS malignancy, infectious processes (e.g., herpes simplex encephalitis and toxoplasmosis), multiple sclerosis, abscess, infarction, and inflammatory processes (13).

The Utility of Body Imaging in Paraneoplastic Syndromes

Paraneoplastic syndromes are rare and heterogeneous disorders that arise when antibodies form due to an underlying malignancy. Such antibodies may target various normal tissues (autoantibodies), thereby producing organ dysfunction via disruption of normal cellular processes. Manifestations can include endocrinologic, rheumatologic, dermatologic, hematologic, and neurologic abnormalities. By definition, the symptoms and signs that develop in the setting of such paraneoplastic syndromes cannot be the consequence of direct tumor invasion. It is estimated that paraneoplastic neurologic syndromes may develop as frequently as 1 in 334 cancers, though for a neuropsychologist, the most pertinent paraneoplastic syndrome is often paraneoplastic encephalitis (14).

The mechanisms by which paraneoplastic autoantibodies form are incompletely understood, and it is important to note that antibodies producing neurologic and non-neurologic syndromes may arise separate from malignancy, often co-occurring with other autoimmune disorders. However, data suggest that antibodies strongly associated with underlying malignancy develop against intracellular neuronal targets and are T-, rather than B-, cell mediated. Paraneoplastic syndromes are more commonly seen with small cell lung, gynecologic, and hematologic cancers, but can be seen with neoplasms of nearly every organ system. Table 4.3 demonstrates paraneoplastic antibodies that have a common ($\geq$50%) association with cancer.

Other less well characterized paraneoplastic antibodies that appear to be highly associated with malignancy and may produce limbic encephalitis include

Table 4.3 Paraneoplastic antibodies, their associations with systemic malignancies, and common neurologic presentations

Antibody	Cancer Likelihood	Malignancy	Presentation(s)
Hu (ANNA-1) (15)	↑↑↑	SCLC, NSCLC, NET, neuroblastoma	Sensory neuropathy, chronic GI pseudo-obstruction, encephalomyelitis, limbic encephalitis
CV2/CRMP5 (16)	↑↑↑	SCLC, thymoma	Encephalomyelitis, sensory neuronopathy
PCA2 (MAP1B) (17)	↑↑↑	SCLC, NSCLC, breast	Sensorimotor neuropathy, rapidly progressive cerebellar syndrome, encephalomyelitis
Amphiphysin (18)	↑↑↑	SCLC, breast	Polyradiculoneuropathy, sensory neuronopathy, encephalomyelitis, stiff-person syndrome
KLHL11 (19, 20,21)	↑↑↑	Testicular	Brainstem/cerebellar syndrome
Ma2/Ma (22)	↑↑	Testicular, NSCLC	Limbic encephalitis, diencephalitis, brainstem encephalitis
Ri (ANNA-2) (23,24)	↑↑	Breast, SCLC, NSCLC	Brainstem/cerebellar syndrome, OMS
AMPA-R (25,26,27)	↑	SCLC, thymoma	Limbic encephalitis
$GABA_B$-R (28,29)	↑	SCLC	Limbic encephalitis
mGluR5 (30)	↑	Hodgkin lymphoma	Encephalitis
CASPR2 (31,32,33)	↑	Thymoma	Morvan syndrome

Abbreviations: NET = neuroendocrine tumor, NSCLC = non-small cell lung cancer, OMS = opsoclonus-myoclonus syndrome, SCLC = small cell lung cancer.

KCTD16, neuronal intermediate filament, ARHGAP26, CRMP3-4, BRSK2, ROCK2, and TRIM46 (34); these are reported in low numbers of patients to date.

Detection of an autoantibody in the setting of a characteristic paraneoplastic syndrome does not necessarily imply that a malignancy will be found on workup at the time of antibody detection or in the future. However, when initial tumor screening is negative, Graus et al., in an update to accepted diagnostic criteria for paraneoplastic neurologic syndromes, have proposed that targeted imaging should be repeated every 4–6 months for 2 years in patients with high-risk phenotypes and positive antibodies (34). Below are screening recommendations by type of cancer suspected as proposed by Graus et al.:

- Breast
    - First line: mammography or ultrasound in young women
    - Second line: FDG-PET/CT
- Hodgkin lymphoma
    - Full-body CT or FDG-PET/CT
- Lung cancer (SCLC and NSCLC) and malignant thymoma
    - CT chest with or without FDG-PET/CT
    - MRI chest may also be considered
- Neuroblastoma
    - CT or MRI; CT identifies calcifications, but MRI is better for thoracic tumors
- Ovarian carcinoma
    - First line: transvaginal ultrasound with or without MRI abdomen/pelvis or FDG-PET/CT
    - Second line: post-menopausal women with anti-Yo antibodies and classical phenotype should be considered for exploratory surgery or prophylactic hysterectomy with bilateral salpingo-oophorectomy
- Ovarian teratoma
    - First line: transvaginal ultrasound with or without MRI abdomen/pelvis or FDG-PET/CT
    - Second line: CT chest to evaluate for extra-pelvic teratomas
    - Nota bene: prophylactic oophorectomy is not recommended for NMDA-R encephalitis if there is no detected ovarian teratoma
- Testicular
    - First line: ultrasound with or without CT pelvis
    - Second line: FDG-PET/CT
- Unknown malignancy
    - Full-body CT with consideration for FDG-PET/CT if negative

Paraneoplastic antibodies that are less often associated with malignancy target cell-surface antigens of both neuronal and glial cells. Examples of these antibodies and their associated malignancies are seen in Table 4.4.

In patients with antibodies less commonly associated with malignancy, comprehensive screening at the time of antibody identification may be sufficient (34). It is reasonable to perform additional screening if a patient develops signs or symptoms of malignancy, demonstrates persistence of symptoms despite appropriate treatment, or experiences recurrence of their syndrome.

Diagnosis of paraneoplastic syndromes can be difficult and often requires high clinical suspicion or a characteristic syndrome, which prompts practitioners to perform autoantibody evaluation of the serum, cerebrospinal fluid, or both. Despite relatively high specificity, the sensitivity of these tests is generally lower and may be less than 50%—particularly in the serum (50,51,52). Hence, while positive findings from autoantibody evaluation are highly predictive of underlying pathology, negative results do not necessarily rule out disease. If imaging has not been obtained, but a serum or CSF paraneoplastic panel has returned with a positive antibody, it is important to consider whether the detected antibody corresponds to the patient's paraneoplastic syndrome as well as the patient's likelihood of an associated malignancy.

Table 4.4 Paraneoplastic antibodies less strongly associated with malignancy, the neoplasms with which they may be associated, and neurologic presentations

Antibody	Cancer Likelihood	Malignancy	Presentation(s)
NMDA-R (35,36,37,38)	↓	Ovarian or teratoma	Encephalitis
$GABA_A$-R (39,40)	↓	Thymoma	Encephalitis
CASPR2 (31,32,33)	↓	Thymoma	Limbic encephalitis, Isaac syndrome
GFAP (41,42)	↓↓	Ovarian teratoma and adenocarcinoma	Meningoencephalitis
GAD65 (43)	↓↓	SCLC, NET, thymoma	Limbic encephalitis, stiff-person syndrome, cerebellar ataxia
LGI1 (44,45,46)	↓↓	Thymoma, NET	Limbic encephalitis
DPPX (47)	↓↓	B-cell neoplasms	Encephalitis with PERM
GlyR (48,49)	↓↓	Thymoma, Hodgkin lymphoma	Limbic encephalitis, PERM

Abbreviations: NET = neuroendocrine tumor, PERM = progressive encephalomyelitis with rigidity and myoclonus, SCLC = small cell lung cancer.

Unfortunately, paraneoplastic serum and CSF studies, which are performed as a panel of tests with reflex testing as appropriate, are labor- and time-intensive with a commensurate, often substantial, financial burden. Because of the turnaround time for testing—typically on the order of weeks—and an often-urgent need to initiate therapy, there is interest in the use of complementary, more readily available diagnostic modalities such as imaging.

The most well described brain imaging findings associated with paraneoplastic neurologic syndromes are those seen with encephalidities. In general, limbic encephalitis is characterized on MRI by cortical thickening with T2 hyperintensity and edema of the temporal lobes; enhancement may be seen (53,54). Most cases demonstrate bilateral, though usually asymmetric, temporal lobe involvement (53). This should be distinguished from herpes simplex virus-associated encephalitis, which spares the basal ganglia and commonly produces hemorrhage; whereas paraneoplastic limbic encephalitis is rarely hemorrhagic and frequently involves the basal ganglia (53). In many cases—particularly early in the disease course—there is scant radiographic change appreciated. Accordingly, the absence of imaging abnormalities does not rule out a paraneoplastic syndrome.

It is important to acknowledge that brain imaging findings for paraneoplastic and non-paraneoplastic limbic encephalitis may be similar and clinical information—such as the presence of psychiatric symptoms—is critical in interpreting imaging results. A reasonable radiographic differential diagnosis for limbic encephalitis would include status epilepticus; primary brain tumors, including of glial or lymphoproliferative origin; autoimmune thyroid encephalopathy; and other inflammatory etiologies such as neurosarcoidosis.

Surgical resection and/or initiation of treatment for an identified systemic malignancy is paramount in the treatment of paraneoplastic syndromes, and there is often a role for immunosuppression. Treatment modalities are beyond the scope of this discussion but adjuncts such as plasma exchange (PLEX), intravenous immunoglobulins (IVIg), and other immunomodulatory therapies are considered.

Positron Emission Tomography and Neuropsychology

Positron emission tomography imaging was first devised in the early 1960s with research applications starting in the mid-1970s using FDG to measure cerebral glucose metabolism (8). Since then, research and clinical applications of PET have expanded broadly within neurodegenerative and movement disorders, cardiology, musculoskeletal imaging, medical oncology, and neuropsychiatric disease.

Fluorodeoxyglucose is the most widely used and studied radioisotope, with FDG-PET scans currently FDA approved for assessment of systemic malignancy within medical oncology; coronary artery disease and left ventricular dysfunction within cardiology; and identification of epileptogenic foci within neurology (55). A large body of research exists for the utility of PET imaging of the brain for neurologic disorders, with compelling characteristic findings for Alzheimer disease, frontotemporal dementia, dementia with Lewy bodies, Creutzfeldt-Jakob disease, Parkinson-associated dementia, atypical parkinsonism, vascular dementia, temporal lobe epilepsy, schizophrenia, and bipolar disorder (56). Systemic PET imaging, which is often combined with CT studies, has a less clear upfront role in the diagnosis of neuropsychiatric disease as compared to brain PET. However, as discussed in prior sections, PET can be used to identify malignancy that has produced either intracranial metastatic disease or a paraneoplastic syndrome; both may manifest with memory deficits, decreased awareness, executive dysfunction, and behavioral change prompting medical evaluation.

Notably, many factors can alter the uptake of FDG including blood glucose levels, endogenous and exogenous insulin, psychoactive medications such as barbiturates and benzodiazepines, and potentially emotional states (57). Optimization of PET scan conditions includes holding metformin—an oral medicine commonly used for diabetes mellitus—for 48 hours; avoiding strenuous activity for 24 hours; eating a low carbohydrate diet for 24 hours; abstaining from caffeine, nicotine, and alcohol for 12 hours; fasting for at least 4–6 hours; and hydrating well in the 2 hours prior to the study (57). When CT scans are obtained with the PET, administration of IV contrast is allowed and has not been shown to affect PET quality. False-negative results may arise in the setting of small lesions; hyperglycemia and hyperinsulinemia; recent chemotherapy, radiation, or corticosteroid use; and metabolically hypoactive tumors such as prostate carcinoma, mucinous neoplasms, medullary thyroid cancer, lobular carcinoma of the breast, hepatocellular carcinoma, and indolent lymphomas (57,58).

Included here is a case study of a patient who presented to our institution with a constellation of symptoms and findings that might be relevant to a practicing neuropsychologist. He underwent neuro-oncologic evaluation incorporating the principles discussed above, leading to his diagnosis.

Case Study

A 55-year-old male patient presented to the emergency department for evaluation of increased pain in the right hip since undergoing surgical repair of a right femur fracture. In addition, the patient also reported intermittent word-finding difficulties and progressively worsening weakness of the right arm, both developing in recent weeks. No fevers, chills, or unexplained weight loss was present, nor was there associated headache, dizziness, vision change, facial weakness, swallowing difficulty, loss of consciousness, unexplained movement or overt seizure-like activity, sensory disturbance, or bowel or bladder dysfunction. Medical history was significant for chronic obstructive pulmonary disease, ongoing tobacco use, and alcohol use disorder.

Physical examination revealed minimal acute distress or discomfort, but labored and mildly rapid breathing was noted (tachypnea). Vital signs were otherwise within normal limits. Cardiopulmonary, abdominal, and extremity evaluations were also normal. A neurologic examination revealed an awake and fully alert patient with orientation to self, location, date, and recent events. Language was fluent during conversation and the patient had normal comprehension of simple and complex commands. However, difficulty with naming of low-frequency objects was elicited. Cranial nerve function was intact, including facial strength and sensation. On motor testing, a slight right arm pronator drift was seen, and strength was diminished in both the right arm and right leg. Sensation was intact to light touch in all limbs. Deep tendon reflexes were symmetric throughout. No incoordination or abnormal movements were noted on cerebellar examination. Given ongoing hip discomfort and weakness, ambulation was not tested (see also Chapter 1 discussion on cranial nerves and motor exam).

Laboratory evaluation, including complete blood counts and basic metabolic profile was unremarkable (see also Chapter 9). A non-contrast head CT was performed, which demonstrated a hyperdense lesion with significant surrounding hypodensity within the left frontal lobe. Representative images are included in Figure 4.1, part A.

Given a lack of infectious signs or symptoms, the patient's ongoing tobacco use, and the progressive nature of neurologic deficits, there was suspicion that the mass represented a neoplastic process. To help characterize the lesion more fully and to assess for additional masses, MRI of the brain without and with intravenous contrast (gadolinium) was undertaken. This study confirmed the presence of a 3.9 cm by 3.9 cm by 3.5 cm lobulated, avidly enhancing tumor in the superior left frontal lobe with significant surrounding edema and mass effect on the ventricular system, as shown in Figure 4.1, part B.

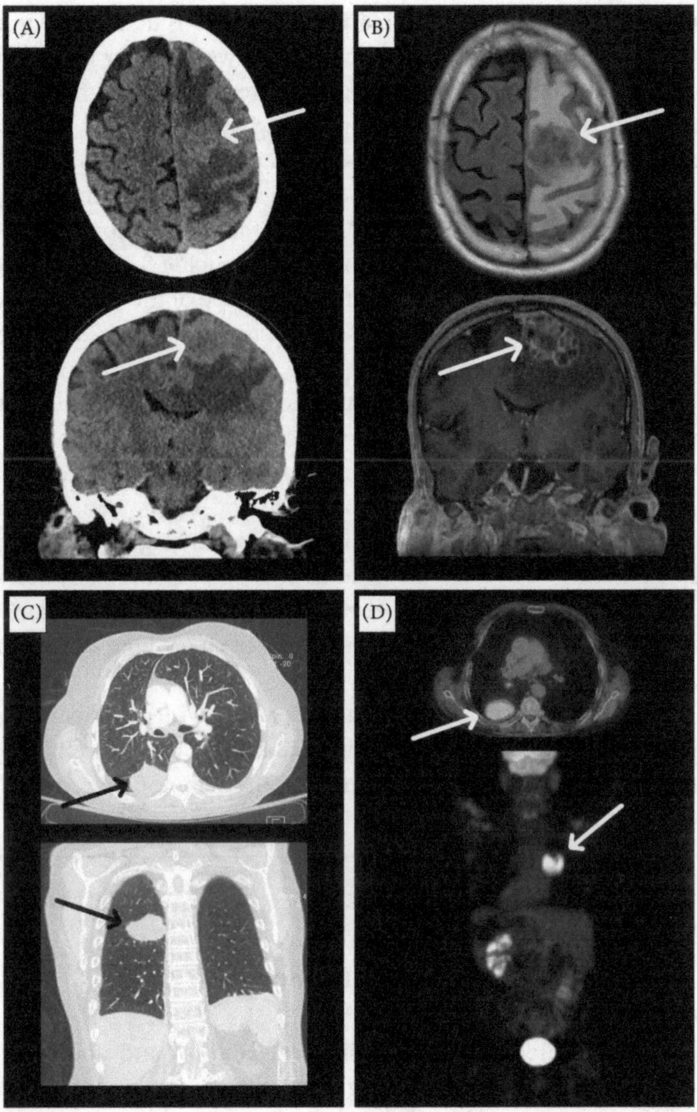

Figure 4.1 (A) Axial and coronal non-contrast head CT imaging demonstrating a moderately sized, iso- to hyperdense left frontal lobe mass with surrounding vasogenic edema. (B) Axial T2/fluid-attenuated inversion recovery (FLAIR) and coronal T1 post-gadolinium sequences delineating the 3.9 cm by 3.9 cm by 3.5 cm lobulated, enhancing mass and surrounding edema (best seen on FLAIR sequence). (C) Axial and coronal T1 post-iodinated contrast CT scans of the chest confirming the 5.2 cm by 3.7 cm by 4.2 cm enhancing right lung mass. (D) Axial and coronal FDG-PET scans performing for systemic malignancy staging. Radiotracer uptake is seen in the right lung, left kidney, and bladder, with renal and bladder contrast normally seen as radiotracer is excreted.

The patient was evaluated by neurosurgery for consideration of resection for both therapeutic and diagnostic benefit. However, prior to any surgical intervention, further systemic imaging was entertained given the possibility of a secondary brain tumor from an underlying systemic cancer. A chest x-ray in the emergency department demonstrated an indeterminate opacity to the posterior right lung; in conjunction with the head imaging findings, suspicion for a metastatic malignancy increased. A CT of the chest, abdomen, and pelvis was performed and confirmed a hypodense enhancing 5.2 cm by 3.7 cm by 4.2 cm mass in the upper segment of the lower lobe of the right lung and an enlarged right hilar lymph node. The lesion can be appreciated on the chest CT imaging in Figure 4.1, part C.

The patient subsequently underwent resection of the left frontal brain mass and intraoperative frozen pathology was consistent with adenocarcinoma. With administration of dexamethasone for reduction of cerebral edema, there was notable improvement in language and right-sided weakness in the postoperative setting. While hospitalized, cancer staging was completed with an FDG-PET scan from the head to mid-thighs. This study confirmed significant tracer uptake in both the right lung mass and in hilar lymph nodes consistent with malignancy and likely primary lung cancer seen in Figure 4.1, part D.

Corticosteroids were tapered off as an outpatient, with persistence of symptom improvement. Final brain pathology confirmed a small cell carcinoma from a primary lung cancer, and treatment with chemotherapy and radiation was pursued thereafter.

Conclusion

Within neuropsychology, there is an intuitive comfort and understanding of the role of brain imaging in the diagnosis and monitoring of our patients. However, the evaluation of a patient with a new brain mass can produce a clinical conundrum, wherein the role of additional imaging is not always certain. In many circumstances, systemic imaging—CT, MRI, or PET—has been obtained reflexively, but it is beneficial to an ever-overloaded healthcare system to consider carefully the indications for and limitations to these studies. By understanding the risk factors and imaging characteristics that point toward a systemic malignancy or paraneoplastic syndrome, neuropsychologists can augment meaningfully the evaluation of and recommendations for patients under their care.

References

1. Schneider W, Bortfeld T, Schlegel W. Correlation between CT numbers and tissue parameters needed for Monte Carlo simulations of clinical dose distributions. Phys Med Biol. 2000;45(2):459.
2. Wilms G, Gryspeerdt S, Bosmans H, Boulanger T, Van Hoe L, Marchal G, et al. Spiral CT of cerebral aneurysms. J Belge Radiol. 1995;78(2):75–78.
3. Rahner N, Steinke V. Hereditary cancer syndromes. Dtsch Ärztebl Int [Internet]. 2008 Oct 10 [cited 2022 May 19]; Available from: https://www.aerzteblatt.de/10.3238/arztebl.2008.0706
4. Nayak L, Abrey LE, Iwamoto FM. Intracranial dural metastases. Cancer. 2009 May 1; 115(9):1947–1953.
5. Lv J, Liu B, Quan X, Li C, Dong L, Liu M. Intramedullary spinal cord metastasis in malignancies: an institutional analysis and review. OncoTargets Ther. 2019 Jun;12: 4741–4753.
6. Davis PC, Hudgins PA, Peterman SB, Hoffman JC. Diagnosis of cerebral metastases: double-dose delayed CT vs contrast-enhanced MR imaging. Am J Neuroradiol. 1991;12(2):293–300.
7. Louis DN, Perry A, Wesseling P, Brat DJ, Cree IA, Figarella-Branger D, et al. The 2021 WHO Classification of Tumors of the Central Nervous System: a summary. Neuro-Oncol. 2021 Aug 2;23(8):1231–1251.
8. Basu S, Kwee TC, Surti S, Akin EA, Yoo D, Alavi A. Fundamentals of PET and PET/CT imaging: fundamentals of PET and PET/CT. Ann N Y Acad Sci. 2011 Jun;1228(1): 1–18.
9. Hoekstra CJ, Paglianiti I, Hoekstra OS, Smit EF, Postmus PE, Teule GJJ, et al. Monitoring response to therapy in cancer using [18 F]-2-fluoro-2-deoxy-d-glucose and positron emission tomography: an overview of different analytical methods. Eur J Nucl Med Mol Imaging. 2000 May 29;27(6):731–743.
10. Karp JS, Viswanath V, Geagan MJ, Muehllehner G, Pantel AR, Parma MJ, et al. PennPET Explorer: design and preliminary performance of a whole-body imager. J Nucl Med. 2020 Jan;61(1):136–143.
11. Levin CS, Hoffman EJ. Calculation of positron range and its effect on the fundamental limit of positron emission tomography system spatial resolution. Phys Med Biol. 1999 Mar 1;44(3):781–799.
12. Erdi Y. Limits of tumor detectability in nuclear medicine and PET. Malecular Imaging Radionucl Ther. 2012 Apr 1;21(2):23–28.
13. Klein JP, Dietrich J. Neuroradiologic pearls for neuro-oncology. Contin Lifelong Learn Neurol. 2017;23(6):1619–1634.
14. Vogrig A, Gigli GL, Segatti S, Corazza E, Marini A, Bernardini A, et al. Epidemiology of paraneoplastic neurological syndromes: a population-based study. J Neurol. 2020 Jan;267(1):26–35.
15. Graus F. Anti-Hu-associated paraneoplastic encephalomyelitis: analysis of 200 patients. Brain. 2001 Jun 1;124(6):1138–1148.
16. Dubey D, Lennon VA, Gadoth A, Pittock SJ, Flanagan EP, Schmeling JE, et al. Autoimmune CRMP5 neuropathy phenotype and outcome defined from 105 cases. Neurology. 2018 Jan 9;90(2):e103–10.
17. Jitprapaikulsan J, Klein CJ, Pittock SJ, Gadoth A, McKeon A, Mills JR, et al. Phenotypic presentations of paraneoplastic neuropathies associated with MAP1B-IgG. J Neurol Neurosurg Psychiatry. 2020 Mar;91(3):328–330.

18. Dubey D, Jitprapaikulsan J, Bi H, Do Campo RV, McKeon A, Pittock SJ, et al. Amphiphysin-IgG autoimmune neuropathy: a recognizable clinicopathologic syndrome. Neurology. 2019 Nov 12;93(20):e1873–80.
19. Mandel-Brehm C, Dubey D, Kryzer TJ, O'Donovan BD, Tran B, Vazquez SE, et al. Kelch-like protein 11 antibodies in seminoma-associated paraneoplastic encephalitis. N Engl J Med. 2019 Jul 4;381(1):47–54.
20. Vogrig A, Péricart S, Pinto AL, Rogemond V, Muñiz-Castrillo S, Picard G, et al. Immunopathogenesis and proposed clinical score for identifying Kelch-like protein-11 encephalitis. Brain Commun. 2021 Jul 1;3(3):fcab185.
21. Maudes E, Landa J, Muñoz-Lopetegi A, Armangue T, Alba M, Saiz A, et al. Clinical significance of Kelch-like protein 11 antibodies. Neurol–Neuroimmunol Neuroinflammation. 2020 May;7(3):e666.
22. Dalmau J. Clinical analysis of anti-Ma2-associated encephalitis. Brain. 2004 Jun 16;127(8):1831–1844.
23. Simard C, Vogrig A, Joubert B, Muñiz-Castrillo S, Picard G, Rogemond V, et al. Clinical spectrum and diagnostic pitfalls of neurologic syndromes with Ri antibodies. Neurol–Neuroimmunol Neuroinflammation. 2020 May;7(3):e699.
24. Pittock SJ, Lucchinetti CF, Lennon VA. Anti-neuronal nuclear autoantibody type 2: paraneoplastic accompaniments. Ann Neurol. 2003 May;53(5):580–587.
25. Hoftberger R, van Sonderen A, Leypoldt F, Houghton D, Geschwind M, Gelfand J, et al. Encephalitis and AMPA receptor antibodies: novel findings in a case series of 22 patients. Neurology. 2015 Jun 16;84(24):2403–2412.
26. Joubert B, Kerschen P, Zekeridou A, Desestret V, Rogemond V, Chaffois MO, et al. Clinical spectrum of encephalitis associated with antibodies against the α-amino-3-hydroxy-5-methyl-4-isoxazolepropionic acid receptor: case series and review of the literature. JAMA Neurol. 2015 Oct 1;72(10):1163.
27. Zhang Z, Fan S, Ren H, Zhou L, Guan H. Clinical characteristics and prognosis of anti-alpha-amino-3-hydroxy-5-methyl-4-isoxazolepropionic acid receptor encephalitis. BMC Neurol. 2021;21(1):1–10.
28. Jeffery OJ, Lennon VA, Pittock SJ, Gregory JK, Britton JW, McKeon A. GABAB receptor autoantibody frequency in service serologic evaluation. Neurology. 2013 Sep 3; 81(10):882–887.
29. Zhu F, Shan W, Lv R, Li Z, Wang Q. Clinical characteristics of anti-GABA-B receptor encephalitis. Front Neurol. 2020 May 21;11:403.
30. Spatola M, Sabater L, Planagumà J, Martínez-Hernandez E, Armangué T, Prüss H, et al. Encephalitis with mGluR5 antibodies: symptoms and antibody effects. Neurology. 2018 May 29;90(22):e1964–72.
31. van Sonderen A, Ariño H, Petit-Pedrol M, Leypoldt F, Körtvélyessy P, Wandinger KP, et al. The clinical spectrum of Caspr2 antibody–associated disease. Neurology. 2016 Aug 2;87(5):521–528.
32. Irani SR, Pettingill P, Kleopa KA, Schiza N, Waters P, Mazia C, et al. Morvan syndrome: clinical and serological observations in 29 cases. Ann Neurol. 2012 Aug;72(2):241–255.
33. Muñiz-Castrillo S, Joubert B, Elsensohn MH, Pinto AL, Saint-Martin M, Vogrig A, et al. Anti-CASPR2 clinical phenotypes correlate with HLA and immunological features. J Neurol Neurosurg Psychiatry. 2020 Oct;91(10):1076–1084.
34. Graus F, Vogrig A, Muñiz-Castrillo S, Antoine JCG, Desestret V, Dubey D, et al. Updated diagnostic criteria for paraneoplastic neurologic syndromes. Neurol–Neuroimmunol Neuroinflammation. 2021 Jul;8(4):e1014.

35. Titulaer MJ, McCracken L, Gabilondo I, Armangué T, Glaser C, Iizuka T, et al. Treatment and prognostic factors for long-term outcome in patients with anti-NMDA receptor encephalitis: an observational cohort study. Lancet Neurol. 2013 Feb;12(2):157–165.
36. Titulaer MJ, McCracken L, Gabilondo I, Iizuka T, Kawachi I, Bataller L, et al. Late-onset anti-NMDA receptor encephalitis. Neurology. 2013 Sep 17;81(12):1058–1063.
37. Bost C, Chanson E, Picard G, Meyronet D, Mayeur ME, Ducray F, et al. Malignant tumors in autoimmune encephalitis with anti-NMDA receptor antibodies. J Neurol. 2018 Oct;265(10):2190–2200.
38. Barry H, Byrne S, Barrett E, Murphy KC, Cotter DR. Anti-N-methyl-d-aspartate receptor encephalitis: review of clinical presentation, diagnosis and treatment. BJPsych Bull. 2015 Feb;39(1):19–23.
39. Spatola M, Petit-Pedrol M, Simabukuro MM, Armangue T, Castro FJ, Barcelo Artigues MI, et al. Investigations in $GABA_A$ receptor antibody-associated encephalitis. Neurology. 2017 Mar 14;88(11):1012–1020.
40. O'Connor K, Waters P, Komorowski L, Zekeridou A, Guo CY, Mgbachi VC, et al. $GABA_A$ receptor autoimmunity: a multicenter experience. Neurol–Neuroimmunol Neuroinflammation. 2019 May;6(3):e552.
41. Fang B, McKeon A, Hinson SR, Kryzer TJ, Pittock SJ, Aksamit AJ, et al. Autoimmune glial fibrillary acidic protein astrocytopathy: a novel meningoencephalomyelitis. JAMA Neurol. 2016 Nov 1;73(11):1297.
42. Kunchok A, Zekeridou A, McKeon A. Autoimmune glial fibrillary acidic protein astrocytopathy. Curr Opin Neurol. 2019 Jun;32(3):452–458.
43. Muñoz-Lopetegi A, de Bruijn MAAM, Boukhrissi S, Bastiaansen AEM, Nagtzaam MMP, Hulsenboom ESP, et al. Neurologic syndromes related to anti-GAD65: clinical and serologic response to treatment. Neurol–Neuroimmunol Neuroinflammation. 2020 May;7(3):e696.
44. Guasp M, Landa J, Martinez-Hernandez E, Sabater L, Iizuka T, Simabukuro M, et al. Thymoma and autoimmune encephalitis: clinical manifestations and antibodies. Neurol–Neuroimmunol Neuroinflammation. 2021 Sep;8(5):e1053.
45. Li W, Wu S, Meng Q, Zhang X, Guo Y, Cong L, et al. Clinical characteristics and short-term prognosis of LGI1 antibody encephalitis: a retrospective case study. BMC Neurol. 2018 Dec;18(1):96.
46. Rodriguez A, Klein CJ, Sechi E, Alden E, Basso MR, Pudumjee S, et al. LGI1 antibody encephalitis: acute treatment comparisons and outcome. J Neurol Neurosurg Psychiatry. 2022 Mar;93(3):309–315.
47. Hara M, Ariño H, Petit-Pedrol M, Sabater L, Titulaer MJ, Martinez-Hernandez E, et al. DPPX antibody–associated encephalitis: Main syndrome and antibody effects. Neurology. 2017 Apr 4;88(14):1340–1348.
48. Carvajal-González A, Leite MI, Waters P, Woodhall M, Coutinho E, Balint B, et al. Glycine receptor antibodies in PERM and related syndromes: characteristics, clinical features and outcomes. Brain. 2014 Aug;137(8):2178–2192.
49. Hinson SR, Lopez-Chiriboga AS, Bower JH, Matsumoto JY, Hassan A, Basal E, et al. Glycine receptor modulating antibody predicting treatable stiff-person spectrum disorders. Neurol–Neuroimmunol Neuroinflammation. 2018 Mar;5(2):e438.
50. Albadareen R, Gronseth G, Goeden M, Sharrock M, Lechtenberg C, Wang Y. Paraneoplastic autoantibody panels: sensitivity and specificity, a retrospective cohort. Int J Neurosci. 2017 Jun 3;127(6):531–538.

51. McKeon A, Pittock SJ, Lennon VA. CSF complements serum for evaluating paraneoplastic antibodies and NMO-IgG. Neurology. 2011 Mar 22;76(12):1108–1110.
52. Budhram A, Yang L, Bhayana V, Mills JR, Dubey D. Clinical sensitivity, specificity, and predictive value of neural antibody testing for autoimmune encephalitis. J Appl Lab Med. 2022 Jan 5;7(1):350–356.
53. Oyanguren B, Sánchez V, González FJ, de Felipe A, Esteban L, López-Sendón JL, et al. Limbic encephalitis: a clinical-radiological comparison between herpetic and autoimmune etiologies. Eur J Neurol. 2013 Dec;20(12):1566–1570.
54. Madhavan AA, Carr CM, Morris PP, Flanagan EP, Kotsenas AL, Hunt CH, et al. Imaging review of paraneoplastic neurologic syndromes. Am J Neuroradiol. 2020 Dec;41(12):2176–2187.
55. Fludeoxyglucose f18 injection [package insert] [Internet]. Columbia, MO: Essential Isotopes LLC; 2021 Oct [cited 2022 May 22]. Available from: https://dailymed.nlm.nih.gov/dailymed/drugInfo.cfm?setid=a7f03847-fa2c-4fbc-8071-3f96500ce68f
56. Schöll M, Damián A, Engler H. Fluorodeoxyglucose PET in neurology and psychiatry. PET Clin. 2014 Oct;9(4):371–390.
57. Waxman AD, Herholz K, Lewis DH, Herscovitch P, Minoshima S, Mountz JM, et al. Society of Nuclear Medicine procedure guideline for FDG PET brain imaging. Society of Nuclear Medicine; 2009 Feb. Report No. 1.0.
58. Kinahan PE, Fletcher JW. Positron emission tomography–computed tomography standardized uptake values in clinical practice and assessing response to therapy. Semin Ultrasound CT MRI. 2010 Dec;31(6):496–505.

5

EEG and Other Neurodiagnostic Procedures

C. J. Brush, Gabriel M. MacAlevy, Jill Z. Stuart, and M. Omar Subei

Electroencephalography

The use of electroencephalography (EEG) can be traced back to its historical roots in 1929 when Hans Berger, a German psychiatrist, first reported a series of experiments documenting electrical activity recorded from the human brain using scalp electrodes. To be considered useful, this electrical activity had to be amplified, and then the recorded voltages plotted over time (Berger, 1929). However, these experiments were not enough to convince other neurophysiologists at the time that this electrical activity reflected useful brain activity, as it was believed that what Berger recorded was some sort of artifact. Fast-forward a few years to 1935, and other respected experts, including Adrian and Matthews (1934), Jasper and Carmichael (1935), and Gibbs et al. (1935) accepted this truth and came to agreement that this electrical activity was indeed real and reflected voltages emanating from the human brain.

Since the 1930s, EEG has been used across a variety of scientific and clinical settings. EEG is a noninvasive neurophysiological tool that provides real-time insight into gross brain activity that reflects underlying neurotransmitter binding and postsynaptic potentials of the cerebral cortex. As one can imagine, this type of real-time information provided by EEG has resulted in many applications in neuroscientific research and clinical settings to better understand the human brain and its clinical manifestations (e.g., epilepsy, sleep abnormalities, and monitoring during surgical interventions).

In this chapter, we focus specifically on applications of EEG, with a focus on how it has been used in clinical settings. Specific clinical conditions will be discussed along with an overview of the procedures involved in using EEG in relation to those conditions. However, first, we will provide a brief background of the neuroscience underlying EEG.

EEG relies on measuring small voltage fluctuations from the brain, which results from the flow of ions across cortical pyramidal neuronal cell membranes

C. J. Brush et al., *EEG and Other Neurodiagnostic Procedures*. In: *Medical and Laboratory Examinations for Neuropsychologists*. Edited by: Thomas J. Farrer and R. Brock Frost, Oxford University Press.
 DOI: 10.1093/9780197619537.003.0006

(Britton, 2016). This ionic flow occurs during a change in postsynaptic potentials from incoming excitatory postsynaptic potentials and inhibitory postsynaptic potentials, which signal to ligand-gated ion channels. It should be noted, however, that EEG *does not typically reflect the measurement of action potentials.* Action potentials are typically too short in duration to be measured with this method; however, early sensory potentials called evoked potentials (EPs) are an exception (Biasiucci, 2019; for an extended commentary, refer to Luck, 2014). The synchronous firing of thousands of pyramidal neurons creates a net dipole, the voltage of which can be measured by electrodes. Besides electrical grounding, two recording electrode sites are necessary to measure voltage potential changes by subtracting the difference between an active measuring electrode and a reference electrode. By convention, an upward deflecting wave is created when the active electrode is more negative than the reference, and a downward deflection is created when the active electrode is more positive than the reference.

The data gathered from these changes in voltage measurements provides a wealth of information that can be analyzed using various methods. Given that the brain is constantly active, whether you are awake or asleep, all of this constant activity leads to ongoing variations that can be recorded and detected as oscillations occurring in specific frequency domains, such as alpha (8–13 Hz), delta (<4 Hz), theta (4–8 Hz), beta (13–30 Hz), and gamma (>30 Hz) bands (Abhang et al., 2016). The location and related activity of these events are summarized in Table 5.1.

Other common techniques include EPs and event-related potentials (ERPs), which reflect electrical activity time-locked to specific events (e.g., brief, transient brain responses to internal and external stimuli; Kappenman & Luck, 2016). We will discuss EPs and how they are applied clinically at the end of this chapter; however, a discussion of ERPs, which are longer-latency brain responses, is outside the scope of this chapter, as they are typically the focus of research initiatives.

Table 5.1 Common EEG frequency bands along with their frequency (Hz), brain state, and location of predominant activity

Frequency Band	Frequency	Brain State	Predominant Location
Gamma	>30 Hz	Cognitive Processing	Various
Beta	13–30 Hz	Conscious Thought, Anxiety, Active	Various
Alpha	8–13 Hz	Relaxed Wakefulness	Occipital Lobe
Theta	4–8 Hz	Deep Relaxation	Frontal Midline
Delta	<4 Hz	Sleep	Frontal Midline

With technological advances in EEG, the brain's electrical activity can be sampled rapidly (e.g., 250 Hz to 1,000 Hz) and recorded by electrodes placed at many sites across the scalp, which are usually embedded in elastic mesh caps worn by individuals. These electrodes are placed in positions corresponding to the International 10/20 system, which provides a standardized mapping of EEG electrode locations using a combination of letters and numbers to define a general brain region (American Encephalographic Society, 1994; Jasper, 1958). For example, F refers to frontal, C refers to central, P refers to parietal, and O refers to occipital. An accompanying number indicates the hemisphere of the electrode position, with odd numbers indicating left hemisphere and even numbers indicating right hemisphere. While a lowercase z, or often referred to as zed, or zero, indicates an electrode placed on the midline, the number corresponding to the electrode can indicate the distance from the midline (i.e., larger numbers mean larger distances from the midline).

Although EEG can provide critical insight into meaningful brain activity, there are some inherent pitfalls to be aware of when using this tool. The first consideration is that EEG is sensitive to various sources of noise in the recording environment, which includes biological electrical potentials produced outside the brain and induced electrical activity from external sources (e.g., lights, powerlines). For example, when recording EEG, a common nuisance is the creation of physiological artifacts such as eye blinks, tongue movements, rapid eye movements, and so forth. Given that the EEG signal is relatively small (i.e., in microvolts plotted over time, which are magnitudes smaller than other biological and external electrical potentials), if precautions are taken to minimize these external sources of noise the EEG signal will be relatively large and will indicated signals of meaningful brain activity. Another major pitfall—due to the superficial electrode placement on the scalp and the low conductivity of the human skull causing smearing—is that it is exceedingly challenging to localize electrical activity from deeper brain structures such as the thalamus and brainstem. The outer cortical region of the brain is the only area known to be readily assessed by this method. Despite these pitfalls, EEG's major strength lies in its temporal resolution in measuring electrical activity as it unfolds in real time as compared to other tools with superior spatial resolution (e.g., functional magnetic resonance imaging [fMRI]; Hajcak et al., 2019; Luck, 2014). A brief overview of the standard procedure to obtain an EEG is described in the next section.

General Procedure Overview for EEG

It is important to note that differences in procedures depend on the recording environment and clinical settings, and the particular need for an EEG assessment; however, as mentioned above, electrodes are typically embedded in an elastic mesh cap worn by individuals. This elastic mesh cap is placed on a participant's head. This cap is sized accordingly to the individual's head to achieve correct electrode placement in accordance with the International 10/20 system. The cap is centered front to back using the area between the eyes and the occipital protuberance (or inion) as anatomical landmarks. The cap is centered at the vertex of the head and uses the ears as landmarks (Yamada, 2017). These basic procedures are critical to allow for comparisons across individuals and patient populations.

To record the EEG, there needs to be a conductive medium, typically gel or liquid, between each electrode and the skin to make a stable electrical connection. Once a stable electrical connection is achieved, recording can proceed, with each electrode recording its own electrical activity. The types of recordings that occur depend on the clinical manifestation. After the recordings are completed, EEG technologists analyze the data, and these results can be shared with a physician or neuropsychologist to inform clinical evaluation.

Clinical Applications

Now that we have covered some background behind EEG and its general procedures, we will discuss how EEG plays an important role in the evaluation and management of common neuropsychiatric conditions.

Seizures

One of the most common uses for EEG is to aid in the detection of seizures. These types of assessments most commonly occur in clinical outpatient settings (e.g., a hospital). When it comes to seizure detection, there are distinct changes in brain activity that occur in the continuous EEG, which appear as synchronized, abnormal waveforms called epileptiform discharges. These epileptiform discharges manifest as sharp waves, spike waves, or spike-and-wave complexes in the EEG record. Identification of these abnormal EEG patterns allows clinicians to detect seizure events and classify their type based on their characteristics, such as location, duration, and frequency. Importantly, to accurately detect the presence of these waves or complexes, these abnormal EEG patterns need to be distinguished from other possible biological potentials (e.g., movement artifacts) and sources of electrical noise that contaminate the signal. As such, EEGs used to assess seizures are often video recorded to correlate observed behavioral changes with suspected abnormalities on the EEG readouts. EEG can help clinicians make informed decisions regarding whether normal or abnormal patterns of electrical activity are present, which can reduce the risk of misdiagnosis.

By observing EEG patterns across different conditions and contexts, clinicians can assess the consistency and reproducibility of abnormal findings, aiding in the differential diagnosis between epileptic and non-epileptic events.

Sleep

EEG is also commonly used to characterize sleep stages and is ordered by a sleep medicine physician to inform the evaluation and management of various sleep disorders, such as sleep apnea, insomnia, parasomnias, and movement disorders during sleep. For instance, in obstructive sleep apnea, disruptions in breathing during sleep lead to changes in EEG patterns, such as arousals and alterations in sleep architecture. Similarly, insomnia may be associated with prolonged periods of wakefulness during sleep, reflected in EEG recordings showing reduced sleep efficiency and increased wakefulness. Parasomnia, such as sleepwalking and REM sleep behavior disorder, often exhibit distinctive EEG abnormalities during episodes, aiding in their diagnosis and differentiation from other sleep disorders.

Sleep can be divided into non-rapid eye movement (NREM) and rapid eye movement (REM) sleep phases; both of which are characterized by distinct EEG patterns. NREM sleep is further subdivided into N1, N2, and N3 stages, with progressively deeper sleep characterized by slower EEG frequencies and higher-voltage delta waves. REM sleep, on the other hand, is associated with low-voltage, fast-frequency EEG activity resembling wakefulness. By analyzing these EEG patterns, clinicians can assess the quality and quantity of each sleep stage, aiding in the diagnosis and management of sleep abnormalities.

Furthermore, EEG is often used in conjunction with other physiological measurements, such as electromyography (EMG) and electrooculography (EOG), to provide a comprehensive assessment of sleep physiology. EMG helps monitor muscle activity, while EOG tracks eye movements, both of which are essential for identifying sleep stages and detecting abnormalities like sleep-related movement disorders and sleep-related breathing disorders.

In a sleep study using EEG, the patient's brain activity is monitored while they are encouraged to sleep. A sleep technologist monitors their brain activity overnight and then evaluates and scores their EEG recordings to inform specific sleep diagnoses, EEG abnormalities, and any other important observations. The sleep medicine physician can then use this information to make a clinical judgment.

Rapid Dementia/Prion Disease (Creutzfeldt-Jakob Disease)

EEG has also been employed to detect prion diseases, and of these diseases, it has played an integral role in the diagnostic process among patients with Creutzfeldt-Jakob disease (CJD). In CJD, EEG findings reveal abnormal electrical activity, which is characterized by generalized periodic sharp wave complexes (PSWCs). These PSWCs manifest as high-amplitude, sharp waveforms that occur at somewhat regular intervals. Their presence within the EEG is used to differentiate CJD from other causes of dementia, as these patterns are not indicative of other neurodegenerative diseases.

Other Monitoring Procedures

It is important to note that there are different monitoring procedures that might be used depending on the specific clinical indication. We highlight some alternative procedures used beyond routine EEG assessments below.

Ambulatory EEG Monitoring

The purpose of ambulatory EEG monitoring is to record brain activity over an extended period of time outside a clinical setting. Oftentimes, brief, routine EEG assessments performed in a clinical setting may not reveal abnormal brain activity due to the short duration of the recording. Therefore, additional testing, such as ambulatory EEG monitoring, is performed while patients engage in their typical activities of daily living.

This allows physicians and EEG technologists to detect the presence of normal-to-abnormal patterns of brain activity that occur across daily contexts, which may aid in the detection of a clinical manifestation. During ambulatory EEG monitoring, a patient will be asked to keep a log of their activities, such as when they eat, sleep, or engage in particular movements, and may also be asked to record any occurrences (e.g., when a seizure was experienced). This type of information, along with a recording of brain activity, will provide refined insight that can be used to make informed clinical judgments. This type of monitoring typically occurs for at least a day (e.g., 24–72 hours).

Long-Term EEG Monitoring

Long-term EEG monitoring is recommended when trying to detect the presence of spikes and seizures (Thompson & Ebersole, 1999) or other episodic disturbances of neurological function (Lagerlund et al., 1996) that are difficult to record during brief, routine EEG assessments. The duration of long-term EEG monitoring is considered to be longer than any brief, routine EEG assessment, but can range anywhere from hours to weeks (American Clinical Neurophysiology Society, 2008). This type of monitoring occurs in in-patient settings (e.g., clinic or hospital setting), and the recording duration is usually up to the discretion of the neurologist. During long-term EEG monitoring, behavior can be self-reported, reported by the observer, or recorded via video. A major objective is for the EEG technologist and neurologist to link behavior with EEG findings to inform their clinical evaluation.

Intraoperative EEG

Intraoperative EEG monitoring, often included as a part of intraoperative neurophysiological monitoring (IONM), is performed in the operating room, especially when there is a risk of neurological damage (Tewari et al., 2018). Before IONM can be performed, baseline recording should occur so comparisons to determine variations in brain activity can be made during the procedure. When the EEG is monitored during the procedure, there is a recording of all events occurring during the procedure to link up with the EEG record. Any loss of or variation in brain activity during an operation could indicate a neural injury and predict postoperative neurological deficits (Ghatol & Widrich, 2023).

Event-Related Potential Tests

Event-related potential (EP) tests provide important information regarding the functional integrity and conductivity in important paths of the central nervous system. EPs can be defined as electrical activity that is time-locked to an event. In the current context, we focus briefly on three types of EPs: somatosensory EPs, auditory EPs, and visual EPs.

Somatosensory EPs are used to study the sensory system, whereby a peripheral nerve, such as the median or tibial nerve, is stimulated by small electrical pulses. Visual EPs assess the functional integrity of optic pathways (i.e., from the retina to the visual cortex in response to visual stimuli). Auditory EPs are often referred to as auditory brainstem responses (ABRs) and are typically labeled with roman numerals (I–VI), with each of these ABRs representing a specific process occurring during auditory processing. Collectively, these EPs are examined during tests to assess the presence of abnormalities by examining the onset latency of specific somatosensory, visual, and/or auditory EPs, which can be used to support a particular diagnosis.

Other Neurodiagnostics

A variety of other neurodiagnostic studies are available to further assist in a comprehensive neurologic evaluation. These tests aid in differential diagnoses of motor neuron and neuromuscular diseases. It is important for neuropsychologists working with these patient populations to be aware of these procedures. In this chapter, we provide an overview of the indications and the limitations of each test, including EMG, NCS, electronystagmography, and autonomic testing. As is the case with most procedures, an appropriately requested test can provide critical clues to the diagnosis and subsequent management of patients, whereas an unnecessarily ordered test can conversely provide misleading data while often being costly and uncomfortable for the patient.

Electromyography and Nerve Conduction Studies

EMG and NCS form the backbone of electroneurodiagnostic studies. However, they do not substitute for the history and clinical neurological examination. When ordered in a targeted manner and with an appropriate differential diagnosis, they have the potential to narrow down the diagnosis and, in some cases, to suggest a specific etiology. On the contrary, it is impractical to do a comprehensive study testing all (or even most) peripheral nerves and muscles available.

EMG and NCS are most useful in diagnosing peripheral nerve disorders, which include disorders affecting the anterior horn gray matter of the spine, dorsal root ganglia, brachial plexus, lumbosacral plexus, peripheral nerves, and neuromuscular junctions. They can also be useful when diagnosing central nervous system diseases, mainly by excluding disorders of the peripheral nervous system.

Broadly speaking, the EMG and NCS can help categorize neuromuscular disorders into one of the three main categories: (1) peripheral neuropathy, (2) neuromuscular junction disorder, or (3) myopathy. If the study is read as normal, then neurologists will likely consider a central nervous system disorder. Therefore, the absence of a positive result still yields valuable diagnostic information. Certainly, more specific localization can help determine the distribution of involvement within the peripheral nervous system, such as within the nerve root, plexus, trunk, division, cord, branch, and peripheral nerve (see Figure 5.1).

In addition to the level of nerve injury, the number of nerves involved (mononeuropathy vs. polyneuropathy), and the size of nerve fibers involved (large, medium, or small fiber neuropathy) can all be critical in arriving at an accurate diagnosis. These studies become even more powerful when coupled with the clinical history such as the clinical symptoms, duration of symptoms, and possible provoking factors.

A variety of neurological disorders can have neuropsychological manifestations, and often these manifestations may be the first clinical symptom of such disorders, leading a patient to medical evaluation. As such, it is of high importance for a clinical neuropsychologist to be familiar with the manifestations

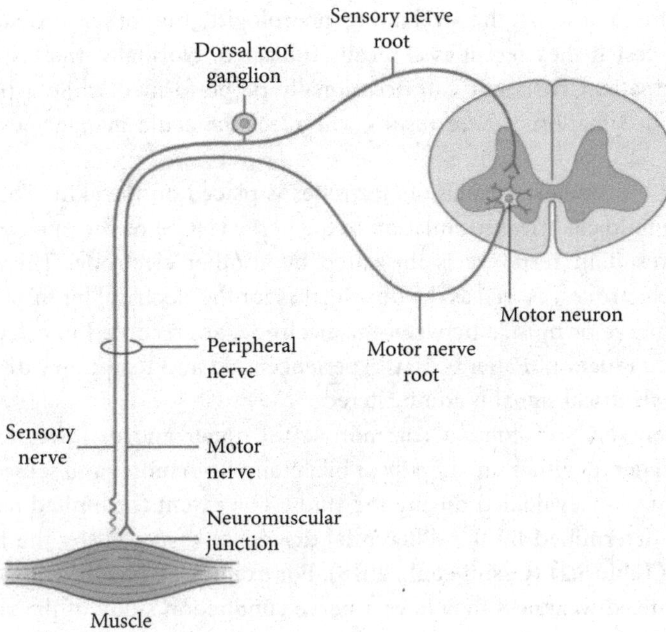

Figure 5.1 Peripheral muscle pathway

From *Electromyography and Neuromuscular Disorders: Clinical-Electrophysiologic Correlations* (3rd ed.), by D. C. Preston & B. E. Shapiro, B. E., 2013, Elsevier Saunders. Copyright 2013 by Elsevier Saunders. Reprinted with permission from Elsevier.

of these disorders. For example, proximal weakness with cognitive decline can be seen in a wide array of neurological disorders, such as amyotrophic lateral sclerosis (ALS) or spinocerebellar ataxia (SCA). In some cases, EMG and nerve conduction studies can be used to differentiate between two or more etiologies of neuropsychological symptoms. For example, an EMG showing myopathic changes (that can be further investigated by a muscle biopsy) can help limit the differential diagnosis to two or three major diagnoses: dermatomyositis, inclusion body myositis, and polymyositis.

Procedure Overview

Accounting for the variety of test structures, NCS and EMG studies together usually take between 30–90 minutes to complete in the EMG lab. The nerve conduction study is performed first, followed by the EMG, as the findings of the former often guide the planning for the latter study. Patients are often referred for a test by the evaluating neurologist, but other providers can order the test if they feel it is clinically indicated. Normally, the test is done on an outpatient basis but can occasionally be performed while a patient is admitted in situations where results will affect the acute management of the patient.

During the studies, a patch of electrodes is placed on the skin. The patient receives a mild electrical stimulation to the nerve (NCS) or the muscle (EMG) and the resultant response is measured by another electrode. The distance between electrodes, as well as the time it takes for the electrical impulse to travel from the nerve or muscle between the electrode, are recorded to calculate the speed of conduction. Patients may experience mild and temporary discomfort while the electrical signal is administered.

Routinely, NCS evaluate a common set of upper and/or lower extremity peripheral nerves either unilaterally or bilaterally. Both motor and sensory pathways are usually evaluated during the study. The extent (or limited nature) of testing is determined by the differential diagnoses suspected by the referring clinician (Table 5.2) (Orsini et al., 2018). For example, a patient with symmetric generalized weakness may have a nerve conduction study of the right arm and the right leg, whereas a patient with asymmetric lower extremity weakness may be limited to testing of the lower extremities. As such, history and clinical exam by the referring provider routinely guide the extent of the study, as crucial segments of the testing may be overlooked if not appropriately requested. Only a limited history and clinical exam are performed prior to the performing the NCS and EMG studies in the lab.

Electronystagmography

Electronystagmography (ENG) is a diagnostic procedure used to evaluate and measure the movements of the eyes, particularly the involuntary and rapid eye movements known as nystagmus. It is a noninvasive test that provides valuable information about the function of the inner ear and the vestibular system (cranial nerve VIII [vestibulocochlear], which is responsible for maintaining balance and spatial orientation. Findings may also be abnormal when cranial nerves III (oculomotor), IV (trochlear), and VI (abducens) are involved (see Chapter 1 for further discussion on cranial nerve function).

ENG is a valuable tool in the field of otolaryngology and neurology, aiding in the diagnosis of a wide range of conditions. It is commonly used to assess patients with vertigo, Ménière's disease, vestibular neuritis, benign paroxysmal positional vertigo (BPPV), or other disorders affecting balance and equilibrium. By providing objective measurements of eye movements, ENG

Table 5.2 Neuromuscular disorders with neuropsychiatric manifestations

Disorder	Initial Presentation	Neuropsychiatric Manifestations	EMG/NCS findings
Duchenne Muscular Dystrophy (DMD)	-Progressive proximal muscle weakness -Global developmental delay -Calf muscle hypertrophy	FSIQ impairment more than one standard deviation below mean. Specifically, verbal memory, visual attention, and processive speed impairment	Spontaneous activity suggesting denervation
Myotonic Dystrophy type 1 (DM 1)	-Distal muscle weakness -Myotonia -Facial muscle weakness	Moderate impairment of cognitive performance in frontal, visuospatial, naming, and memory abilities.	Spontaneous (myotonic) discharges; Normal vs decreased CMAP
Mitochondrial Encephalomy-opathy, Lactic Acidosis, and Stroke-Like Episodes (MELAS)	-Pediatric onset -Recurrent attacks of migraine, seizures -Proximal muscle weakness	Specific cognitive deficits: visual construction, attention, and abstraction/flexibility	Normal (important to rule out other etiologies)
Amyotrophic Lateral Sclerosis (ALS)	-Bulbar dysfunction -Spasticity -Slurred speech	Fulfill criteria for frontotemporal dementia. Cognitive impairment without executive dysfunction.	Active denervation with reinnervation in multiple body segments
Parkinson's disease; Lewy body dementia	-Resting tremor -Muscle rigidity -Masked facies -Bradykinesia	Depression Cognitive impairment (Parkinson's dementia)	Normal

CMAP, compound muscle action potential

assists healthcare professionals in formulating accurate diagnoses and designing effective treatment plans to alleviate symptoms and improve patients' quality of life.

In recent years, advanced diagnostic techniques such as videonystagmography (VNG) and rotary chair testing have gained popularity, offering additional benefits and improved accuracy compared to traditional ENG. These newer methods often incorporate video-based eye-tracking technology and provide more detailed information about eye movements and vestibular function. However, ENG remains a valuable tool in many clinical settings and continues to play a significant role in the assessment and management of balance disorders.

Procedure Overview

During an ENG test, electrodes are placed around the eyes to detect and record eye movements. The test typically involves a series of different stimuli, such as visual patterns, cold or warm air (or water in a caloric test) stimulation in the ear canals, or changes in body position. These stimuli help to provoke nystagmus and provide insights into the underlying causes of balance disorders, vertigo, or dizziness. The recorded eye movements are then analyzed to assess various aspects of vestibular function. This includes measuring the intensity and duration of nystagmus, the presence of spontaneous or positional nystagmus, and the response of the eyes to different stimuli (Baloh & Kerber, 2011). The test results can help clinicians determine if there are abnormalities in the vestibular system, identify the affected ear or structure, and guide appropriate treatment strategies.

ENG was found to be helpful in differentiating organic psychiatric disorders from functional symptoms. Patients with organic psychiatric disorders and nystagmus almost always exhibit abnormal eye movements and nystagmus patterns on ENG due to underlying structural or functional abnormalities in the brain or inner ear (Kiyomizu et al., 2011). These abnormalities may be indicative of conditions such as brain lesions, neurodegenerative diseases, or vestibular disorders.

In contrast, patients with functional symptoms may show normal ENG results, indicating that their symptoms are not caused by an underlying organic condition. This can be helpful in ruling out specific organic causes and guiding clinicians toward considering psychiatric factors as the primary contributors to the symptoms. It is important to note that ENG alone cannot definitively diagnose or exclude organic psychiatric disorders, but it can provide valuable information that complements the overall clinical assessment. The use of ENG, in conjunction with a thorough clinical evaluation, psychiatric assessments, and other diagnostic tests, allows healthcare professionals to make more informed decisions in the differential diagnosis of patients with psychiatric symptoms. This helps ensure appropriate management strategies, whether they involve medical interventions, psychological therapies, or a combination of both, leading to improved patient care and outcomes.

Autonomic Testing

Autonomic testing plays an important role in the evaluation and management of neuropsychiatric disorders. The autonomic nervous system (ANS) regulates involuntary functions in the body, such as heart rate, blood pressure, digestion, and sweating. Autonomic testing involves a range of assessments designed to evaluate the integrity and functioning of the ANS. One commonly used technique is heart rate variability (HRV) analysis, which measures the variation in time intervals between successive heartbeats. HRV provides insights into the balance between sympathetic and parasympathetic activity, with reduced HRV indicating autonomic dysfunction. It has been shown to be altered in conditions like anxiety disorders and depression, reflecting increased sympathetic tone and reduced parasympathetic activity.

Another important autonomic test is the tilt-table test, which assesses the body's cardiovascular response to postural changes. During this test, the individual lays flat on a table that is then tilted to simulate standing up. Tilt-table testing is particularly useful in conditions such as orthostatic hypotension, where individuals experience a sudden drop in blood pressure upon standing, often leading to dizziness or fainting. This can be seen in patients with postural orthostatic tachycardia syndrome (POTS) or other forms of dysautonomia, as well as neurodegenerative conditions that often have abnormal ANS function, such as multiple system atrophy (MSA).

Furthermore, sudomotor function testing measures sweat production and distribution across the body. This evaluation helps assess the integrity of the sympathetic nervous system, which controls sweating. It is important for neuropsychologists to be aware of sudomotor dysfunction, as it is frequently observed in neurodegenerative disorders, such as Parkinson's disease and multiple system atrophy (Jarczok et al., 2013).

Relevance to Neuropsychology

Although neuropsychological evaluations are primarily requested to characterize cognitive functioning, understanding a patient's medical history provides important clinical and historical context for interpreting the results of the neuropsychological evaluation. Increasingly, neuropsychology is being integrated into multidisciplinary clinics, where basic familiarity with various neurologically related tests and procedures becomes critical to differential diagnosis, case conceptualization, and treatment planning. Neuropsychologists are likely to encounter patients who have had EMGs or autonomic testing. It is important to understand the patient's physical symptoms and functioning as this may impact factors relevant to testing, such as fatigue, stamina, or motor skills needed for certain tests. For example, if the patient has weakness in their dominant upper extremity, the neuropsychologist may want to minimize tests that have demands for written responses. It is also helpful to know about that symptomatology when interpreting tests. The neuropsychologist may want to know if their patient has a risk of falling when considering how they will ambulate to the testing room.

In general, understanding the presentation of a patient's sensorimotor symptoms can help the neuropsychologist mitigate discomfort and maximize patient safety during the neuropsychological evaluation.

Case Example: Myasthenia Gravis

Myasthenia gravis is a neuromuscular disorder that can cause drooping of the eyelids (ptosis) or weakness of eye muscles, blurred or double vision (diplopia), changes in facial expression, dysarthria, difficulty swallowing, and muscle weakness. Many patients will experience a medical emergency known as a myasthenic crisis, which occurs when the muscles that control breathing weaken to the point that the person requires a ventilator to breathe. Typically, myasthenia gravis is thought to be primarily a neuromuscular disorder and not directly associated with cognitive impairment. However, patients with myasthenia gravis can experience a number of non-motor symptoms which can secondarily impact their cognitive functioning and overall quality of life.

Ms. Jane Doe is a 67-year-old woman with a medical history of type 2 diabetes, hypertension, hyperlipidemia, and seronegative generalized myasthenia gravis diagnosed 8 years ago. The diagnosis of myasthenia gravis was based on medical history (including response to prior therapy) and a single-fiber EMG. At the time of her EMG, she presented with fluctuating generalized weakness, diplopia, and bilateral ptosis starting approximately 3 years prior. Single fiber studies of the left extensor digitorum communis (EDC), left frontalis, left orbicularis oculi, and left deltoid muscle were performed. Per the report, the left EDC demonstrated blocking in 10% of fibers with normal jitter in 50% of pairs. The other muscles were unremarkable. Ms. Doe was treated with pyridostigmine, an acetylcholinesterase inhibitor that works at the neuromuscular junction, and a course of IVIG with marked symptomatic relief. She returned to the Neurology Clinic after a few years. She has not taken pyridostigmine for many months. She told the neurologist that she continues to experience diplopia and ptosis on a daily basis. The neurologist noted that the patient's speech and language were normal. Pupils were equal, round, and reactive to light. There was moderate bilateral ptosis that did not fatigue. Eye closure, cheek puff, and tongue protrusion were strong. Her palate elevated symmetrically. With red lens testing, diplopia on right gaze was elicited, localizing to the left medial rectus muscle. There was also reduced upgaze with mild exophoria in the left eye. Neck flexion was strong. The remainder of the motor exam was normal. There was a stocking-and-glove distribution decrement in pinprick and vibratory sensation; joint position was normal. The neurologist described extensive distal lower extremity pitting edema and weeping, as well as serous lesions over the distal anterior calves bilaterally. Reflexes were reportedly absent in the lower extremities and noted as trace in the upper extremities. The neurologist ordered a nerve conduction study and repeat EMG to evaluate for polyneuropathy and myasthenia gravis. The report noted electrophysiologic evidence of: (1) a diffuse sensorimotor polyneuropathy (the study was technically difficult given the patient's body habitus and lower extremity edema, but the abnormal sensory responses in the upper extremity support an underlying neuropathy), and (2) a disorder of neuromuscular transmission compatible with myasthenia gravis.

Ms. Doe is referred for neuropsychological evaluation after continuing to report memory concerns to her doctor following a myasthenia gravis crisis for which she was hospitalized and placed on a ventilator for several weeks. She reports that her memory seems to be getting worse and that she cannot find words, despite reportedly having an excellent vocabulary. Ms. Doe states she has experienced decline in her attention, concentration, and ability to see tasks to completion. She feels she cannot multitask. She reports some trouble following recipes and leaving the stove on. She has been managing her medications, but her regimen is quite complex and she has found it difficult to organize and remember all of her medications correctly. She continues to manage finances, but occasionally has had trouble due to change in her organizational skills. Additionally, poor stamina and physical fatigue also impact her ability to manage complex activities of daily living.

Ms. Doe's neurocognitive profile reveals relative weakness in aspects of attention/executive functioning, such as working memory and processing speed. Learning and memory is secondarily impacted and characterized by inefficient encoding of rote information (word list or visual display); however, retention and recall of information over time is commensurate with expectation. Ms. Doe demonstrates impaired motor dexterity in her dominant (left) hand. Verbal and nonverbal reasoning skills, as well as problem-solving and word retrieval are generally intact. Ms. Doe's emotional functioning is notable for mild depressive symptoms, and moderate anxiety.

Overall, Ms. Doe is doing fairly well from a cognitive perspective. The most likely etiology for decline in manual motor dexterity, as well as her fatigue, is her history of myasthenia gravis. The inefficiency with memory encoding is likely attributable to cerebrovascular risk factors. There is no evidence of an amnestic memory impairment to suggest mesial temporal damage or a secondary neurodegenerative process. It is also noteworthy that Ms. Doe currently endorses moderate symptoms of anxiety and some mild symptoms of depression. On interview she describes feelings of frustration around her subjective cognitive difficulties, as well as medically related fears, which has started to affect her sleep. She would likely benefit from discussing these concerns with a therapist who could help her cope with her functional changes, as well as provide strategies which may improve her sleep.

Conclusion

This chapter has reviewed the foundational role of electroencephalography (EEG), electromyography (EMG), and nerve conduction studies (NCS) in diagnosing and managing neurological and neuropsychiatric conditions. These tools provide critical insights into brain and peripheral nerve activity, allowing for the assessment of disorders ranging from seizures and sleep disturbances to neuromuscular and neurodegenerative diseases. As demonstrated through specific applications, EEG, EMG, and NCS aid clinicians in forming precise diagnoses by analyzing distinct electrical patterns, helping to tailor interventions. Their diagnostic precision, despite inherent limitations, emphasizes their value in clinical practice, supporting improved patient outcomes across a spectrum of neurological conditions.

References

Abhang, P. A., Gawali, B. W., & Mehrotra, S. C. (2016). *Introduction to EEG-and speech-based emotion recognition*. Academic Press.

Adrian, E. D., & Matthews, B. H. C. (1934). The Berger rhythm: Potential changes from the occipital lobes in man. *Brain, 57*(4), 355–385. https://doi.org/10.1093/brain/57.4.355

American Clinical Neurophysiology Society. (2008). Guidelines twelve: Guidelines for long-term monitoring for epilepsy. *Journal of Clinical Neurophysiology, 25*(3), 170–180.

American Electroencephalographic Society. (1994). Guidelines in electroencephalography, evoked potentials, and polysomnography. *Journal of Clinical Neurophysiology, 11*(1), 1–147.

Baloh, R. W., & Kerber, K. A. (2011). *Clinical neurophysiology of the vestibular system*. Oxford University Press.

Berger, H. (1929). Ueber das elektrenkephalogramm des menschen. *Archives for Psychatrie Nervenkrankheiten, 87*, 527–580.

Biasiucci, A., Franceschiello, B., & Murray, M. M. (2019). Electroencephalography. *Current Biology, 29*(3), R80–R85.

Britton, J. W., Frey, L. C., Hopp, J. L., et al. (2016). *Electroencephalography (EEG): An introductory text and atlas of normal and abnormal findings in adults, children, and infants*. American Epilepsy Society.

Ghatol, D., & Widrich, J. (2024). Intraoperative neurophysiological monitoring. *StatPearls*. https://www.ncbi.nlm.nih.gov/books/NBK563203/

Gibbs, F. A., Davis, H., & Lennox, W. G. (1935). The electro-encephalogram in epilepsy and in conditions of impaired consciousness. *Archives of Neurology & Psychiatry*, *34*(6), 1133–1148. https://doi.org/10.1001/archneurpsyc.1935.02250240002001

Hajcak, G., Klawohn, J., & Meyer, A. (2019). The utility of event-related potentials in clinical psychology. *Annual Review of Clinical Psychology*, *15*, 71–95.

Jarczok, M. N., Jarczok, M., Mauss, D., Koenig, J., Li, J, Herr, R. M., & Thayer, J. F. (2013). Autonomic nervous system activity and workplace stressors—a systematic review. *Neuroscience & Biobehavioral Reviews*, *37*(8), 1810–1823. https://doi.org/10.1016/j.neubiorev.2013.07.004

Jasper, H. H., & Carmichael, L. (1935). Electrical potentials from the intact human brain. *Science*, *81*, 51–53.

Jasper, H. H. (1958). Ten-twenty electrode system of the international federation. *Electroencephalogr Clin Neurophysiol*, *10*, 371–375.

Kappenman, E. S., & Luck, S. J. (2016). Best practices for event-related potential research in clinical populations. *Biological Psychiatry: Cognitive Neuroscience and Neuroimaging*, *1*(2), 110–115.

Kiyomizu, K., Matsuda, K., Torihara, K., Nakayama, M., Ishida, Y., Yoshida, K., & Tono, T. (2011). Neuro-otological findings in psychiatric patients with nystagmus. *European Archives of Oto-Rhino-Laryngology*, *268*(12), 1713–1719. https://doi.org/10.1007/s00405-011-1524-z

Lagerlund, T. D., Cascino, G. D., Cicora, K. M., & Sharbrough, F. W. (1996, October). Long-term electroencephalographic monitoring for diagnosis and management of seizures. *Mayo Clinic Proceedings*, *71*(10), 1000–1006.

Luck, S. J. (2014). *An introduction to the event-related potential technique*. MIT Press.

Orsini, M., Ferreira, A. de, Damm de Assis, A. C., Magalhães, T., Teixeira, S., Bastos, V. H., Marinho, V., Oliveira, T., Fiorelli, R., Oliveira, A. B., & De Freitas, M. R. (2018). Cognitive impairment in neuromuscular diseases: A systematic review. *Neurology International*, *10*(2). https://doi.org/10.4081/ni.2018.7473

Preston, D. C., & Shapiro, B. E. (2013). *Electromyography and neuromuscular disorders: Clinical-electrophysiologic correlations*. Elsevier/Saunders.

Tewari, A., Francis, L., Samy, R. N., Kurth, D. C., Castle, J., Frye, T., & Mahmoud, M. (2018). Intraoperative neurophysiological monitoring team's communiqué with anesthesia professionals. *J Anaesthesiol Clin Pharmacol*, *34*(1), 84–93.

Thompson, J. L., & Ebersole, J. S. (1999). Long-term inpatient audiovisual scalp EEG monitoring. *Journal of Clinical Neurophysiology*, *16*(2), 91–99.

Yamada, T., & Meng, E. (2017). *Practical guide for clinical neurophysiologic testing: EEG* (2nd ed.). Lippincott Williams & Wilkins.

6
Intraoperative Mapping and Monitoring

Barbara Tomasino, Jill Z. Stuart, and Miran Skrap

Background

Awake surgery (e.g., Duffau et al., 2003; Duffau et al., 2005; Ojemann, 1993; Ojemann et al., 1989) is a widely used technique (e.g., Krieg, 2022) in neurosurgery units where patients with brain glioma or language hemisphere dominant medication refractory epilepsy are treated surgically. For these patients, surgical intervention is necessary to avoid disease progression and/or mitigate the potential for morbidity. Use of neurocognitive assessment to map, and hopefully spare, eloquent tissue during these surgeries reduces cognitive deficits (e.g., Morshed et al., 2021). This is particularly beneficial for language functions, including speech, comprehension, and reading skills (e.g., DeWitt Hammer et al., 2012; Morshed et al., 2021).

Patients with brain glioma are generally young, and thanks to neurosurgical and neuro-oncological efforts, along with more recent technical advances, their survival is increasing. A brain glioma often invades cortical areas involved in cognitive processing and the white matter underneath. In cases requiring surgical treatment, resection in these areas is not without risk to the patient's cognitive status (Sanai & Berger, 2008). Often in brain tumor surgeries, a key factor in a patient's survival is the extent of resection, with maximal resection typically resulting in best outcomes in terms of disease progression/re-emergence (e.g., Duffau et al., 2003; Ius et al., 2012; Keles et al., 2004; Sanai et al., 2008; Sanai, Polley, & Berger, 2010; Sanai & Berger, 2011). That said, surgeons aim to balance the preservation of cognitive functioning, while being as surgically aggressive as possible, to ensure and maintain the highest possible quality of life for the patient.

Case Example

As an example of how a low-grade glioma invades the eloquent cortex and white matter, we show below the MRI images of a 24-year-old right-handed

Barbara Tomasino, Jill Z. Stuart, and Miran Skrap, *Intraoperative Mapping and Monitoring*. In: *Medical and Laboratory Examinations for Neuropsychologists*. Edited by: Thomas J. Farrer and R. Brock Frost, Oxford University Press. © Oxford University Press (2026). DOI: 10.1093/9780197619537.003.0007 young man with a low-grade glioma involving the left temporal cortex and the adjacent subcortical white matter (Figure 6.1). Symptom presentation was characterized by seizures consisting of the perception of a sweet "Christmas-like" smell, pleasant but difficult to identify, initially occurring multiple times a day (6–7 times). At the time of examination, the smell sensation was occurring about once a month, with episodes lasting for approximately 1 minute. Not infrequently, these episodes were accompanied by emotional lability and crying. After experiencing one of these spells while he was at work, the patient decided to meet a psychiatrist who recommended an MRI examination. These episodes were eventually determined to be seizures.

As part of a comprehensive workup, the patient underwent a neuropsychological evaluation (Table 6.1) and an fMRI examination. The neuropsychological evaluation showed that cognitive functions were within the normal range. The purpose of pre-intervention neuropsychological assessment is multifold: (1) to collect the patient's history and current cognitive and psychiatric symptoms; (2) to assess suitability for awake surgery based on cognitive and psychiatric status; (3) to administer in-depth cognitive assessment to establish a presurgical cognitive baseline, including any deficits present before surgery; and (4) to establish comprehension and baseline performance for tasks repeatable during an awake surgery. Understanding the neuroanatomy around the area of the tumor allows the neuropsychologists to effectively select tasks to monitor important linguistic or neurocognitive functions in and around the area being considered for resection.

The figures below show the patient's T2-FLAIR and T1 MRI clinical images (Figure 6.1), in which the low-grade glioma is highlighted in red.

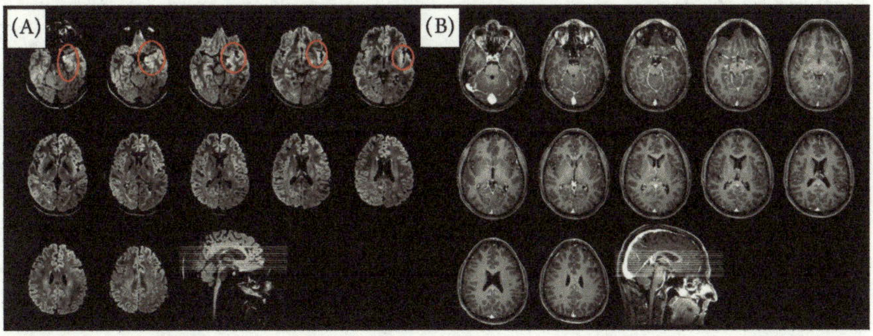

Figure 6.1 T2 Flair and T1 MIR images of glioma for case 1. The T2-flair and T1 MRI clinical images, in which the low-grade glioma is highlighted in red on the T2-flair image.

Table 6.1 Case 1 neuropsychological status

Cognitive Domain	Function	Test	Reference		Pre-Surg	Imm. Post Surg	Follow-Up
Fluid intelligence	Logic reasoning	Raven's colored progressive matrices	(Basso et al., 1987)	RS	33	-	-
				CS	30.5	-	-
				ES	3	-	-
Executive functions	Strategy use	Verbal fluency	(Novelli et al., 1996)	RS	51	38	46
				CS	45	32	40
				ES	4	4	4
	Digit span	Backward test	(Monaco et al., 2015)	RS	4	4	4
				CS	3.42	3.42	3.42
				ES	2	2	2
Short-term memory	Digit span	Forward test	(Monaco et al., 2015)	RS	6	5	7
				CS	5.44	4.44	6.44
				ES	3	2	4
Praxis	Upper limb imitation	Ideomotor apraxia	(De Renzi et al., 1980)	RS	70	69	72
				PE	4	4	4
	Buccofacial imitation	oral apraxia	(Spinnler & Tognoni, 1987)	RS	20	20	20
				CS	19.75	19.75	19.75
				ES	4	4	4
Language	Comprehension	Token test	(De Renzi & Faglioni, 1978)	RS	34	35	35
				CS	31.5	32.5	32.5
				ES	2	3	3

	Objects comprehension	(Miceli et al., 1994)	Cutoff 39	40	40	40
	Verbs comprehension	(Miceli et al., 1994)	Cutoff 19	20	20	20
Semantics	Semantic fluency	(Novelli et al., 1996)	RS	53	50	66
			S	47	44	60
			ES	4	4	4
	Object naming	(Miceli et al., 1994)	Cutoff 28	39	30	30
	Verb naming	(Miceli et al., 1994)	Cutoff 26	28	26	28
Semantic memory	Pyramids and palm trees test	(Gamboz et al., 2009)	RS	52	52	52
			CS, cutoff 40.15	49.69	49.69	49.69
Writing	Word and pseudoword writing	(Luzzatti et al., 1998)	cutoff 16.675	180	179	180
Reading	Word reading	(Ripamonti et al., 2014)	cutoff 78	79	79	79
	Pseudoword reading	(Ripamonti et al., 2014)	cutoff 18	20	20	20
	Phrase reading	(Miceli et al., 1994)	cutoff 6	6	6	6
	Three-syllabic word reading	(Toraldo et al., 2006)	cutoff 24	24	24	24
Repetition	Word repetition	(Miceli et al., 1994)	cutoff 80	80	80	80
	Pseudoword repetition	(Miceli et al., 1994)	cutoff 20	20	20	20
Lexical processing	Visual lexical decision	(Ripamonti et al., 2016)	cutoff 136	140	139	1387
Lexical processing	Auditory lexical decision	(Ripamonti et al., 2016)	cutoff 121	127	125	125

RS (raw score), CS (corrected score), ES (equivalent score)

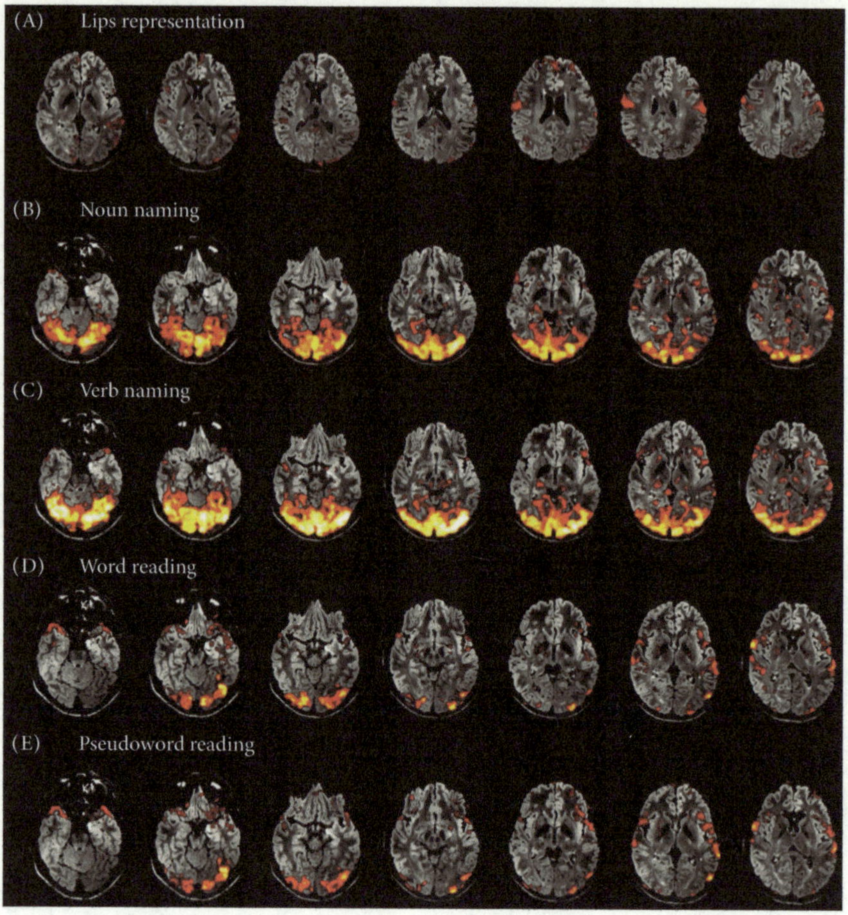

Figure 6.2 fMRI maps by task. fMRI maps obtained for each task we used (tongue movements, A; object naming, B; verb naming, C; word reading, D; pseudoword reading, E).

We acquired the fMRI images showing the fMRI maps obtained for each task we used (tongue movements, object naming, verb naming, word reading, pseudoword reading (Figure 6.2).

To better understand key language areas that could be affected by the patient's brain tumor or the resection of the tumor, we acquired diffusion tensor images (DTI), showing the inferior fronto-occipital fasciculus (IFOF), the inferior longitudinal fasciculus (ILF), the superior longitudinal fasciculus (SLF), and the uncinate fasciculus (Figure 6.3). These areas are typically tracked when the glioma invades the language-related areas in the left hemisphere given possible mass effect pressing on these fasciculi. In this example, the fMRI maps related to the language task show that there was part of the activation localized in the left inferior frontal cortex close to the low-grade glioma for naming nouns and verbs (Figures 6.2B and 6.2C) and for reading (Figures 6.2D and 6.2E). As to the white matter tracts, the DTI showed that a part of the IFOF is traced inside the hyperintensity (Figure 6.3A); the ILF passes externally to the lesion (Figure 6.3B); the anterior part of the SLF, ending in the inferior frontal gyrus, passes internally to the hyperintensity (Figure 6.3C); and the uncinate fasciculus is traced anterior to the hyperintensity (Figure 6.3D).

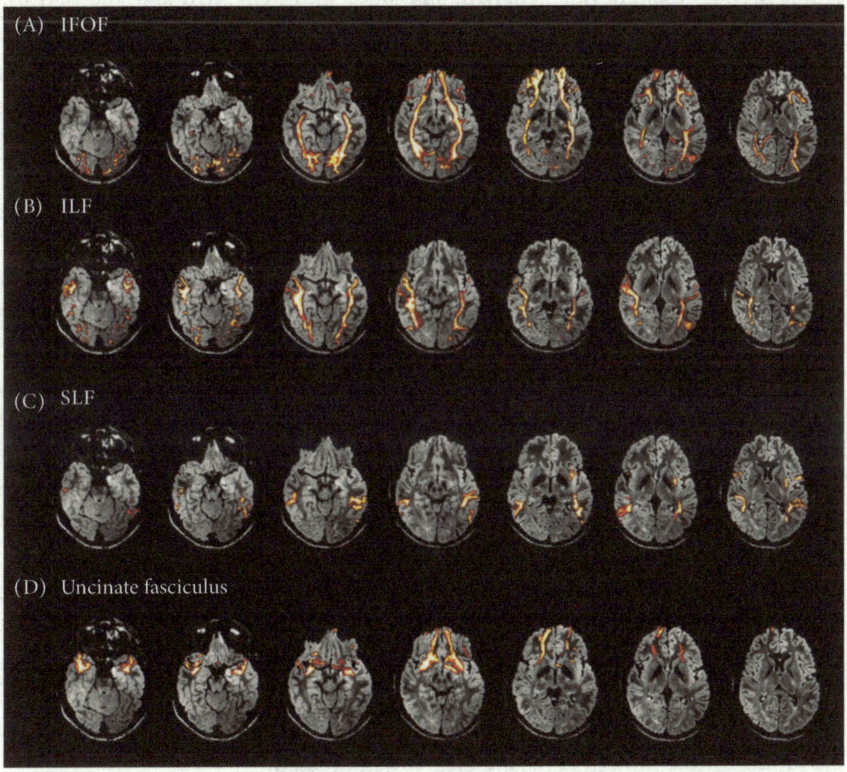

Figure 6.3 DTI images showing the inferior frontal-occipital fasciculus (IFOF), inferior longitudinal fasciculus (ILF), superior longitudinal fasciculus (SLF), and the uncinate fasciculus

Understanding Intraoperative Mapping and Monitoring

During awake brain surgery, the patient is usually anesthetized during the initial craniotomy and awoken when the resection cavity is exposed. As noted above, the goal of awake surgery is for the surgeon to tailor the resection to achieve maximal resection of the tumor while maintaining the functional tissue. To do so, the gold standard, routinely used, is direct electrical stimulation (DES) (e.g., Berger, 1994, 1996; Berger & Ojemann, 1992; Duffau et al., 2000; Duffau et al., 2005). DES is designed to inhibit or excite the function of the nodule or tissue that is being probed during a cognitive task, which allows for a pseudo-lesional approach to functional assessment of underlying tissue.

Intraoperative monitoring (IOM) is a set of monitoring techniques such as electroencephalography, electromyography, and evoked potentials that are applied continuously during resection. The aim is to provide functional guidance to the surgeon to reduce the risk of postsurgery damage to the nervous system. IOM provides a continuous neurophysiological recording reflecting the status and functional integrity of certain neural structures during surgery. IOM allows monitoring of sensorimotor and visual functions.

DES is a mapping technique that is applied cortically and subcortically, before resection and intermittently during the resection phase (e.g., Manfield et al., 2021). The result of DES is a spatial map showing the areas to be preserved or avoided during resection because they represent eloquent or functional tissue (e.g., those sites identified as "positive" during stimulation). DES is performed in conjunction with IOM. For example, intracranial electroencephalography (EEG) whereby electrodes are placed directly on the surface of the brain. The EEG records activity from the cortex to determine if the amount of stimulation delivered (in milliamps) is sufficient to disrupt function, yet not so much that the patient runs the risk of having a seizure. During DES, the patient performs a cognitive task (Figure 6.4). The patient's response during stimulation, or lack thereof, allows the neuropsychologist and neurosurgeon to determine whether the tissue in and around the tumor is functional (e.g., Axelson et al., 2022). For example, by stimulating Broca's area, a speech arrest is obtained. This indicates that this portion of the tissue is a functional component contributing to expressive language. This process is repeated in multiple cortical areas around the tumor until eloquent areas are clearly demarcated.

If cognitive functioning is disrupted with DES, the result is considered a "positive" result. In contrast, a "negative" result means that stimulation did not cause any interference. However, a negative result does not allow the conclusion that this portion of the tissue is not functional. There are multiple factors that could contribute to a negative mapping result. First, due to the focal nature of stimulation exerted, DES can electrically condition only a small and circumscribed

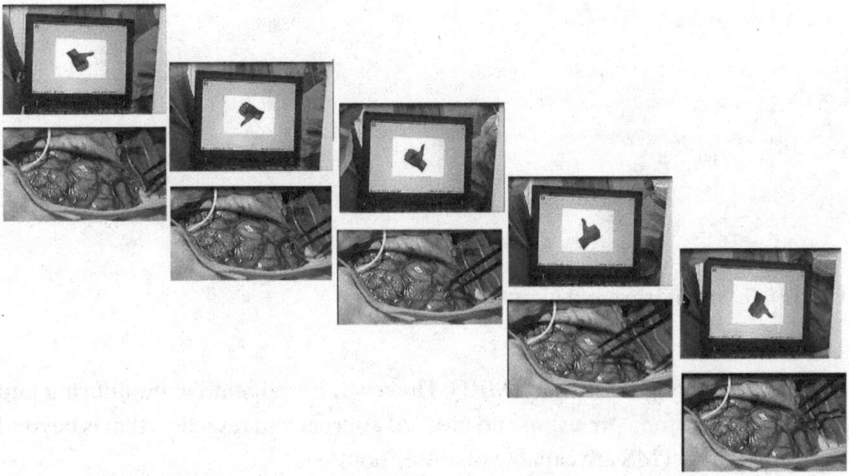

Figure 6.4 Example of stimuli sequence paired with DES. An example of DES along with the stimuli sequence presentation occurring paired with every stimulation. The probe stimulates different portions of the cortex.

area. Second, a mismatch between the cognitive test and the cognitive functions supported by the tested area. Third, higher cognitive functions are organized in network of areas, and DES could have insufficient local effects to disrupt the entire network (e.g., Mandonnet et al., 2021). Fourth, due to time constraints, the number of tasks that can be administered during DES is limited. Fifth, the risk of causing seizure when using DES can limit the number of stimulations in a particular area. Finally, it is also important to consider patient-specific variance in the anatomical representation of a particular cognitive function. As in, there can be some individual differences in functional neuroanatomy across patients. Although DES allows for precise determination of the functional eloquence of the tissue being probed, it has the disadvantage of requiring many repetitions of the stimulations, which can be time consuming (e.g., Bertani et al., 2009).

Real-Time Neuropsychological Testing

The causal effect of brain change on cognition has long been investigated by correlating patients' impaired behavior to identified lesions. Even for virtual lesions created by using transcranial magnetic stimulation (TMS), which induces an electrical temporary reversible-lesion, effects can be measured in real-time simultaneously with stimulation delivery. There are other real-time applications within neuroscience, such as with real-time functional magnetic

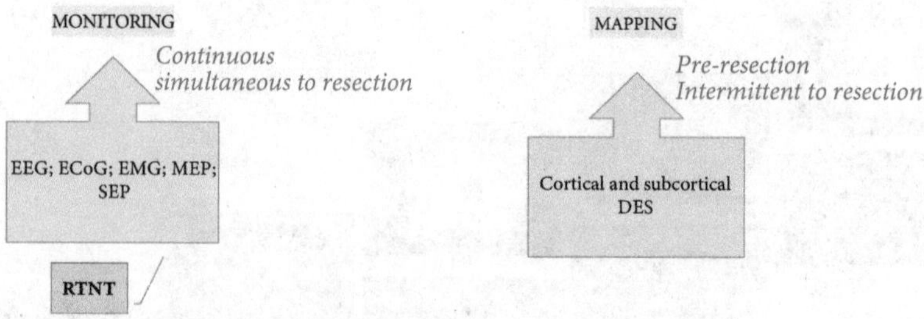

Figure 6.5 Monitoring and mapping techniques

resonance imaging (real-time fMRI). However, intraoperative monitoring and mapping allows for a precision and tailored approach to resection that is beyond what fMRI and TMS are capable of at this point.

Our approach, real-time neuropsychological testing (RTNT), is a monitoring technique that is applied continuously during resection (Figure 6.5). The goal of RTNT is to guide surgical resection to allow the surgeon to achieve maximum extension of resection with minimal risk for functional deficits. As such, the neuropsychologist directly contributes to the outcome of the clinical interventions (Ng et al., 2020). When using DES, an anatomical site is considered functional if the effect is obtained on at least two out of three stimulations. When using RTNT, if the patient continues performing the task without error, resection continues; errors are communicated to the surgeon immediately.

Selection of Neurocognitive Tasks for Intraoperative Mapping

Real-time neuropsychological testing (RTNT) allows for monitoring many neuropsychological functions, including executive functions, sensorimotor abilities, language-related functions, visuospatial, or emotional-related abilities. The most common language-based tasks utilized during intraoperative mapping include auditory and visual naming paradigms (Hamberger et al., 2003). However, it is also common to assess additional cognitive function typically located within the area being considered for surgical resection. In this section, we describe several neuropsychological tasks by region of interest. A selection of these tasks is utilized depending on the areas of interest for a particular surgical case. It is not clinically necessary, or efficient, to utilize an exhaustive test battery during awake brain surgery.

This section is meant to give the reader an idea of skills that could be sampled, as well as corresponding tasks that have been used by the authors. For the case presented in this chapter, neurocognitive tasks were selected from published

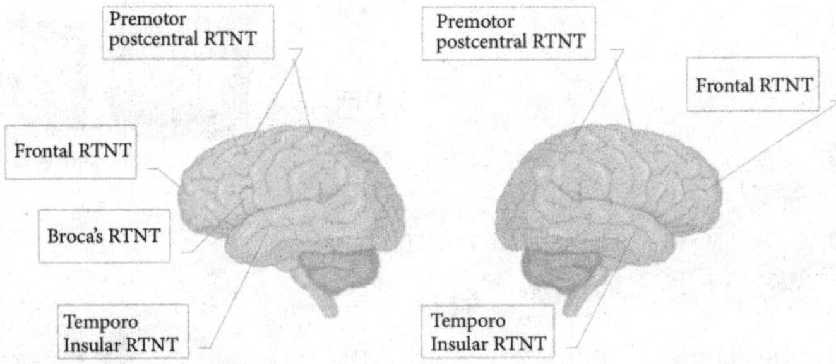

Figure 6.6 Areas of interest for RTNT mapping and monitoring

neuropsychological batteries available in the Italian norming data, many of which will be familiar to English-based neuropsychologists. The authors created a fixed task order to capture skills relevant for functions in the following areas: left frontal, left Broca's area, left premotor/postcentral, left temporo-insular, right frontal, right premotor/postcentral, and right temporo-insular areas (Figure 6.6). A certain task might be included when assessing more than one area, as that particular cognitive function may be supported by a network of areas. Stimuli have to be tailored to patients' presurgery ability, as some patients may exhibit selective naming difficulties at the baseline (presurgical) assessment (Tomasino, Tronchin, et al., 2018). A subset of stimuli is selected for each task, using items from the neuropsychological baseline assessment.

Several surgery programs utilize a dynamic assessment approach with RTNT, assessing several skills in succession before applying electrical stimulation to a new location. For example, the Italian authors of this chapter utilize a RTNT that allows for the assessment of object naming, word reading, and pseudoword repetition, among other skills within a relatively circumscribed area (Figure 6.7). The rationale for using multiple tasks is that if a single task is used (e.g., a visual naming task) there is the possibility, due to mismatch between an area and the function of the tissue being assessed, of obtaining a false negative. For example, in the accompanying clinical case, only by testing the patient with multiple tasks was a deficit identified.

If there is concern for postoperative conduction aphasia, repetition tasks for words, pseudowords, and sentences may be included in the selected tests (Miceli et al., 1994; Tomasino et al., 2015b). Reading tasks of words and pseudowords (Miceli et al., 1994) are included for some patients due to the circumscribed nature of some lesions, such as those affecting repetition and writing, but not reading aloud (Tomasino et al., 2015b). Reading could be very useful for resection in the temporo-occipital areas and the white matter beneath such as the inferior fronto-occipital fasciculus (Tomasino et al., 2015a). In order to monitor comprehension and phonological input processing, RTNT protocols may include phonemic discriminations (Miceli et al., 1994), auditory object

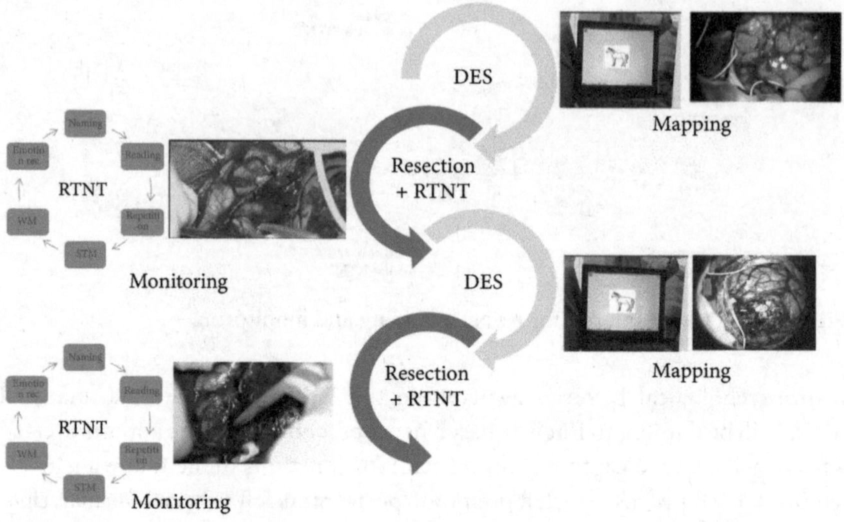

Figure 6.7 Continuous mapping and monitoring with a variety of tasks

comprehensions (Miceli et al., 1994), lexical decisions (Miceli et al., 1994), sentence comprehension (Miceli et al., 1994), and the comprehension of syntax (De Renzi & Faglioni, 1978). Tasks tapping semantic associations (Gamboz et al., 2009) and phonological fluency (Novelli et al., 1996) are also included in some batteries. For a detailed description of the Italian approach to RTNT of language in left temporal region see Tomasino et al. (2021), where the authors extensively describe the left temporal RTNT protocol.

In addition to language-related tasks, attention and working memory are assessed through digit span forward (Orsini et al., 1987) and digit span backward (Orsini et al., 1987), respectively. Assessment of executive functions includes inhibitory control via the Stroop task (Barbarotto et al., 1998), phonological fluency, and semantic fluency (Novelli et al., 1996). To control for information processing speed, which can be influenced by sustained attention and selective attention, we included also the oral version of the Symbol Digit Modalities Test (Nocentini et al., 2006). In addition, for surgeries involving the frontal lobe or the temporo-insular region, RTNT also involves tasks tapping social cognition such as a theory of mind task (Prior et al., 2003), as well as evoking emotions upon visual presentation of pictures (International Affective Picture Stimuli) (Lang et al., 2008). For a detailed description of RTNT in insular regions, see (Tomasino, Marin, et al., 2018), where the authors extensively describe the RTNT insular protocol.

For surgeries involving the right hemisphere, the RTNT protocols were designed to assess functions of right frontal, right premotor, right temporo-insular, and right parietal areas. Visuospatial processing is assessed by using the landmark task (Capitani et al., 2000), a visual scanning task requiring reading a list of words and of sentences (Miceli et al., 1994) appearing in different positions and different quadrants on the screen, and a picture description task (Paradis, 1986) in which patients are asked to detect all the relevant target stimuli appearing in different positions and different quadrants on the PC screen. Visual processing is tested by using an object decision task in which patients decide whether a stimulus is a real object or a chimera (Rusconi, 1998). The ability to create and manipulate mental visual images is tested through a clock test (Antonietti et al., 2008); the patient is asked to imagine a pair of clock faces indicating different times, proposed verbally, and to indicate where the clock hands appear. This task is useful also to monitor neglect. For surgeries involving right temporal areas RTNT also includes a face recognition task (Bizzozero et al., 2005). The ability to comprehend figurative language is assessed by using a metaphor comprehension task (Papagno et al., 1995).

For resections in both the left and the right hemisphere, especially around sensorimotor areas, tasks involve handedness decisions (Tomasino et al., 2003) since patients may exhibit an inability to predict and simulate action plans, or hand and arm movements (Tomasino et al., 2004; Tomasino et al., 2011). The RTNT also includes the Florida praxis imagery test (Ochipa et al., 1997), kissing and dancing test (Bak & Hodges, 2003), assessment of finger agnosia (Spinnler & Tognoni, 1987), oral apraxia (De Renzi et al., 1966), and ideomotor apraxia tasks (De Renzi et al., 1980). The RTNT in sensorimotor areas is intermixed with neurophysiological tasks tapping motor and sensory functions. For a detailed description of RTNT in the pre- and postcentral gyrus region, see Tomasino et al. (2022), where the authors extensively describe the pre- and postcentral gyrus RTNT protocol.

Diagnostic and Clinical Implications of Intraoperative Monitoring and Mapping

From a clinical perspective, as mentioned above, medical outcomes, especially in neuro-oncology, are related to the maximum extent of resection. Fortunately, when there is a real-time deterioration of the patient's accuracy on cognitive testing, it typically occurs gradually, allowing the neuropsychologist to inform the surgeon of the functional change to prevent lasting deficits. The RTNT method, which incorporates trials during both neurocognitive mapping and neurocognitive monitoring during resection, significantly enhanced the extent of resection compared to that achieved in a series of patients operated on using only mapping with DES (Skrap et al., 2016). Simply put, RTNT, or monitoring during resection, allows for an increase in the extent of resection while maintaining the patient's cognitive status.

As to the diagnostic implications of the RTNT during intraoperative mapping and monitoring, we find that the RTNT performance recorded in the operating room correlates with postsurgery cognitive status. In a validation study (Skrap et al., 2016), we defined three groups of neuropsychological tests based on the patient's performance during surgery and their performance at follow-up: (1) those who showed no decrements or minimal decreases in accuracy at RTNT and complete recovery at the end of the procedure; (2) those who showed a decrement at RTNT, a lower performance at the immediate postsurgery assessment (but a recovery at follow-up); and (3) those who at follow-up still show only a partial recovery. In the latter, we found that most of the tasks were related to language processing, confirming the importance of monitoring and maintenance of language functions for the patient's postsurgery recovery and return to function.

Importance to Neuropsychology

Real-time neurocognitive assessment during awake brain surgery introduced a new approach of applying neuropsychology to the intrasurgery monitoring. Designing new methods for intraoperatory neuropsychological monitoring is a challenge. The RTNT battery and method described in this chapter is counted among the neuropsychological techniques available to the neuropsychologist (Tomasino & Rumiati, 2020). In general, RTNT requires a different approach from traditional neuropsychological assessment. For starters, the test setting is a major difference, in that the patient is not sitting at a desk in front of the neuropsychologist. The patient is on an operating table with their head attached to intraoperative instrumentation, with movements restricted, and forced to manage distraction from several unusual environmental stimuli (e.g., sounds, smells, sensory experiences). Nonetheless, it is possible to administer neuropsychological testing in this setting with reliability. Neuropsychologists are well versed in establishing rapport in any condition and facilitating the patient's best performance. Neuropsychologists are required to be flexible both in the timing of the administration of tests, as well as with the condensed time to complete the evaluation. The extent of neuropsychological monitoring depends on a number of variables, including the timing of the resection, lesion size, lesion location, and other clinical variables. A resection may be as brief as 20 minutes but could take more than 1 hour. Test instruments must be fast and dynamic as the surgery proceeds. Tests are repeated at different stages of resection. For this reason, it can be defined as a neuropsychological *interventional* method.

The neuropsychologist must demonstrate clear, concise, and timely communication with the neurosurgeon.

Intraoperative testing may also reveal sudden reduction or decline in performance, which may completely recover to normal levels after several minutes. We defined this behavior as a *transitory neuropsychological dysfunction.* The dysfunction may be related to mechanical tissue manipulation that occurs during surgery. When this dysfunction occurs, the surgeon is informed and can adopt a more dynamic surgical strategy. The neuropsychologist is in a unique position to observe and communicate with the surgical team, so that patient factors are appropriately interpreted.

The RTNT gains importance in relation to anatomical functional correlations. Namely, when the patient shows a decrement in neurocognitive performance that is stable, it is often possible to correlate this finding with MRI T1 imaging (see Chapter 2 on structural imaging techniques). In other words, neuroimaging can corroborate the neuroanatomical changes related to the decrease in task performance. This is the quintessential demonstration of brain–behavior relationships. For example, we previously showed that the distribution of the site coordinates corresponding to a decrement in reading accuracy span from the posterior inferior and middle temporal gyrus to the temporoparietal areas and the precentral gyrus (Tomasino et al., 2020).

Review of Case Examples

Case 1, presented at the beginning of this chapter, showed a slight decrement in short-term memory (STM) and working memory (Figure 6.8). At the immediate postsurgery assessment, he showed a slight decrease of STM span, which recovered at follow-up assessment 4 months after surgery (Table 6.1).

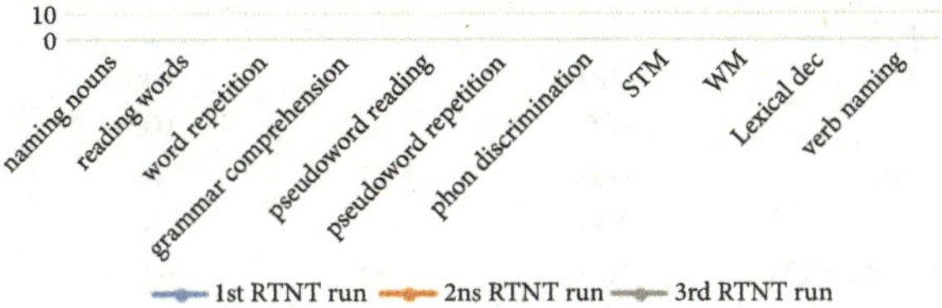

Figure 6.8 RTNT performance of case 1, patient with left temporal lesion

We now present a second case of a 26-year-old, right-handed male with 13 years of education. The patient was diagnosed with a larger low-grade glioma in the left temporo-insular cortex. His symptoms were pain localized behind the eye and episodes of difficulty finding some words lasting a few minutes. In this case, the RTNT tests used during awake surgery were selected to focus on areas important for language processing, including, but not limited to, noun naming (see Figure 6.9), repetition, lexical and semantic decision-making, reading, and phonological skills.

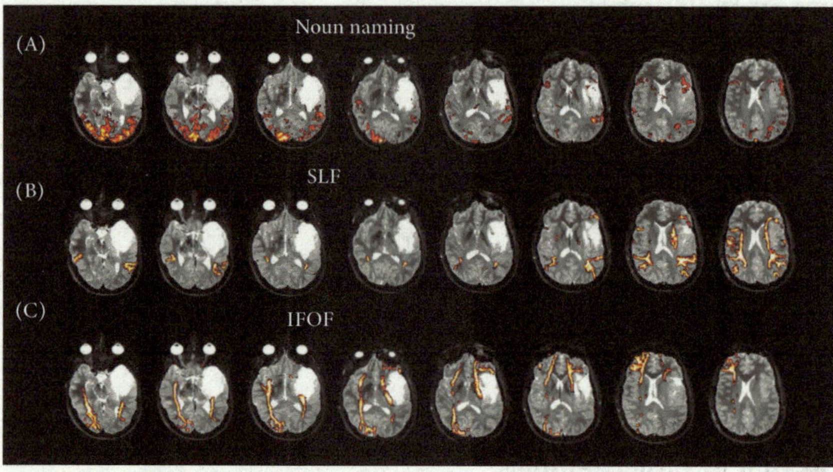

Figure 6.9 Major white matter tracts involved in language processing. A case example of a patient with a low-grade glioma volume in the left temporo-insular cortex. The fMRI maps for noun naming and the major white matter fascicles involved in language processing are shown.

Table 6.2 RTNT performance for case 2

	1st RTNT	2nd RTNT	3rd RTNT	4th RTNT
Verbal Comprehension	80	100	100	
Noun Naming	100	100	80	60
Verb Naming	90	80	80	
Pseudoword Repetition	100	90	80	
Word Repetition	100	100	100	100
Word Reading	100	100	100	80
Pseudoword Reading	100	100	70	
Lexical Decision	100	80	70	
Phonologic Discrimination	100	100	80	
Grammatical Comprehension	90	90	100	

Table 6.3 Case 2 neuropsychological status

Cognitive Domain	Function	Test	Reference		Pre-Surg	Imm. Post Surg	Follow-Up
Fluid intelligence	Logic reasoning	Raven's colored progressive matrices	(Basso et al., 1987)	RS	36	-	-
				CS	33.5	-	-
				ES	4	-	-
Executive functions	Strategy use	Verbal fluency	(Novelli et al., 1996)	RS	29	14	42
				CS	23	8	36
				ES	2	0	4
	Digit span	Backward test	(Monaco et al., 2015)	RS	6	5	4
				CS	5.42	4.42	3.42
				ES	4	4	2
Short term memory	Digit span	Forward test	(Monaco et al., 2015)	RS	7	6	5
				CS	6.44	5.44	4.44
				ES	4	3	1
Praxis	Upper limb imitation	Ideomotor apraxia	(De Renzi et al., 1980)	RS	72	68	72
				PE	4	3	4
	Buccofacial imitation	Oral apraxia	(Spinnler and Tognoni, 1987)	RS	20	20	20
				CS	19.75	19.75	19.75
				ES	4	4	4
Language	Comprehension	Token test	(De Renzi and Faglioni, 1978)	RS	36	31	36
				CS	33.5	28.5	33.5
				ES	4	1	4

continued

Table 6.3 *continued*

Cognitive Domain	Function	Test	Reference		Pre-Surg	Imm. Post Surg	Follow-Up
		Objects Comprehension	(Miceli et al., 1994)	Cut-off 39	40	39	40
		Verbs comprehension	(Miceli et al., 1994)	Cutoff 19	20	20	20
	Semantics	Semantic fluency	(Novelli et al., 1996)	RS	48	45	56
				S	42	39	50
				ES	4	4	4
		Object naming	(Miceli et al., 1994)	Cut-off 28	29	29	29
		Verb naming	(Miceli et al., 1994)	Cutoff 26	28	25	27
	Semantic memory	Pyramids and palm trees test	(Gamboz et al., 2009)	RS	52	49	52
				CS, cutoff 40.15	49.69	49.69	49.69
	Writing	Word and pseudoword writing	(Luzzatti et al., 1994)	cutoff 16.675	174	177	176
	Reading	Word reading	(Ripamonti et al., 2014)	cutoff 78	79	77	79
		Pseudoword reading	(Ripamonti et al., 2014)	cutoff 18	20	17	20
		Phrase reading	(Miceli et al., 1994)	cutoff 6	6	6	6
		Three-syllabic word reading	(Toraldo et al., 2006)	cutoff 24	24	24	24
	Repetition	Word repetition	(Miceli et al., 1994)	cutoff 80	80	80	80
		Pseudoword repetition	(Miceli et al., 1994)	cutoff 20	20	20	20
	Lexical processing	Visual lexical decision	(Ripamonti et al., 2016)	cutoff 136	141	136	143
	Lexical processing	Auditory lexical decision	(Ripamonti et al., 2016)	cutoff 121	125	125	125

RS (raw score); CS (corrected score), ES (equivalent score)

During surgery, he showed a decrease in naming nouns, reading, and lexical decisions (Table 6.2). At follow-up (Table 6.3), he demonstrated recovery and showed a performance within the normal range.

Conclusion

The integration of real-time neuropsychological testing (RTNT) with awake brain surgery offers a vital advancement in maximizing tumor resection success, while preserving essential cognitive functions. Through continuous neurocognitive monitoring, RTNT allows neurosurgeons to achieve more extensive resections safely, minimizing postsurgery cognitive deficits and enhancing overall patient outcomes. By providing immediate feedback on functional changes during surgery, neuropsychologists play a pivotal role in guiding surgical decisions, especially when an operation occurs near eloquent cortical and subcortical areas. This method demonstrates the profound utility of neuropsychology in surgical contexts, embodying the essential brain–behavior relationship and underscoring the importance of interdisciplinary collaboration for successful patient care in neuro-oncology.

References

Axelson, H. W., Latini, F., Jemstedt, M., Ryttlefors, M., & Zetterling, M. (2022) Continuous subcortical language mapping in awake glioma surgery. *Frontiers in Oncology*, 12, 947119. https://doi.org/10.3389/fonc.2022.947119

Antonietti, A., Bartolomeo, P., Colombi, A., Incorpora, C., & Olivieri, S. (2008). *Batteria immaginazione e percezione (BIP) per la valutazione della cognizione visuo-spaziale. Traduzione e adattamento italiano.* ISU–Università Cattolica del Sacro Cuore.

Bak, T., & Hodges, J. R. (2003). Kissing and dancing—a test to distinguish the lexical and conceptual contributions to noun/verb and action/object dissociation: Preliminary results in patients with frontotemporal dementia. *Journal of Neurolinguistics, 16*, 169–181.

Barbarotto, R., Laiacona, M., Frosio, R., Vecchio, M., Farinato, A., & Capitani, E. (1998). A normative study on visual reaction times and two Stroop colour-word tests. *Italian Journal of Neurological Science, 19*, 161–170.

Basso, A., Capitani, E., & Laiacona, M. (1987). Raven's coloured progressive matrices: normative values on 305 adult normal controls. *Functional Neurology, 2*(2), 189–194.

Berger, M. S. (1994). Lesions in functional ("eloquent") cortex and subcortical white matter. *Clinical Neurosurgery, 41*, 444–463.

Berger, M. S. (1996). Minimalism through intraoperative functional mapping. *Clinical Neurosurgery, 43*, 324–337.

Berger, M. S., & Ojemann, G. A. (1992). Intraoperative brain mapping techniques in neuro-oncology. *Stereotactic Functional Neurosurgery, 58*, 153–161.

Bertani, G., Fava, E., Casaceli, G., et al. (2009). Intraoperative mapping and monitoring of brain functions for the resection of low-grade gliomas: technical considerations. *Neurosurg Focus, 27*(4), E4.

Bizzozero, I., Ferrari, F., Pozzoli, S., Saetti, M. C., & Spinnler, H. (2005). Who is who: Italian norms for visual recognition and identification of celebrities. *Neurological Science, 26*, 296.

Capitani, E., Neppi-Modona, M., & Bisiach, E. (2000). Verbal-response and manual-response versions of the Milner landmark task: Normative data. *Cortex, 36*, 593–600.

De Renzi, E., & Faglioni, P. (1978). Normative data and screening power of a shortened version of the token test. *Cortex, 14*, 41–49.

De Renzi, E., Motti, F., & Nichelli, P. (1980). Imitating gestures: A quantitative approach to ideomotor apraxia. *Archives of Neurology, 37*, 6–10.

De Renzi, E., Piezcuro, A., & Vignolo, L. A. (1966). Oral apraxia and aphasia. *Cortex, 2*, 50–73.

DeWitt Hamer, P. C., Robles, S. G., Zwinderman, A. H., Duffau, H., & Berger, M. S. (2012). Impact of intraoperative stimulation brain mapping on glioma surgery outcome: A meta-analysis. *Journal of Clinical Oncology, 30* (20), 2559–2565.

Duffau, H., Capelle, L., Denvil, D., Sichez, N., Gatignol, P., Taillandier, L., et al. (2003). Usefulness of intraoperative electrical subcortical mapping during surgery for low-grade gliomas located within eloquent brain regions: Functional results in a consecutive series of 103 patients. *Journal of Neurosurgery, 98*(4), 764–778.

Duffau, H., Capelle, L., Sichez, J. P., Bitar, A., Faillot, T., Arthuis, F., et al. (2000). Surgery of the central nervous system: Value of preoperative functional brain mapping by direct electric stimulation. *Presse Medicale, 29*, 421–422.

Duffau, H., Lopes, M., Arthuis, F., Bitar, A., Sichez, J. P., Van Effenterre, R., & Capelle, L. (2005). Contribution of intraoperative electrical stimulations in surgery of low grade gliomas: A comparative study between two series without (1985–96) and with (1996–2003) functional mapping in the same institution. *Journal of neurology, neurosurgery, and psychiatry, 76*, 845–851. https://doi.org/10.1136/jnnp.2004.048520

Gamboz, N., Coluccia, E., Iavarone, A., & Brandimonte, M. A. (2009). Normative data for the pyramids and palm trees test in the elderly Italian population. *Neurological Sciences, 30*(6), 453–458.

Hamberger, M. J, Seidel, W. T., Goodman, R. R., Perrine, K., & McKhann, G. M. (2003). Temporal lobe stimulation reveals anatomic distinction between auditory naming processes. *Neurology, 60* (9), 1478–1483.

Ius, T., Isola, M., Budai, R., Pauletto, G., Tomasino, B., Fadiga, L., et al. (2012). Low-grade glioma surgery in eloquent areas: Volumetric analysis of extent of resection and its impact on overall survival. A single-institution experience in 190 patients: Clinical article. *Journal of Neurosurgery, 117*, 1039–1052.

Keles, G. E., Lundin, D. A., Lamborn, K. R., Chang, E. F., Ojemann, G., & Berger, M. S. (2004). Intraoperative subcortical stimulation mapping for hemispherical perirolandic gliomas located within or adjacent to the descending motor pathways: Evaluation of morbidity and assessment of functional outcome in 294 patients. *Journal of Neurosurgery, 100*, 369–375.

Krieg, S. M. (2022). Awake craniotomy as a mandatory part of the armamentarium of surgical neuro-oncologists. *Lancet Oncology*, 23(6), 698–699.

Lang, P. J., Bradley, M. M., & Cuthbert, B. N. (2008). *International Affective Picture System (IAPS): Affective ratings of pictures and instruction manual* (Technical Report A-8). University of Florida, Gainesville, FL.

Luzzatti, C., Laiacona, M., Allamano, N., De Tanti, A., & Inzaghi, M. G. (1998). Writing disorders in Italian aphasic patients. A multiple single-case study of dysgraphia in a language with shallow orthography. *Brain: A Journal of Neurology*, 121(Pt 9), 1721–1734. https://doi.org/10.1093/brain/121.9.1721

Mandonnet, E. (2021). Should complex cognitive functions be mapped with direct electrostimulation in wide-awake surgery? A commentary. *Frontiers in neurology*, *12*, 721038.

Manfield, J., Waqar, M., Mercer, D., Ehsan, S., Bambrough, J., Ibrahim, N., Sivarajan, K., Bailey, M., Karabatsou, K., Coope, D., Ponnusamy, A., Phang, I., & D'Urso, P. I. (2023). Multimodal mapping and monitoring is beneficial during awake craniotomy for intra-cranial tumours: Results of a dual centre retrospective study. *British Journal of Neurosurgery*, *37*(2), 182–187. https://doi.org/10.1080/02688697.2021.2016622

Miceli, G., Laudanna, A., Burani, C., & Capasso, R. (1994). *Batteria per l'analisi dei deficit afasici. B. A. D. A. [B. A. D. A.: A Battery for the assessment of aphasic disorders.]*. CEPSAG.

Monaco, M., Costa, A., Caltagirone, C., & Carlesimo, G. A. (2015). Erratum to: Forward and backward span for verbal and visuo-spatial data: standardization and normative data from an Italian adult population. *Neurological Sciences: Official Journal of the Italian Neurological Society and of the Italian Society of Clinical Neurophysiology*, *36*(2), 345–347. https://doi.org/10.1007/s10072-014-2019-7

Morshed, R. A., Young, J. S., Lee, A. T., Berger, M. S., & Hervey-Jumper, S. L. (2021). Clinical pearls and methods for intraoperative awake language mapping. *Neurosurgery*, *89*(1), 1–10. https://doi.org/10.1093/neuros/nyaa400.

Ng, S., Herbet, G., Lemaitre, A. L., Cochereau, J., Moritz-Gasser, S., & Duffau, H. (2020). Neuropsychological assessments before and after awake surgery for incidental low-grade gliomas. *Journal of Neurosurgery*, *135*(3), 871–880. https://doi.org/10.3171/2020.7.JNS201507

Nocentini, U., Giordano, A., Di Vincenzo, S., Panellaa, M., & Pasqualetti, P. (2006). The symbol digit modalities test–oral version: Italian normative data. *Functional Neurology*, *21*, 93–96.

Novelli, G., Papagno, C., Capitani, E., Laiacona, M., Vallar, G., & Cappa, S. F. (1996). Tre test clinici di ricerca e produzione lessicale. Taratura su soggetti normali. *Archivio di Psicologia, Neurologia e Psichiatria*, *47*, 477–505.

Ochipa, C., Rapcsak, S. Z., Maher, L. M., Rothi, L. J. G., Bowers, D., & Heilman, K. M. (1997). Selective deficit of praxis imagery in ideomotor apraxia. *Neurology*, *49*, 474–480.

Ojemann, G. (1993). Functional mapping of cortical language areas in adults: Intraoperative approaches. *Advances in neurology*, *63*, 155–163.

Ojemann, G., Ojemann, J., Lettich, E., & Berger, M. (1989). Cortical language localization in left, dominant hemisphere: An electrical stimulation mapping investigation in 117 patients. *Journal of Neurosurgery*, *71*, 316–326.

Orsini, A., Grossi, D., Capitani, E., Laiacona, M., Papagno, C., & Vallar, G. (1987). Verbal and spatial immediate memory span: Normative data from 1355 adults and 1112 children. *Italian Journal of Neurological Science*, *8*, 539–548.

Papagno, C., Cappa, S. F., Forelli, A., Garavaglia, G., Laiacona, M., & Capitani, E. (1995). Non-literal language comprehension: Standardization of a metaphors and idiom comprehension test. *Archivio di Psicologia, Neurologia e Psichiatria, 56*, 402–420.

Paradis, M. (1986). *Bilingual aphasia test (Italian version)*. EMS.

Prior, M., Marchi, S., & Sartori, G. (2003). *Cognizione sociale e comportamento* (Vol. 1). Domeneghini Editore UPSEL.

Ripamonti, E., Aggujaro, S., Molteni, F., Zonca, G., Frustaci, M., & Luzzatti, C. (2014). The anatomical foundations of acquired reading disorders: a neuropsychological verification of the dual-route model of reading. *Brain and Language, 134*, 44–67. https://doi.org/10.1016/j.bandl.2014.04.001

Ripamonti, E., Cattani, B., Molteni, F., & Luzzatti, C. (2016). Toward an empirical basis for theories of lexical access: Evidence from Italian lexical decision and word naming. *Frontiers in Psychology, 7*(960). https://doi.org/10.3389/fpsyg.2016.00960

Rusconi, M. L. (1998). Test delle figure chimeriche nella valutazione dell'eminegligenza spaziale. *Archivio di psicologia neurologia e psichiatria, 59*(6), 810–823.

Sanai, N., & Berger, M. S. (2008). Glioma extent of resection and its impact on patient outcome. *Neurosurgery, 62*, 753–766.

Sanai, N., & Berger, M. S. (2011). Extent of resection influences outcomes for patients with gliomas. *Rev Neurol (Paris), 167*, 648–654.

Sanai, N., Polley, M. Y., & Berger, M. S. (2010). Insular glioma resection: Assessment of patient morbidity, survival, and tumor progression. *Journal of Neurosurgery, 112*, 1–9.

Skrap, M., Marin, D., Ius, T., Fabbro, F., & Tomasino, B. (2016). Brain mapping: A novel intraoperative neuropsychological approach. *Journal of Neurosurgery, 125*(4), 877–887.

Spinnler, M., & Tognoni, G. (1987). Standardizzazione e taratura italiana di test neuropsicologici. *Italian Journal of Neurological Science, 6*(Suppl. 8), 1–120.

Tomasino, B., Guarracino, I., Ius, T., Budai, R., & Skrap, M. (2022). Real-time neuropsychological testing of sensorimotor cognition during awake surgery in precentral and postsomatosensory areas. *World Neurosurgery, 164*, e599–e610. https://doi.org/10.1016/j.wneu.2022.05.018

Tomasino, B., Guarracino, I., Ius, T., Maieron, M., & Skrap, M. (2021). Real-time neuropsychological testing protocol for left temporal brain tumor surgery: A technical note and case report. *Frontiers in human neuroscience*,15, 760569.

Tomasino, B., Ius, T., Skrap, M., & Luzzatti, C. (2020). Phonological and surface dyslexia in individuals with brain tumors: Performance pre-, intra-, immediately post-surgery and at follow-up. *Human Brain Mapping*, 41(17), 5015–5031.

Tomasino, B., Marin, D., Ius, T., & Skrap, M. (2018). Neuropsychology in insular lesions prior, during, and after brain surgery. In M. Turgut, C. Yurtta, & S. Tubbs (Eds.), *Island of Reil (Insula) in the human brain: Anatomical, functional, clinical and surgical aspects* (pp. 281–292). (Springer).

Tomasino, B., Marin, D., Maieron, M., D'Agostini, S., Fabbro, F., Skrap, M., et al. (2015a). Double-letter processing in surface dyslexia and dysgraphia following a left temporal lesion: A multimodal neuroimaging study. *Cortex, 73*, 112–130.

Tomasino, B., Marin, D., Maieron, M., D'Agostini, S., Medeossi, I., Fabbro, F., et al. (2015b). A multimodal mapping study of conduction aphasia with impaired repetition and spared reading aloud. *Neuropsychologia, 70*, 214–226.

Tomasino, B., Skrap, M., & Rumiati, R. I. (2011). Causal role of the sensorimotor cortex in action simulation: Neuropsychological evidence. *Journal of Cognitive Neuroscience, 23*(8), 2068–2078.

Tomasino, B., Toraldo, A., & Rumiati, R. I. (2003). Dissociation between the mental rotation of visual images and motor images in unilateral brain-damaged patients. *Brain and Cognition, 51*, 368–371.

Tomasino, B., Tronchin, G., Marin, D., Maieron, M., Fabbro, F., Cubelli, R., et al. (2018). Noun–verb naming dissociation in neurosurgical patients. *Aphasiology, 33*(12), 1418–1440. https://doi.org/10.1080/02687038.2018.1542658

Tomasino, B., Vorano, L., Skrap, M., Gigli, G., & Rumiati, R. I. (2004). Effects of strategies of mental rotation performed by unilateral brain damaged patients. *Cortex, 40*, 197–199.

Toraldo, A., Cattani, B., Zonca, G., Saletta, P., & Luzzatti, C. (2006). Reading disorders in a language with shallow orthography: A multiple single-case study in Italian. *Aphasiology, 20*(9), 823–850. https://doi.org/10.1080/02687030600738838

van Kessel, E., Schuit, E., Huenges Wajer, I. M. C., et al. (2021). Added value of cognition in the prediction of survival in low and high grade glioma. *Frontiers in neurology, 12*, 773908.

7

Wada Testing

Assessment of Language and Memory Dominance

Amanda Gooding and Seth A. Margolis

Abbreviations That Appear in the Text

ATL	Anterior Temporal Lobectomy
EEG	Electroencephalogram
ESL	English Second Language
ESM	Electrocortical Stimulation Mapping
FDG-PET	Fluorodeoxyglucose Positron Emission Tomography
fMRI	Functional Magnetic Resonance Imaging
IAP	Intracarotid Amobarbital Procedure
ICA	Internal Carotid Artery
IQ	Intelligence Quotient
MCA	Middle Cerebral Artery
MEG	Magnetoencephalogram
MRI	Magnetic Resonance Imaging
MTS	Mesial Temporal Sclerosis
PCA	Posterior Cerebral Artery
PCOM	Posterior Communicating Artery
PET	Positron Emission Tomography
SPECT	Single Photon Emission Tomography
TLE	Temporal Lobe Epilepsy
TMS	Transcranial Magnetic Stimulation

History and Purpose

The Wada test—also known as the intracarotid amobarbital procedure (IAP)—was first reported on by an epileptologist named Juhn Wada (Wada, 1949; Wada, 2008; Wada & Rasmussen, 1960). It was subsequently adopted by Brenda Milner and Theodore Rassmussen at the Montreal Neurological Institute, where it was routinely administered in all left-handed or ambidextrous patients with

Amanda Gooding and Seth A. Margolis, *Wada Testing*. In: *Medical and Laboratory Examinations for Neuropsychologists*. Edited by: Thomas J. Farrer and R. Brock Frost, Oxford University Press.
 DOI: 10.1093/9780197619537.003.0008

medically refractory focal epilepsy under consideration for neurosurgical interventions, as well as all right-handed surgical candidates whenever there was suspicion that speech functions might be atypically lateralized to the right hemisphere (Rasmussen & Milner, 1977). Initially developed to determine cerebral language representation (Wada & Rasmussen, 1960), it was soon modified to include assessment of memory (Milner et al., 1962).

Historically, the Wada test has been most common in the presurgical workup of candidates with temporal lobe epilepsy (TLE), though it may be used in other populations (e.g., brain tumor). The general purpose of Wada testing is to provide information on lateralization of language dominance, functional adequacy of the hippocampal formation thought to be epileptogenic or unhealthy, and the functional reserve of the contralateral hippocampal formation. Though less commonly considered with advances in electrophysiology and structural and functional neuroimaging, some centers employ the Wada test to help verify the side of seizure onset and prognosticate the extent of postsurgical seizure control (Alpherts et al., 2000).

Participants and Setup

The Wada procedure takes place in an interventional radiology or angiography suite and involves a multidisciplinary team working together to ensure patient safety and procedure success. A neurointerventional radiologist or endovascular neurosurgeon is responsible for performing the angiogram, placing the catheter, and dosing the anesthetic. An electroencephalogram (EEG) technician and an epileptologist/neurologist are responsible for monitoring the EEG and overseeing the procedures. EEG monitoring should run throughout the duration of the procedure to identify potential seizures and monitor dosing effect. The neuropsychologist assesses language and memory. Several other members of the interventional radiology technological and nursing staff are present to assist with the procedure and ensure patient stability.

Although cognitive testing is relatively brief, typically a minimum of 2 hours in the angiography suite is required, and patients often remain in postoperative recovery for ≥6-hours afterward to ensure the puncture wound heals and stability for discharge. Notably, there is no single Wada test protocol in use. Across centers, neuropsychologists design protocols that fit their needs. Several well-researched protocols have been discussed elsewhere (Lee, 2010; Loring & Meador, 2015; Pearson et al., 2019). The description to follow highlights general procedures that most centers employ.

Cerebral Angiography

Prior to dosing and test administration, cerebral angiography is performed in each hemisphere. Catheterization is done via the right femoral artery, with a contrast medium used to determine the precise location of the catheter as it traverses toward the internal carotid artery (ICA). Angiography allows the team to visualize the vasculature to evaluate the degree of cross-fill into the contralateral hemisphere, as well as perfusion in the ipsilateral posterior circulation. If mild, the cross-fill can sometimes be alleviated by applying less force upon injection. Examination of the posterior circulation is particularly important to ascertain whether a persistent fetal circulation is present (i.e., a posterior communicating artery [PCOM] larger than the P1 segment of the posterior cerebral artery [PCA]), which if large enough could supply dosing of the anesthetic into brainstem areas and cause respiratory distress. Examples of interhemispheric cross-fill and persistent fetal circulation are presented in Figures 7.1 and 7.2, respectively.

In cases of normal vasculature, the catheter typically remains placed in the ICA for dosing, allowing perfusion of the anesthetic into both the anterior and middle cerebral artery (MCA) territories. In cases of abnormal vasculature, a more selective approach can be taken—for example, by placing the catheter into the MCA distal to the PCOM and proximal to the anterior choroidal artery to ensure appropriate dosing to the hippocampus, while reducing the probability of perfusing the posterior circulation (Brassel et al., 1996).

Anesthetizing Agents and Dosing

Two types of anesthetics are typically used. The most common and historically used is amobarbital (also known as amylobarbitone or sodium amytal, or by the brand name Amytal). The other is methohexital (also known by the brand name Brevital), which is gaining in popularity across centers for several reasons including supply availability, cost, and efficiency. Other drugs that have been utilized include sodium secobarbital (Oguro et al., 2008), propofol (Loddenkemper et al., 2004; Masters et al., 2000; Mikuni et al., 2005; McCleary et al., 2018), and etomidate (Andelman et al., 2013; Jones-Gotman et al., 2005). Both amobarbital and methohexital are barbiturate derivatives with sedative-hypnotic properties, and while both cause similar clinical effects (Andelman et al., 2006; Buchtel et al., 2002; Loddenkemper et al., 2009; Loddenkemper et al., 2007), the shorter half-life of methohexital allows repeated injections and a shorter testing interval. Also, it allows for shorter durations of motor, speech, and EEG recovery after each injection, perhaps with relatively fewer behavioral side-effects such as agitation and somnolence (Conradi et al., 2020).

The initial Amytal dose is typically administered at 75–125 mg (diluted at 25 mg/cc) with an average dose of 100 mg (Bajammal et al., 2020). This initial dose varies based on numerous factors including age, body size, gender, and type and dose of antiepileptic drug(s) prescribed (Lee, 2010). Depending on quality and duration of the clinical effect, additional syringes of 25 mg each should be available for follow-up injections before or during cognitive testing. These boluses can be used either to achieve complete flaccid hemiparesis or for subsequent injections should the anesthetic effect wear off prior to test completion. By contrast, methohexital is typically used at much smaller doses—often starting with an initial dose of 3–5 mg (diluted at 1 mg/cc) with a median dose of 3 mg, and additional boluses of 2–5 mg with a median bolus of 2 mg available for follow-up injections as needed (Conradi et al., 2020). It is typically recommended that either drug be injected at a rate of 1–2 cc/sec by hand.

Regardless of which anesthetic is used, the catheter should remain in place until the neuropsychologist indicates that all test stimuli have been presented. This allows additional dosing should the effect of the initial injection wear off prematurely. Although the relative safety and risk/benefit ratio must be considered on a case-by-case basis, withdrawal of the catheter before all testing procedures are completed may lead to uninterpretable results and perhaps the need for a repeat injection.

Cognitive Testing

Testing often begins with the patient holding both hands above their head with palms turned upward toward the examiner who is standing behind them. The patient is asked to count aloud while an initial injection of the drug is administered by the interventional radiologist. With adequate dosing, a contralateral arm drop will occur with continuous hemiparesis for the duration of the drug effect. A visual field deficit (typically a contralateral homonymous hemianopia) is expected; thus, caution should be taken to ensure that all stimuli are presented in the ipsilateral visual field.

If the language-dominant hemisphere has been sufficiently anesthetized, there will be abrupt speech arrest (e.g., patient stops counting) for the duration of maximal drug effect. Basic language functions are assessed to verify the expected aphasia. Whereas language task procedures, test stimuli, scoring, and psychometric criteria used to operationalize left, mixed, or atypical right hemisphere language dominance vary across centers, a study of seven epilepsy centers found high interrater reliability when determining hemispheric language dominance

Table 7.1a Examples of Wada test procedure stimuli

Function	Tasks	Examples
Language Expression	State own name	
	Object and/or picture naming	Battery Eraser Pen Glasses
	Sentence repetition	"No ifs, ands, or buts" "The spy fled to Greece" "Methodist Episcopal"
Language Comprehension	Follow midline commands	"Stick out your tongue" "Close your eyes"
Oral Reading	Numbers	
	Single words	
	Sentences	"The rabbit hopped down the lane." "The car backed over the curb."
Non-language stimuli presentation (for memory testing)	Faces Line drawings Patches of color Abstract designs Printed words and sentences	

Note: Centers vary in terms of whether they use all or a selection of these tasks and whether language test items are also employed for memory assessment purposes.

(Haut et al., 2002). Most Wada protocols assess whether or not speech arrest has occurred, the ability to name to visual confrontation, repetition, reading, and the ability to follow simple midline commands (see Table 7.1a). Once language is assessed, memory stimuli are presented during the duration of drug effect. Drug effects are expected to wear off by 8–10 minutes or 1–3 minutes after injection of sodium amytal and methohexital, respectively.

To guide interpretation of wearing off effects, grip strength is tested throughout to determine when motor strength returns to the affected limb. At many centers, EEG technologists or epileptologists provide updates on resolution of EEG slowing in the affected hemisphere as well. As the patient returns to their neurologic baseline, language functions are re-assessed to determine resolution of aphasia. Once the anesthetic has worn off, language abilities tend to recover quickly along a predictable course with successive resolution of reading, naming, comprehension, then repetition (Ravdin et al., 1997). Paraphasic errors are considered strong signs of language representation in the affected hemisphere.

Table 7.1b Memory assessment

Methodologies		
	Free Recall	
	Yes/No Recognition	
	Forced Choice Recognition	
	Multiple Choice Recognition	
Scores		**Equations**
	Total Score	
	Percent Correct	Total Correct/Total Items
	Discriminability Index	1—[(Total Misses + Total False Positives)/Total Items]
Interpretive Values	Absolute cut-points Interhemispheric Memory Asymmetry score	

Assessing grammatical ability during the postinjection recovery phase may have added lateralizing value (Polczynska et al., 2014).

Once language returns to baseline, full strength is regained in the affected limb, and EEG slowing normalizes, memory retention is tested. There are various methods for assessing memory retention (see Table 7.1b). Regardless of methodology, at least 10 minutes must pass between initial injection and assessment to allow for drug effects to wear off. Additionally, timing of memory stimuli presentation is important to track; for example, if drug effects wear off while memory stimuli are presented (e.g., grip strength returns, language impairment resolves to approximate baseline levels), it must be considered that memory items presented during that time frame are less valid because the hippocampal formation was insufficiently anesthetized during stimulus exposure (Loring et al., 1997).

After the initial hemisphere is tested, an analogous procedure using different stimuli is performed with a drug administered to the contralateral hemisphere. This second injection can occur in as little as 30 minutes to help ensure effects of the first injection completely wore off. Some centers opt to perform this second injection on a different day; Grote et al. (1999) demonstrated that Wada memory scores were lower for the second injection when performed on the same day but were similar for the first and second injections if conducted on separate days. Either way, testing of the contralateral side should be done using the same anesthetic and same dose as the initial side to ensure comparable clinical conditions.

Interpretation of Results

Hemispheric Language Dominance

Determining hemispheric language dominance is important for gauging risk of language impairment following surgery. Language dominance can serve as a predictor of postoperative memory decline in standard resection cases (Chelune & Najm, 2000). Thus, determining the side(s) of language dominance helps guide the extent of resection, whether additional language localization will be necessary prior to intervention via cortical language mapping, functional magnetic resonance imaging (fMRI), or magnetoencephalogram (MEG), and aid in clinical decision-making about the type of surgical options presented to the patient. When eloquent language cortex overlaps with epileptogenic zones, alternative techniques that may offer less seizure freedom but spare language function may be advised, such as selective amygdalohippocampectomy and/or palliative neuromodulation approaches.

Definitions of hemispheric language dominance vary across centers. Speech arrest—when ongoing automatic speech (e.g., counting, reciting alphabet) comes to an abrupt and sustained halt in the presence of drug effect (Benbadis et al., 1998; Rosenow et al., 1997)—is a common hallmark of language dominance. In most cases, the patient exhibits speech arrest with approximately 3–3.5 minutes of aphasia following sufficient sodium amytal dosing (Morris et al., 1998). However, more detailed language assessments are encouraged (Loring, Meador, Lee, et al., 1990), since the validity and reliability of using speech arrest alone to characterize language dominance has been questioned (Benbadis et al., 2018). Additionally, cases of mixed language representation (i.e., language disruption after both left and right hemisphere injections) are challenging to confirm and categorize but are best captured with a comprehensive language investigation, as programs reporting lower rates of mixed language representation tend to utilize less detailed language assessment protocols (Snyder et al., 1990).

Continued speech is common/anticipated following nondominant hemisphere injection. So, when this does not occur atypical language organization should be considered (Helmstaedter, 2003). Rasmussen and Milner's (1977) early investigation of speech lateralization using the Wada found that 96% of non-lesional right-handers and 70% of left- or mixed-handers have left hemisphere language dominance, and only left- or mixed-handers have bilateral (15%) or right (15%) hemisphere language representation. When examining patients with definite clinical evidence of early-life left hemisphere lesion, rates of bilateral or right hemisphere language dominance were higher in both right- (19%) and left-handers (72%).

Later research showed that mixed or reversed language dominance is most common in left-handers (Knecht et al., 2000), among patients who sustained early-life brain injuries, with lesions near eloquent language cortex, and/or in persons with seizures of childhood onset (usually those <6-years of age though the upper age limit may extend as high as 16) (Boatmann et al.,1999; Hertz-Pannier et al., 2002; Loddenkemper et al., 2003; Telfeian et al., 2002). Transhemispheric transfer of language occurs frequently in left mesial TLE (DeVos et al., 1995; Pataraia et al., 2004). Later-onset disease in childhood and left-sided congenital lesions tend to maintain left hemisphere language representation (DeVos et al., 1995; Duchowny et al., 1996; Saltzman-Benaiah et al., 2003). Reorganization of cortical function is known to result from abnormal neuronal/glial proliferation (cortical dysplasia) but not abnormal cortical organization (polymicrogyria) (Burneo et al., 2004).

When Wada results suggest bilateral or right hemisphere language representation, usual assumptions about localization and lateralization of material-specific memory functions and other cognitive abilities cannot be made as confidently (Loring, Lee, Meador, et al., 1990). Some patients may experience a "crowding effect" when hemispheric language networks are reorganized to the nondominant hemisphere or both hemispheres, such that nonverbal aspects of cognition appear underdeveloped by comparison (Strauss et al., 1990).

Hemispheric Memory Support

Most programs utilize the Wada to investigate functionality of the hippocampal formation for learning/memory within the hemisphere where proposed surgery would occur and the contralateral hemisphere, which would be presumed healthy. An optimal Wada memory result would demonstrate good performance in the hemisphere contralateral to the to-be-resected lesion/epileptogenic area and poor memory ipsilateral to that area. In effect, this optimal outcome suggests that the tissue intended for resection is already dysfunctional from a memory standpoint and its removal would do little functional harm. Relative functional sparing of the contralateral hemisphere suggests the possibility of decent compensatory memory support from the unresected side. Conversely, good memory performance in the ipsilateral hemisphere and poor contralateral memory performance suggests high functional adequacy of the to-be-resected tissue and low functional reserve of the contralateral tissue (Chelune, 1995). Such an unexpected asymmetry warrants concern for postoperative memory decline (Baxendale et al., 2007; Hermann, 2005; Kirsch et al. 2005; Lee et al., 2005; Sabsevitz et al., 2001).

When memory performance is similarly poor bilaterally, these results may be equivocal and reflect bilateral hippocampal dysfunction or complications associated with the procedure (e.g., oversedation, antiepileptic drug burden, unresponsiveness, cross-fill of drug; Lozano-García et al., 2021; Mani et al., 2008). In such instances, there is a decision to be made regarding whether to repeat the Wada on the same or a different day. Although patients for whom a correctable technical problem may benefit from a repeat study, overall test-retest reliability of the Wada is high and so there is little benefit to repeating the procedure in patients with discordant or unexpected results (i.e., results are unlikely to change) (Whitman et al., 2012). An additional consideration is whether more selective Wada testing via the anterior choroidal artery or PCA is warranted. However, these more selective procedures should be used judiciously given general absence of language effects and increased risk of vasospasm and inadvertent perforation of the vessel wall associated with catheterization and anesthetization of smaller/narrower vessels (Brazzel et al., 1996).

Notably, reports suggest that type and quantity of memory stimulus items can affect memory results. For example, Loring et al. (1997) found that using real objects (as opposed to line drawings) was superior in differentiating lateralized temporal dysfunction. That study also suggested that using a smaller number of memory items may decrease the likelihood of obtaining memory asymmetries simply because the scores offer less reliable performance measures, potentially resulting in a higher number of indeterminate findings. Hamberger and Hirsch (1999) report that incorporating memory confidence ratings into Wada protocols may increase the likelihood of obtaining asymmetry scores that more accurately reflect memory dysfunction associated with the epileptogenic hemisphere.

Prediction of Postoperative Outcomes

Although many centers avoid resecting lesions that are clearly in eloquent language cortex to avoid inducing an iatrogenic aphasia, Wada memory results are not always as clear-cut. The original purpose of Wada memory testing was to predict the risk for *global* amnesia (Jones-Gotman, 1987; Milner, 1962). Results vary across centers, with differences in Wada memory protocols resulting in differential outcomes. One group reported showing memory stimuli to patients while anesthetized and testing recall/recognition once the drug effect wore off with a resulting 48% predictive value for postoperative amnesia (Ojemann & Kelley, 2002). However, the Dodrill Seattle protocol, which involves showing test stimuli prior to injection and continuously showing items following the injection, boasts a 76% predictive value for postoperative memory states (Dodrill, 1993; Dodrill & Ojemann, 1997).

Use of Wada testing with other patient-level factors has also been used to predict outcomes with variable effects. For instance, Fatoorechi, Westerveld, and

Lee (2020) note that patients with atypical language dominance have *better* verbal memory outcomes following left hemisphere resections than those with left hemisphere language dominance due to inferred reorganization of typical memory substrates along with language. In cases of typical language lateralization, some report that poor Wada memory results contralateral to the lesion/to-be-resected side predicted those with postoperative verbal memory decline at a sensitivity of 100% and a specificity of 78.6% (Jokeit et al., 1997). However, others found that preoperative verbal learning determined via neuropsychological testing and magnetic resonance imaging (MRI) ratings of mesial temporal sclerosis (MTS) severity predicted postoperative verbal memory declines in left TLE patients who underwent anterior temporal lobectomy (ATL), while Wada memory results did not (Elhorst et al., 2009).

In attempting to disentangle the added benefit of Wada memory asymmetries, some showed that worse performance ipsilateral to seizures predicted lower risk for postoperative neuropsychological deficits (Loring et al., 2009) and worse performance of the contralateral side portended verbal memory decline (Sever et al., 2018). However, these reports did not comment on those who "failed" the Wada due to paradoxically reversed memory profiles (i.e., high functional adequacy in the ipsilateral hemisphere with low functional reserve in the contralateral side) presumably because many of these patients do not progress to surgery. Additionally, the base rate of Wada test failure without apparent technical complications is only 13% (Kemp et al., 2008).

Lacruz et al. (2004) examined 12 patients who had temporal lobectomy despite having failed the Wada and found that such individuals had more deficits than those who passed or who had a less extensive selective amygdalohippocampectomy, showed higher visual memory deficits after surgery than those who passed, and there was a strong trend suggesting a verbal memory deficit following temporal lobectomy compared to patients with selective amygdalohippocampectomy. Additionally, stronger correlations between worse contralateral Wada memory scores and verbal deficits were observed following temporal lobectomy compared to the general sample or those who underwent selective amygdalohippocampectomy. By contrast, Rathore et al., (2015) studied 29 patients with left MTS who failed the Wada but underwent ATL and found no reliable change in memory scores pre- to postsurgery compared to those who passed. Additionally, both groups showed no difference in seizure outcomes or quality of life.

Similar variability has been reported in how well Wada results predict postsurgical seizure freedom with some groups reporting no effects (Rathore et al., 2013) and others highlighting encouraging results (Perrine et al., 1995; Waseem et al., 2015). Given variability across outcome studies, some have questioned the utility of the Wada. This has especially been the case among right temporal patients for whom there is less concern about inducing verbal memory decline (Baxendale et al., 2006; Lineweaver et al., 2006).

Safety Considerations

As with all clinical procedures, it is imperative to weigh potential benefits against risks. Although the Wada procedure is generally well tolerated, patients will occasionally have catastrophic emotional reactions that can interfere with testing (Paola et al., 2004). Detailed evaluation of 5/81 Wada-tested patients who developed bizarre behavior failed to identify predictive factors of such behaviors. The authors suggested that thorough explanation of the procedure and simulation prior to initiating Wada testing might help lessen this risk.

Adverse effects can be as minor as pain at the site of catheter insertion or transient shivering (Shah et al., 2010), or as serious as stroke given potential for endovascular complications. Generally, the risk for stroke is higher in older adults or those with atherosclerosis or vascular abnormalities. A retrospective study of Cleveland Clinic patients (Loddenkemper et al., 2008) found that 10.9% had complications. These included encephalopathy (7.2%), seizures (1.2%), strokes (0.6%), transient ischemic attacks (0.6%), localized hemorrhage at the catheter insertion site (0.6%), carotid artery dissections (0.4%), allergic reaction to contrast (0.3%), bleeding from the catheter insertion site (0.1%), and infection (0.1%). However, another retrospective study performed by the University of Michigan (Beimer et al., 2015) found only 5.8% of patients had notable clinical events. Two percent had events meeting criteria for complication, which included seizures, status epilepticus, internal carotid artery vasospasm, inadvertent injection of anesthetic in the external carotid artery, and transient encephalopathy. No complications were associated with significant morbidity or mortality. Risks and adverse events appear to be similar regardless of which anesthetic is used, including propofol (Szantroch et al., 2019).

Considerations for Special Populations

Clinical experience shows that the traditional Wada procedure can be performed effectively with presurgical patients who do not speak English using translators including speakers of American Sign Language (Miranda et al, 2022). Similarly, there are literature reports documenting successful completion of the Wada test in individuals who are primary speakers of Spanish (Maestú at el., 2002), Chinese (Rapport et al., 1983), Arabic (Alsallom et al., 2014), and French (Lehéricy et al., 2000), among others.

Wada testing has been validated in pediatric populations, though special considerations may be needed to determine which children are suitable, which stimuli are appropriate, and how to interpret results. Modifications can include pre-test instruction and practice, simplified test items, and reduced dosing (Hinz et al., 1994; Szabó & Wyllie, 1993; Williams & Rausch, 1992). Perhaps one of the most important considerations is the child's age and maturity, as the procedure can be frightening and physically uncomfortable (particularly the femoral artery incision) (Lee, 2010). Results of Wada language and memory testing have been validated in children as young as 7 years, though the reliability of findings in those under age 10 is lower than adults (Schevon et al., 2007).

An additional consideration for children and adults is general level of cognitive development or intellectual functioning. One study suggested an intelligence quotient (IQ) of 70 as a potential cut-point for when to proceed with Wada testing, as those with lower IQs had higher rates of failure to complete language and memory testing in a satisfactory manner (Szabo & Wyllie, 1993). Still, though, in a small clinical sample from the Netherlands information on motor, language, and memory functioning was successfully obtained in respectively 13/15, 9/13, and 5/11 children under 10 years of age—some with IQ scores below 50 and/or who were critically ill and/or behaviorally disturbed (Jansen et al., 2002). Regardless of age or IQ, attention should be paid to patients' neurobehavioral baseline, as individuals with preexisting hemiparesis, blindness, visual field deficits, or aphasia may require additional modifications. Similarly, case reports and series suggest that some antiepileptic drugs, such as topiramate, may significantly alter Wada test results (Kipervasser et al., 2004; McCabe & Eslinger, 2000).

Alternatives to the Wada

A retrospective case review from Cleveland Clinic between 1997 and 2007 found that the frequency of Wada testing declined, such that patients with left, likely dominant, TLE and those with subdural evaluations were more likely to undergo Wada testing than other patients (Das et al., 2010). In 2008, an international survey indicated that 87% of responding epilepsy centers were using the Wada test (Baxendale et al., 2008). However, there appeared to be wide variations in test deployment within individual centers, with a study from that same year (Haag et al., 2008) identifying frequencies ranging from 8% to 110% of presurgical candidates (the latter figure presumably accounting for retests of an initial unreliable result). This same survey was updated in 2014, at which time 85% of respondents indicated that the Wada test was unnecessary for TLE surgery patients since intracranial electrodes are often still needed to localize motor-sensory and language functions prior to surgery. There were differences based on residency, with 100% of those from Europe indicating that the Wada test was unnecessary compared with 75% of those from North America (Mathern et al. 2014). Although opposing positions emerged afterward about the field's readiness to retire the Wada (Loring, 2014; Papanicolaou, 2014), these findings suggest that there has been a significant shift in its perceived role.

Some of the most common alternatives include fMRI, MEG (Papanicolaou et al., 2004), and electrocortical stimulation mapping (ESM; Massot-Tarrús et al., 2019). Other less commonly used techniques include transcranial magnetic stimulation (TMS), single photon emission tomography (SPECT), and positron emission tomography (PET). Case studies have also highlighted the potential utility of electrical stimulation via depth electrodes (i.e., "Electric Wada") to predict postoperative memory performance (Kaestner et al., 2022) and language laterality (Alonso et al., 2016).

Notably, fMRI, MEG, SPECT, and PET are activation techniques designed to elicit or invoke a neural and behavioral response. By contrast, TMS, ESM, and Wada are disruption techniques designed to temporarily interrupt or impede neural communication. Thus, the behavioral response being measured is fundamentally different depending on the type of procedure used. While all of these techniques have been utilized specifically for the purpose of language lateralization/localization, fewer have the ability to adequately measure hemispheric memory support, though some studies suggest that lateralization of language is sufficiently correlated with lateralization of verbal memory and that fMRI can in fact be used to predict postoperative verbal memory decline (Binder et al., 2008). Careful consideration should be used when selecting the technique most appropriate for an individual patient given age, medical history, behavioral concerns, and general tolerability of the procedure.

While procedures such as fMRI and MEG are less invasive than Wada, some argue that disruption techniques are more sensitive to identifying the true functional consequences of a surgical procedure. Further, these methods and Wada testing have had less than stellar interrater reliability (Massot-Tarrús et al., 2020); for example, fMRI studies comparing Wada results show the correlation with activation asymmetry in the frontal but not temporal lobes (Lehéricy et al., 2000) and laterality activation can differ between frontal and temporal regions. Additionally, meta-analysis found that Wada testing is still indicated in cases where fMRI fails to show unilateral left-lateralized language dominance (Bauer et al., 2014). Furthermore, Killgore et al. (1999) suggested that fMRI and Wada testing provide complementary data that may not correlate, but both account for non-overlapping unique variance in predicting surgical outcome. Even in situations where language discordance emerges between fMRI and Wada, fMRI does not outperform the Wada's ability to predict postsurgical naming declines in all cases (Janecek et al., 2013).

Qadri et al. (2021) provide an updated review of available evaluation techniques and recommended workflow to limit Wada use to only the most essential cases. They propose limiting Wada testing to instances of: atypical language lateralization, bilateral language representation, inconclusive language lateralization, significant bitemporal dysfunction on EEG or suggested by neuropsychological testing. The authors cautioned readers to apply this framework on a case-by-case basis. See also Wada and Kosaka (2009) for a detailed review of alternative techniques, including their utility and limitations.

Summary

The Wada test has a rich history in the evaluation of language and memory among presurgical patients with and without epilepsy. Care must be taken to ensure that appropriate procedures are followed with consideration given to type/dosage of medicine, stimuli, scoring, and testing conditions, as well as populations for whom the procedure is most/least indicated. The foregoing information is intended to provide clinicians with detailed considerations to keep in mind when proposing, conducting, and interpreting Wada results in their practice.

Case Examples

The following two case examples illustrate how Wada testing was used to inform the risk of potential language and memory declines in two adults with epilepsy with similar cognitive profiles and seizures. As seen in Case 1, resective surgery was performed regardless of cautions raised following Wada testing because the patient weighted the treatment team's suspicion for low-grade glioma more heavily than concern for potential postoperative cognitive decline. In Case 2, the patient was declined resective/ablative surgery on the basis of Wada results and was instead offered additional studies for consideration of a palliative neuromodulation device. Methohexital was used in both cases but Case 1 had abnormal vasculature and the medication regimen included topiramate, each of which may have complicated results. Though similar language protocols were utilized in each case, psychometric classification of hemispheric memory support differed. In Case 1, a recognition memory cutoff of 6/10 was used to indicate a passing score with score differences >2 reflecting asymmetry. In Case 2, a recognition discriminability index was calculated, with a score difference >15% reflecting asymmetry.

Case 1

Demographics and Clinical History

This case describes a 54-year-old, right-handed, English-speaking, white woman who left school in the 9th grade. Her clinical history includes neurofibromatosis–I, depression, and drug-resistant epilepsy with seizures for the preceding 5 years. Baseline medications included levetiracetam, topiramate, amitriptyline, and escitalopram.

Neurodiagnostic Workup

Neurodiagnostic workup revealed a low-grade glioma versus cortical dysplasia within the left hippocampus. A fluorodeoxyglucose positron emission tomography (FDG-PET) revealed decreased uptake in the left mesial temporal lobe. Video EEG monitoring showed left greater than right bitemporal slowing and spike wave discharges but no clinical seizures were captured.

Neuropsychological evaluation was non-lateralizing. Overall intellectual functioning was within the below average range (full scale IQ = 74), with below average to exceptionally low scores on all other tasks except for low average to average learning/recall of a short story and recognition of a word list. Together with her older age at seizure onset, broad cognitive impairment, and low functional baseline, concern was raised that she might experience verbal memory declines with a standard left temporal resection.

Wada Procedure

Left Hemisphere Injection

A total of 5 mg of methohexital was injected through the left ICA. Angiogram revealed mild cross-over into the right hemisphere (Figure 7.1) and a mild fetal distribution of the PCOM (Figure 7.2). There was evidence of left hemisphere EEG slowing and right arm weakness ~20 seconds post-injection. She showed almost immediate speech arrest, which persisted for ~4.5 minutes. She was unable to state her name, repeat phrases, or follow commands. She was shown 10 stimuli and was unable to name/read any despite remaining alert and visually tracking them. EEG returned to baseline at ~5 minutes. Right-hand grip strength returned at ~5.5 minutes. She exhibited impaired naming, reading, and repetition until ~6.5 minutes post-injection. She remained dysarthric and continued to make numerous language errors until ~10 minutes post-injection. She was amnestic for the entire procedure. She freely recalled none of the items she was shown. She correctly identified 3/10 stimuli with a recognition format and stated that she was guessing on most responses.

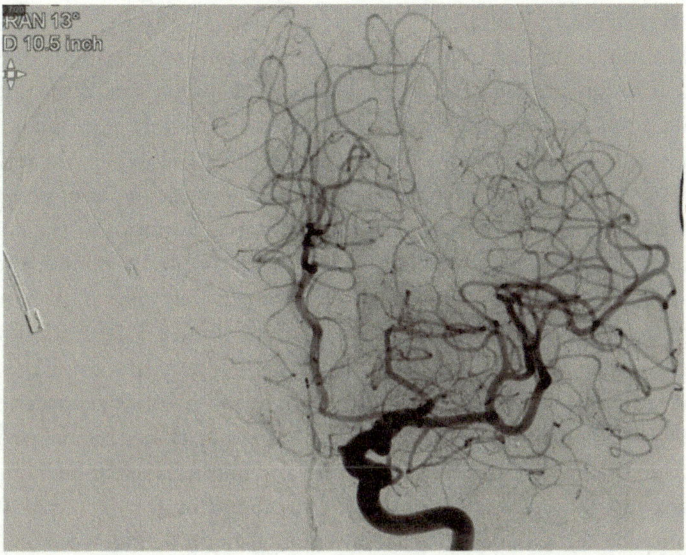

Figure 7.1 CT angiogram showing left to right hemisphere cross-fill during Wada procedure

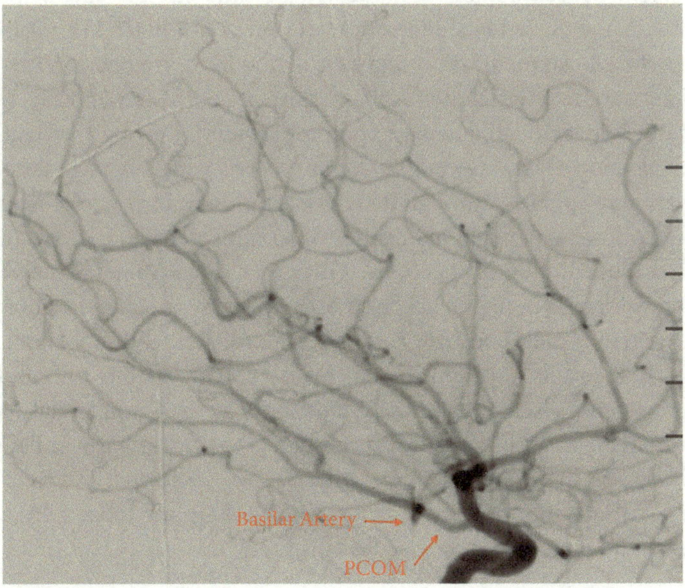

Figure 7.2 CT angiogram showing a fetal-type variant of the left posterior communicating artery (PCOM)

Note: Red arrows denote a mild fetal (origin of the) left PCOM which results in a PCOM larger than the P1 segment of the posterior cerebral artery and a significantly smaller basilar artery than normal. When a fetal circulation is large enough, anesthetics could be supplied to brainstem areas and cause respiratory distress during Wada testing.

Right Hemisphere Injection

Thirty minutes after the first injection, the patient was given 5 mg of methohexital through the right ICA. No medication bolus was given after the fact given the longer than expected duration of her left hemisphere injection. There was evidence of right hemisphere EEG slowing and right arm weakness ~17 seconds post-injection. She was alert throughout and able to state her name, repeat phrases, and follow simple commands. She correctly named/read 8/8 to-be-remembered stimuli and assigned a name to the additional 2 abstract designs she was asked to remember. Her right-hand grip strength returned to 50% ~2 minutes post-injection and the EEG started to normalize at 2.5 minutes, which coincided with administration of the final 2 memory stimuli. Ten minutes after the conclusion of stimulus presentation, she freely recalled none of the new items but spontaneously named one of the items from the set shown to her during the previous left hemisphere injection. She correctly endorsed 3/8 stimuli with a recognition format (or 4/10 items if you include those items she was shown after methohexital wore off).

Impressions

The current procedure was complicated by mild left-to-right hemisphere cross-fill during the left-sided injection, as well as a mild fetal PCOM variant in a patient whose medication regimen included topiramate. Nevertheless, with inactivation of the left hemisphere, she was completely aphasic, whereas she retained speech and language comprehension following the right hemisphere injection. These findings were consistent with language dominance in the left hemisphere. Although there was a more rapid return of neurologic function following the right hemisphere injection, memory performance was impaired following both injections. There was no asymmetry between the two hemispheres (even including items shown to the patient after medication wear off, each hemisphere's performance was within 1-point of the other). These results were considered consistent with her bitemporal slowing and sharp wave discharges on EEG and both verbal- and visual-memory impairment on presurgical neuropsychological testing despite a unilateral left mesial temporal lesion. Therefore, concern was raised that the right hemisphere may not support memory following left temporal lobectomy, and the care team and patient were advised accordingly.

Surgical Decision-Making

The patient opted to undergo bilateral mesial temporal depth electrode placement with electrocorticography to further guide surgical decision-making. Despite holding her antiseizure medications to increase the likelihood of capturing seizures, only one seizure occurred within several days of monitoring. As this seizure originated from the left temporal lobe, the family decided to move forward with a left ATL with amygdalohippocampectomy.

Postoperative Outcome

The patient was seen for neuropsychological re-evaluation 2 years post-surgery. Seizure frequency was reduced from multiple events per week to one every 3–4 weeks with shorter durations (~30 seconds). Seizure reduction notwithstanding, she and her significant other reported worsened memory and word-finding. She also reported increased anxiety, depression, and mood lability, with a hospitalization for suicidal ideation the summer following surgery (no prior psychiatric hospitalizations on record). At the time of re-evaluation, she was in outpatient counseling and working with a psychiatrist. She remained on levetiracetam, topiramate, and amitriptyline with slight changes in dosages and was also started on lorazepam. Consistent with her presurgical evaluation, cognitive testing revealed broad cognitive impairments that were generally stable. However, she showed clinically meaningful declines in verbal memory and confrontation naming (see Table 7.2).

Table 7.2 Pre- and postsurgical naming and verbal memory scores for Case 1

Test	Baseline	Baseline	Follow-up	Follow-up
	Raw Score	T Score	Raw Score	T Score
Boston Naming Test	29	<20	14	<20
California Verbal Learning Test, 2nd edition				
Total Learning	33	32	12	<20
Short Delay Free Recall	5	30	3	25
Long Delay Free Recall	7	35	0	<20
Recognition Discriminability	2.5	45	1.2	25
Neuropsychological Assessment Battery Story Learning Subtest				
Immediate Recall	50	41	35	31
Delayed Recall	30	47	8	31

Case 2

Demographics and Clinical History

This case describes a 43-year-old, right-handed, Mexican American, bilingual (English and Spanish) woman, for whom English was her second language (ESL), with drug-resistant epilepsy since infancy. She had breech presentation at birth and hypoxia, with forceps delivery. She had a vagal nerve stimulator placed at age 30, which was efficacious at first, but she experienced adverse side-effects. It was ultimately deemed nonfunctional and turned off at age 38. Her seizures were managed via sodium valproate and oxcarbazepine.

Neurodiagnostic Workup

Brain MRI showed left MTS with accompanying asymmetric diminished white matter in the left anterior temporal region. FDG-PET suggested a correlating focal area of decreased metabolism within the medial left temporal lobe (in the expected area of the hippocampus). Video EEG monitoring captured seven of her habitual clinical events, which were lateralized to the left hemisphere and maximal in the left temporal region. Interictal epileptiform discharges were also localized to the left anterior temporal region.

Neuropsychological testing was conducted in English and considered to be non-lateralizing. Her overall intellectual functioning was within the below average range (full scale IQ = 71), with diffuse impairments in multiple domains including language, attention/working memory, visual/verbal memory, and fine motor abilities. Her memory profile was characterized by impairments in initial acquisition/learning for both visual and verbal material, but with relatively better recognition memory and no evidence of rapid forgetting. Language testing may have been confounded by her bilingual/ESL status.

Wada Procedure

Left Hemisphere Injection

The patient was administered 4 mg of methohexital into the left ICA. She experienced left hemisphere EEG slowing, total loss of right-hand grip strength (0/5), and speech arrest after ~13 seconds. She was aphasic throughout stimulus presentation, unable to name visually presented objects, read sentences aloud, or follow orally presented commands. However, she remained alert and was observed to be tracking items visually. Grip strength increased to 2/5 at ~60 seconds following the initial injection, and so an additional 2 mg of methohexital was administered. Stimulus presentation concluded at ~108 seconds following initial injection. She remained aphasic until ~180 seconds. Her speech returned to baseline, she regained full 5/5 grip strength, and the EEG normalized at ~240 seconds. Ten minutes after the conclusion of stimulus presentation, recognition memory testing was performed. She correctly identified 1/12 stimulus items with 0 false positive errors, placing her recognition discriminability at 54.2%.

Right Hemisphere Injection

The patient was administered 4 mg of methohexital into the right ICA. She experienced total loss of left-hand grip strength (0/5), as well as EEG slowing of the right hemisphere. She correctly named 2/2 objects (1 in English and 1 in Spanish) and followed 2/2 orally presented commands. When reading sentences aloud, she became briefly distracted by extraneous stimuli but was able to redirect her attention and read aloud with minimal errors. An additional 2 mg of methohexital was administered midway through stimulus presentation (~69 seconds) due to gradual return of left-hand grip strength (2/5). Stimulus presentation concluded at ~144 seconds. She regained full 5/5 grip strength ~210 seconds post-injection, after which the EEG normalized. Ten minutes after the conclusion of the stimulus presentation, recognition memory testing was performed. She correctly identified 5/12 stimulus items with 0 false positive errors, placing her recognition discriminability at 70.8%.

Impressions

Angiography showed no abnormal vasculature and the procedure was considered successful with no complications. Results suggested clear left hemisphere language dominance, as she was completely aphasic following the left ICA injection, and she retained speech and language comprehension following the right ICA injection. In terms of memory, the left hemisphere injection produced a 54% discriminability index, and the right hemisphere injection produced a 71% discriminability index. Despite this patient's unilateral left MTS, left temporal seizures, and left temporal hypometabolism, this 17% memory asymmetry score suggested paradoxically greater left hemisphere memory support.

Surgical Decision-Making

Prior to the Wada, the consensus was to offer a left mesial temporal laser ablation. After Wada results were reviewed at the multidisciplinary surgical epilepsy conference, all treatment team members shared concerns about risk for postsurgical memory decline. The final consensus recommendation was to offer an intracranial stereo-EEG study to confirm more precise seizure localization and possible responsive neurostimulator placement thereafter, which the patient is currently contemplating.

Acknowledgments

Drs. Gooding and Margolis contributed equally to this chapter. Both authors extend sincere thanks to Allyson Goldstein for her formatting, referencing, and editorial assistance.

References

Alsallom, F. A., Sinha, S., Alsenani, F. M., Bamogaddam, F. A., Wahass, S. H., Canillo, A. B., & Siddiqui, K. A. (2014). Tolerability of propofol in Wada testing. *Neurosciences Journal, 19*(3), 218–223. https://nsj.org.sa/content/19/3/218.short

Alonso, F., Sweet, J., & Miller, J. (2016). Speech mapping using depth electrodes: The "electric Wada." *Clinical Neurology and Neurosurgery, 144*, 88–90. https://doi.org/10.1016/j.clineuro.2016.03.017

Alpherts, W. C. J., Vermeulen, J., & Van Veelen, C. W. M. (2000). The Wada test: Prediction of focus lateralization by asymmetric and symmetric recall. *Epilepsy Research, 39*(3), 239–249. https://doi.org/10.1016/S0920-1211(00)00097-8

Andelman, F., Kipervasser, S., Maimon, S., Fried, I., Parmet, Y., & Neufeld, M. Y. (2013). A revised intracarotid etomidate memory (Wada) procedure. *Acta Neurologica Scandinavica, 127*(2), 97–102. https://doi.org/10.1111/j.1600-0404.2012.01685.x

Andelman, F., Kipervasser, S., Reider-Groswasser, I. I., Fried, I., & Neufeld, M. Y. (2006). Hippocampal memory function as reflected by the intracarotid sodium methohexital Wada test. *Epilepsy & Behavior, 9*(4), 579–586. https://doi.org/10.1016/j.yebeh.2006.08.001

Bajammal, S., Babtain, F., Alqadi, K., Baeesa, S., Kurdi, K., Madani, N., & Al Said, Y. (2020). The optimal dose of amobarbital in the Wada test for the presurgical evaluation of patients with temporal lobe epilepsy. *Clinical Neuropharmacology, 43*(6), 185–190. https://doi.org/10.1097/WNF.0000000000000411

Bauer, P. R., Reitsma, J. B., Houweling, B. M., Ferrier, C. H., & Ramsey, N. F. (2014). Can fMRI safely replace the Wada test for preoperative assessment of language lateralisation? A meta-analysis and systematic review. *Journal of Neurology, Neurosurgery & Psychiatry, 85*(5), 581–588. http://dx.doi.org/10.1136/jnnp-2013-305659

Baxendale, S., Thompson, P., & Duncan, J. (2008). The role of the Wada test in the surgical treatment of temporal lobe epilepsy: An international survey. *Epilepsia, 49*, 715–719.

Baxendale, S., Thompson, P., Harkness, W., & Duncan, J. S. (2006). Predicting memory decline following epilepsy surgery: A multivariate approach. *Epilepsia, 47*, 1887–1894. https://doi.org/10.1111/j.1528-1167.2006.00810.x

Baxendale, S., Thompson, P., Harkness, W., & Duncan, J. (2007). The role of the intracarotid amobarbital procedure in predicting verbal memory decline after temporal lobe resection. *Epilepsia, 48*(3), 546–552. https://doi.org/10.1111/j.1528-1167.2006.00940.x

Beimer, N. J., Buchtel, H. A., & Glynn, S. M. (2015). One center's experience with complications during the Wada test. *Epilepsia, 56*(8), e110–e113. https://doi.org/10.1111/epi.13046

Benbadis, S. R., Binder, J. R., Swanson, S. J., Fischer, M., Hammeke, T. A., Morris, G. L., Frost, J. A., & Springer, J. A. (1998). Is speech arrest during Wada testing a valid method for determining hemispheric representation of language? *Brain and Language, 65*(3), 441–446. https://doi.org/10.1006/brln.1998.2018

Binder, J. R., Sabsevitz, D. S., Swanson, S. J., Hammeke, T. A., Raghavan, M., & Mueller, W. M. (2008). Use of preoperative functional MRI to predict verbal memory decline after temporal lobe epilepsy surgery. *Epilepsia, 49*(8), 1377–1394. https://doi.org/10.1111/j.1528-1167.2008.01625.x

Boatman, D., Freeman, J., Vining, E., Pulsifer, M., Miglioretti, D., Minahan, R., Carson, B., Brandt, J., & McKhann, G. (1999). Language recovery after left hemispherectomy in children with late-onset seizures. *Annals of Neurology: Official Journal of the American Neurological Association and the Child Neurology Society, 46*(4), 579–586. https://doi.org/10.1002/1531-8249(199910)46:4<579::AID-ANA5>3.0.CO;2-K

Brassel, F., Weissenborn, K., Rückert, N., Hussein, S., & Becker, H. (1996). Superselective intra-arterial Amytal (Wada test) in temporal lobe epilepsy: Basics for neuroradiological investigations. *Neuroradiology, 38*(5), 417–421. https://doi.org/10.1007/BF00607264

Buchtel, H., Passaro, E., Selwa, L., Deveikis, J., Gomez-Hassan, D. (2002). Sodium methohexital (Brevital) as an anesthetic in the Wada test. *Epilepsia, 43*(9), 1056–1061. https://doi.org/10.1046/j.1528-1157.2002.00902.x.

Burneo, J. G., Kuzniecky, R. I., Bebin, M., & Knowlton, R. C. (2004). Cortical reorganization in malformations of cortical development: A magnetoencephalographic study. *Neurology, 63*(10), 1818–1824. https://doi.org/10.1212/01.WNL.0000144179.87918.2F

Chelune, G. J. (1995). Hippocampal adequacy versus functional reserve: Predicting memory functions following temporal lobectomy. *Archives of Clinical Neuropsychology, 10*(5), 413–432. https://doi.org/10.1093/arclin/10.5.413

Chelune, G. J., & Najm, I. M. (2000). Risk factors associated with postsurgical decrements in memory. In H. O. Luders & Y. Comair (Eds.), *Epilepsy surgery* (2nd ed, pp. 497–504). Lippincott-Raven.

Conradi, N., Rosenberg, F., Biermann, L., Haag, A., Hermsen, A., Gorny, I., Podewils, V., Gurschi, M., Keil, F., Hattingen, E., Menzler, K., Bauer, S., Schubert-Bast, S., Knake, S., Rosenow, F., & Strzelczyk, A. (2020). Advantages of methohexital over amobarbital in determining hemispheric language and memory lateralization in the Wada test—A retrospective study. *Epilepsy & Behavior, 113*. https://doi.org/10.1016/j.yebeh.2020.107551.

Das, R. R., Alexopoulos, A. V., & Loddenkemper, T. (2010). Incidence and clinical decision making for the Wada test over one decade: 1997–2007. *Epileptic Disorders, 12*(4), 249–254. https://doi.org/10.1684/epd.2010.0332

DeVos, K. J., Wyllie, E., Geckler, C., Kotagal, P., & Comair, Y. (1995). Language dominance in patients with early childhood tumors near left hemisphere language areas. *Neurology, 45*(2), 349–356. https://doi.org/10.1212/WNL.45.2.349

Dodrill, C. B. (1993). Preoperative criteria for identifying eloquent brain: Intracarotid Amytal for language and memory testing. *Neurosurgery Clinics of North America*, *4*(2), 211–216. https://doi.org/10.1016/s1042-3680(18)30587-4 PMID: 8467207

Dodrill, C. B., & Ojemann, G. A. (1997). An exploratory comparison of three methods of memory assessment with the intracarotid amobarbital procedure. *Brain and Cognition*, *33*(2), 210–223. https://doi.org/10.1006/brcg.1997.0893

Duchowny, M., Jayakar, P., Harvey, A. S., Resnick, T., Alvarez, L., Dean, P., & Levin, B. (1996). Language cortex representation: Effects of developmental versus acquired pathology. *Annals of Neurology: Official Journal of the American Neurological Association and the Child Neurology Society*, *40*(1), 31–38. https://doi.org/10.1002/ana.410400108

Elshorst, N., Pohlmann-Eden, B., Horstmann, S., Schulz, R., Woermann, F., & McAndrews, M. P. (2009). Postoperative memory prediction in left temporal lobe epilepsy: The Wada test is of no added value to preoperative neuropsychological assessment and MRI. *Epilepsy & Behavior*, *16*(2), 335–340. https://doi.org/10.1016/j.yebeh.2009.08.003

Fatoorechi, S., Westerveld, M., & Lee, G. P. (2020). Atypical language representation as a protective factor against verbal memory decline following epilepsy surgery. *Epilepsy & Behavior: E&B*, *112*, 107451. https://doi.org/10.1016/j.yebeh.2020.107451

Grote, C., Wierenga, C., Smith, M., Kanner, A. M., Bergen, D. C., Geremia, G., Greenlee, W. (1999). Wada difference a day makes: Interpretative cautions regarding same-day injections. *Neurology*, *52*, 1577–1582. https://doi.org/10.1212/WNL.52.8.1577

Haag, A., Knake, S., Hamer, H. M., Boesebeck, F., Freitag, H., Schulz, R., Baum, P., Helmstaedter, D., Wellmer, J., Urbach, H., Hopp, P., Mayer, T., Hufnagel, A., Jokeit, H., Lerche, H., Uttner, I., Meencke, J., Meierkord, H., Paul, E., . . . Rosenow, F. (2008). The Wada test in Austrian, Dutch, German, and Swiss epilepsy centers from 2000 to 2005: A review of 1421 procedures. *Epilepsy & Behavior*, *13*(1), 83–89. https://doi.org/10.1016/j.yebeh.2008.02.012

Hamberger, M. J., & Hirsch, L. J. (1999). Effects of incorporating memory confidence ratings and language handicap modifications on intracarotid amobarbital procedure (Wada test) memory asymmetry scores. *Epilepsia*, *40*(9), 1286–1291. https://doi.org/10.1111/j.1528-1157.1999.tb00859.x

Haut, S. R., Berg A. T., Shinnar, S., Cohen, H. W., Bazil, C. W., Sperling, M. R., Langfitt, J. T., Pacia, S. V., Walczak, T. S., & Spencer, S. S. (2002). Interrater reliability among epilepsy centers: Multicenter study of epilepsy surgery. *Epilepsia*, *43*, 1396–1401. https://doi.org/10.1046/j.1528-1157.2002.20902.x

Helmstaedter, C. (2003). Neuropsychological aspects of epilepsy surgery. *Epilepsy and Behavior*, *5*, S45–S55. https://doi.org/10.1016/j.yebeh.2003.11.006

Hermann, B. (2005). Wada test failure and cognitive outcome. *Epilepsy Currents*, *5*(2), 61–62. https://doi.org/10.1111/j.1535-7597.2005.05206.x

Hertz-Pannier, L., Chiron, C., Jambaque, I., Renaux-Kieffer, V., Moortele, P. F. V. D., Delalande, O., Fohlen, M., Brunelle, F., & Bihan, D. L. (2002). Late plasticity for language in a child's non-dominant hemisphere: A pre-and post-surgery fMRI study. *Brain*, *125*(2), 361–372. https://doi.org/10.1093/brain/awf020

Hinz, A. C., Berger, M. S., Ojemann, G. A., & Dodrill, C. (1994). The utility of the intracarotid Amytal procedure in determining hemispheric speech lateralization in pediatric epilepsy patients undergoing surgery. *Child's Nervous System*, *10*(4), 239–243. https://doi.org/10.1007/BF00301161

Janecek, J. K., Swanson, S. J., Sabsevitz, D. S., Hammeke, T. A., Raghavan, M., Mueller, W., & Binder, J. R. (2013). Naming outcome prediction in patients with discordant Wada and fMRI language lateralization. *Epilepsy & Behavior*, *27*(2), 399–403. https://doi.org/10.1016/j.yebeh.2013.02.030

Jansen, F. E., Jennekens-Schinkel, A., Van Huffelen, A. C., Van Veelen, W. M., Van Rijen, C. P., Alpherts, W. C., Vermeulen, J., & Van Nieuwenhuizen, O. (2002). Diagnostic significance of Wada procedure in very young children and children with developmental delay. *European Journal of Paediatric Neurology: EJPN: Official Journal of the European Paediatric Neurology Society*, *6*(6), 315–320. https://doi.org/10.1016/s1090-3798(02)90619-7

Jokeit, H., Ebner, A., Holthausen, H., Markowitsch, H. J., Moch, A., Pannek, H., Schulz, R., & Tuxhorn, I. (1997). Individual prediction of change in delayed recall of prose passages after left-sided anterior temporal lobectomy. *Neurology*, *49*(2), 481–487. https://doi.org/10.1212/wnl.49.2.481

Jones-Gotman, M. (1987). Commentary: Psychological evaluation: Testing hippocampal function. In J. Engel Jr. (Ed.), *Surgical treatment of the epilepsies* (pp. 68–76). Raven Press.

Jones-Gotman, M., Sziklas, V., Djordjevic, J., Dubeau, F., Gotman, J., Angle, M., Tampieri, D., Olivier, A., & Andermann, F. (2005). Etomidate speech and memory test (eSAM): A new drug and improved intracarotid procedure. *Neurology*, *65*(11), 1723–1729. https://doi.org/10.1212/01.wnl.0000187975.78433.cb

Kaestner, E., Rubino, A., Liu, K. D., Alving, L. I., Kuzniecky, R. I., & Devinsky, O. (2022). Electrical Wada for pre-surgical memory testing: A case report. *Epileptic Disorders*, *24*(3), 527–533. https://doi.org/10.1684/epd.2021.1390

Kemp, S., Wilkinson, K., Caswell, H., Reynders, H., & Baker, G. (2008). The base rate of Wada test failure. *Epilepsy & Behavior*, *13*(4), 630–633. https://doi.org/10.1016/j.yebeh.2008.07.013

Killgore, W. D., Glosser, G., Casasanto, D. J., French, J. A., Alsop, D. C., & Detre, J. A. (1999). Functional MRI and the Wada test provide complementary information for predicting post-operative seizure control. *Seizure*, *8*(8), 450–455. https://doi.org/10.1053/seiz.1999.0339

Kipervasser, S., Andelman, F., Kramer, U., Nagar, S., Fried, I., & Neufeld, M. Y. (2004). Effects of topiramate on memory performance on the intracarotid amobarbital (Wada) test. *Epilepsy & Behavior*, *5*(2), 197–203. https://doi.org/10.1016/j.yebeh.2003.11.033

Kirsch, H. E., Walker, J. A., Winstanley, F. S., Hendrickson, R., Wong, S. T., Barbaro, N. M., Laxer, K. D., & Garcia, P. A. (2005). Limitations of Wada memory asymmetry as a predictor of outcomes after temporal lobectomy. *Neurology*, *65*(5), 676–680. https://doi.org/10.1212/01.wnl.0000174440.31387.65. PMID: 16157898

Knecht, S., Dräger, B., Deppe, M., Bobe, L., Lohmann, H., Flöel, A., Ringelstein, E. B., & Henningsen, H. (2000). Handedness and hemispheric language dominance in healthy humans. *Brain: A Journal of Neurology*, *123*(12), 2512–2518. https://doi.org/10.1093/brain/123.12.2512

Lacruz, M. E., Alarcón, G., Akanuma, N., Lum, F. C., Kissani, N., Koutroumanidis, M., Adachi, N., Binnie, C. D., Polkey, C. E., & Morris, R. G. (2004). Neuropsychological effects associated with temporal lobectomy and amygdalohippocampectomy depending on Wada test failure. *Journal of Neurology, Neurosurgery, and Psychiatry*, *75*(4), 600–607. https://doi.org/10.1136/jnnp.2003.020248

Lee, G. P. (2010). Other neuropsychological procedures in epilepsy surgery. In S. McPherson, I. S. Baron, R. Kaplan, S. Koffler, G. J. Lamberty, & J. Sweet (Eds.), *Neuropsychology of epilepsy and epilepsy surgery* (pp. 183–210). Oxford University Press.

Lee, G. P., Westerveld, M., Blackburn, L. B., Park, Y. D., & Loring, D. W. (2005). Prediction of verbal memory decline after epilepsy surgery in children: Effectiveness of Wada memory asymmetries. *Epilepsia, 46*(1), 97–103.

Lehéricy, S., Cohen, L., Bazin, B., Samson, S., Giacomini, E., Rougetet, R., Hertz-Pannier, D., Le Bihan, D., Marsault, M., & Baulac, M. (2000). Functional MR evaluation of temporal and frontal language dominance compared with the Wada test. *Neurology, 54*(8), 1625–1633. https://doi.org/10.1212/wnl.54.8.1625

Lineweaver, T. T., Morris, H. H., Naugle, R. I., Najm, I. M., Diehl, B., & Bingaman, W. (2006). Evaluating the contributions of state-of-the-art assessment techniques to predicting memory outcome after unilateral anterior temporal lobectomy. *Epilepsia, 47*(11), 1895–1903. https://doi.org/10.1111/j.1528-1167.2006.00807.x

Loddenkemper, T., Möddel, G., Dinner, D. S., Kim, H., Schuele, S. U., Alexopoulos, A. V., Kotagal, P., & Lüders, H. O. (2009). Language assessment in Wada test: Comparison of methohexital and amobarbital. *Seizure, 18*(9), 656–659. https://doi.org/10.1016/j.seizure.2009.08.002

Loddenkemper, T., Möddel, G., Schuele, S. U., Wyllie, E., & Morris, H. H., III. (2007). Seizures during intracarotid methohexital and amobarbital testing. *Epilepsy & Behavior, 10*(1): 49–54. https://doi.org/10.1016/j.yebeh.2006.09.001

Loddenkemper, T., Morris, H. H., & Moddel, G. (2004, April). Complications of the intracarotid amobarbital test. *Neurology, 62*(7), A248–A249.

Loddenkemper, T., Morris, H. H., & Möddel, G. (2008). Complications during the Wada test. *Epilepsy & Behavior, 13*(3), 551–553. https://doi.org/10.1016/j.yebeh.2008.05.014

Loddenkemper, T., Wyllie, E., Lardizabal, D., Stanford, L. D., & Bingaman, W. (2003). Late language transfer in patients with Rasmussen encephalitis. *Epilepsia, 44*(6), 870–871.

Loring D. (2014). Commentary: *Epilepsia*'s Wada survey. *Epilepsia, 55*(12), 1891. https://doi.org/10.1111/epi.12836

Loring, D. W., Bowden, S. C., Lee, G. P., & Meador, K. J. (2009). Diagnostic utility of Wada memory asymmetries: Sensitivity, specificity, and likelihood ratio characterization. *Neuropsychology, 23*(6), 687–693. https://doi.org/10.1037/a0016528

Loring, D. W., Lee, G. P., Meador, K. J., Flanigin, H. F., Smith, J. R., Figueroa, R. E., & Martin, R. C. (1990). The intracarotid amobarbital procedure as a predictor of memory failure following unilateral temporal lobectomy. *Neurology, 40*(4), 605–610. https://doi.org/10.1212/WNL.40.4.605

Loring, D. W., & Meador, K. J. (2015). The Wada test: Current perspectives and applications. In W. B. Barr & C. Morrison (Eds.), *Handbook on the neuropsychology of epilepsy* (pp. 123–137). Springer.

Loring, D. W., Meador, K. J., Lee, G. P., Murro, A. M., Smith, J. R., Flanigin, H. F., Gallagher, B. B., & King, D. W. (1990). Cerebral language lateralization: Evidence from intracarotid amobarbital testing. *Neuropsychologia, 28*(8), 831–838. https://doi.org/10.1016/0028-3932(90)90007-B

Loring, D. W., Meador, K. J., Lee, G. P., Nichols, M. E., King, D. W., Murro, A. M., Parka, Y. D., & Smith, J. R. (1997). Wada memory and timing of stimulus presentation. *Epilepsy Research, 26*, 461–464. https://doi.org/10.1016/S0920-1211(96)01000-5

Lozano-García, A., Hampel, K. G., Garcés-Sánchez, M., Aparici-Robles, F., Rubio-Sánchez, P., González-Bono, E., Cano-López, I., & Villanueva, V. (2021). Drug load and memory during intracarotid amobarbital procedure in epilepsy. *Acta Neurologica Scandinavica, 144*(5), 585–591. https://doi.org/10.1111/ane.13491. Epub 2021 Jun 16. PMID: 34132388.

Maestú, F., Ortiz, T., Fernandez, A., Amo, C., Martin, P., Fernández, S., & Sola, R. G. (2002). Spanish language mapping using MEG: A validation study. *Neuroimage, 17*(3), 1579–1586. https://doi.org/10.1006/nimg.2002.1235

Mani, J., Busch, R., Kubu, C., Kotagal, P., Shah, U., & Dinner, D. (2008). Wada memory asymmetry scores and postoperative memory outcome in left temporal epilepsy. *Seizure, 17*(8), 691–698. https://doi.org/10.1016/j.seizure.2008.04.009

Massot-Tarrús, A., White, K., & Mirsattari, S. M. (2019). Comparing the Wada test and functional MRI for the presurgical evaluation of memory in temporal lobe epilepsy. *Current Neurology and Neuroscience Reports, 19*(6), 1–14. https://doi.org/10.1007/s11910-019-0945-8

Massot-Tarrús, A., White, K. P., Mousavi, S. R., Hayman-Abello, S., Hayman-Abello, B., & Mirsattari, S. M. (2020). Concordance between fMRI and Wada test for memory lateralization in temporal lobe epilepsy: A meta-analysis and systematic review. *Epilepsy & Behavior, 107*, 107065. https://doi.org/10.1016/j.yebeh.2020.107065

Masters, L. T., Perrine, K., Devinsky, O., & Nelson, P. K. (2000). Wada testing in pediatric patients by use of propofol anesthesia. *American Journal of Neuroradiology, 21*(7), 1302–1305.

Mathern, G. W., Beninsig, L., & Nehlig, A. (2014). From the editors: *Epilepsia*'s survey on the necessity of the Wada test and intracranial electrodes. *Epilepsia, 55*(12), 1887–1889. https://doi.org/10.1111/epi.12815

McCabe, P. H., & Eslinger, P. J. (2000). Abnormal Wada and neuropsychological testing results due to topiramate therapy. *Epilepsia, 41*(7), 906–908. https://doi.org/10.1111/j.1528-1157.2000.tb00262.x

McCleary, K., Barrash, J., Granner, M., Manzel, K., Greider, A., & Jones, R. (2018). The safety and efficacy of propofol as a replacement for amobarbital in intracarotid Wada testing of presurgical patients with epilepsy. *Epilepsy & Behavior, 78*, 25–29. https://doi.org/10.1016/j.yebeh.2017.10.037

Mikuni, N., Takayama, M., Satow, T., Yamada, S., Hayashi, N., Nishida, N., Taki, J., Enatsu, R., Ikeda, A., Miyamoto, S., & Hashimoto, N. (2005). Evaluation of adverse effects in intracarotid propofol injection for Wada test. *Neurology, 65*(11), 1813–1816. https://doi.org/10.1212/01.wnl.0000176988.87336.ff

Milner, B., Branch C., & Rasmussen T. (1962). Study of short term memory after intracarotid injection of sodium amytal. *Transactions of the American Neurological Association, 87*, 224–226.

Miranda, M., Arias, F., Arain, A., Newman, B., Rolston, J., Richards, S., Peters, A., & Pick, L. H. (2022). Neuropsychological evaluation in American Sign Language: A case study of a deaf patient with epilepsy. *Epilepsy & Behavior Reports, 19*, 100558. https://doi.org/10.1016/j.ebr.2022.100558. PMID: 35856041; PMCID: PMC9287772

Morris, R. R., Polkey, C. C., & Cox, T. T. (1998). Independent recovery of memory and language functioning during the intracarotid sodium amytal test. *Journal of Clinical and Experimental Neuropsychology, 20*(4), 433–444. https://doi.org/10.1076/jcen.20.4.433.1467

Oguro, K., Yokota, H., & Watanabe, E. (2008). Non-invasive evaluation of language function by optical topography (Japanese). *Rinsho-Noha (Clinical Electroencephalography), 50*, 110–117.

Ojemann, J. G., & Kelley, W. M. (2002). The frontal lobe role in memory: A review of convergent evidence and implications for the Wada memory test. *Epilepsy & Behavior, 3*(4), 309–315. https://doi.org/10.1016/S1525-5050(02)00039-2

Paola, L. D., Mäder, M. J., Germiniani, F., Coral, P., Zavala, J. A., Watzo, D. J., Kanegusuku, J., Silvado, C., & Werneck, L. C. (2004). Bizarre behavior during intracarotid sodium amytal testing (Wada test): Are they predictable? *Arquivos de neuro-psiquiatria, 62*, 444–448. https://doi.org/10.1590/S0004-282X2004000300012

Papanicolaou A. C. (2014). Commentary: *Epilepsia*'s Wada survey. *Epilepsia, 55*(12), 1890. https://doi.org/10.1111/epi.12835

Papanicolaou, A. C., Simos, P. G., Castillo, E. M., Breier, J. I., Sarkari, S., Pataraia, E., Billingsley, R. L., Buchanan, S., Wheless, J., Maggio, V, & Maggio, W. W. (2004). Magnetocephalography: A noninvasive alternative to the Wada procedure. *Journal of Neurosurgery, 100*(5), 867–876. https://doi.org/10.3171/jns.2004.100.5.0867

Pataraia, E., Simos, P. G., Castillo, E. M., Billingsley-Marshall, R. L., McGregor, A. L., Breier, J. I., Sarkari, S., & Papanicolaou, A. C. (2004). Reorganization of language-specific cortex in patients with lesions or mesial temporal epilepsy. *Neurology, 63*(10), 1825–1832. https://doi.org/10.1212/01.WNL.0000144180.85779.9A

Pearson, C. M., Parks, A., & Landazuri, P. (2019). Wada testing and neurosurgical patients. In C. M. Pearson, E. Ecklund-Johnson, & S. D. Gale (Eds.), *Neurosurgical neuropsychology: The practical application of neuropsychology in the neurosurgical practice* (pp. 99–112). Academic Press.

Perrine, K., Westerveld, M., Sass, K. J., Devinsky, O., Dogali, M., Spencer, D. D., Luciano, D. J., & Nelson, P. K. (1995). Wada memory disparities predict seizure laterality and postoperative seizure control. *Epilepsia, 36*(9), 851–856. https://doi.org/10.1111/j.1528-1157.1995.tb01627.x

Połczyńska, M., Curtiss, S., Walshaw, P., Siddarth, P., Benjamin, C., Moseley, B. D., Vigil, C., Jones, M., Eliashiv, D., & Bookheimer, S. (2014). Grammar tests increase the ability to lateralize language function in the Wada test. *Epilepsy Research, 108*(10), 1864–1873. https://doi.org/10.1016/j.eplepsyres.2014.09.014

Qadri, S., Dave, H., Das, R., & Alick-Lindstrom, S. (2021). Beyond the Wada: An updated approach to pre-surgical language and memory testing: An updated review of available evaluation techniques and recommended workflow to limit Wada test use to essential clinical cases. *Epilepsy Research, 174*, 106673. https://doi.org/10.1016/j.eplepsyres.2021.106673

Rapport, R. L., Tan, C. T., & Whitaker, H. A. (1983). Language function and dysfunction among Chinese-and English-speaking polyglots: Cortical stimulation, Wada testing, and clinical studies. *Brain and Language, 18*(2), 342–366. https://doi.org/10.1016/0093-934X(83)90024-X

Rasmussen, T., & Milner, B. (1977). The role of early left-brain injury in determining lateralization of cerebral speech functions. *Annals of the New York Academy of Sciences, 299*, 355–369.

Rathore, C., Alexander, A., Sarma, P. S., & Radhakrishnan, K. (2015). Memory outcome following left anterior temporal lobectomy in patients with a failed Wada test. *Epilepsy & Behavior: E&B, 44*, 207–212. https://doi.org/10.1016/j.yebeh.2015.02.011

Rathore, C., Kesavadas, C., Sarma, S. P., & Radhakrishnan, K. (2013). Usefulness of Wada test in predicting seizure outcome following anterior temporal lobectomy. *Epilepsy Research, 107*(3), 279–285. https://doi.org/10.1016/j.eplepsyres.2013.09.003

Ravdin, L. D., Perrine, K., Haywood, C. S., Gershengorn, J., Nelson, P. K., & Devinsky, O. (1997). Serial recovery of language during the intracarotid amobarbital procedure. *Brain & Cognition, 33*, 151–160.

Rosenow, F., Morris, H. H., III, Foldvary, N., & Bingaman, W. (1997). Focal temporal burst-suppression during Wada-test correlated with speech arrest. *Journal of Epilepsy, 10*(6), 270–272. https://doi.org/10.1016/S0896-6974(97)00067-4

Sabsevitz, D. S., Swanson, S. J., Morris, G. L., III, Mueller, W. M., & Seidenberg, M. (2001). Memory outcome after left anterior temporal lobectomy in patients with expected and reversed Wada memory asymmetry scores. *Epilepsia, 42*(11), 1408–1415. https://doi.org/10.1046/j.1528-1157.2001.38500.x. PMID: 11879343.

Saltzman-Benaiah, J., Scott, K., & Smith, M. L. (2003). Factors associated with atypical speech representation in children with intractable epilepsy. *Neuropsychologia, 41*(14), 1967–1974. https://doi.org/10.1016/S0028-3932(02)00323-8

Schevon, C. A., Carlson, C., Zaroff, C. M., Weiner, H. J., Doyle, W. K., Miles, D., Lajoie, J., Kuzniecky, R., Pacia, S., Vazquez, B., Luciano, D., Najjar, S., & Devinsky, O. (2007). Pediatric language mapping: Sensitivity of neurostimulation and Wada testing in epilepsy surgery. *Epilepsia, 48*(3), 539–545. https://doi.org/10.1111/j.1528-1167.2006.00962.x

Sever, R. W., Vivas, A. C., Vale, F. L., & Schoenberg, M. R. (2018). Wada asymmetry in patients with drug-resistant mesial temporal lobe epilepsy: Implications for postoperative neuropsychological outcomes. *Epilepsia Open, 3*(3), 399–408. https://doi.org/10.1002/epi4.12250

Shah, A. K., Atkinson, M. D., Gupta, P., Zak, I., Watson, C. E., Rothermel, R., Asano, E., & Fuerst, D. (2010). Transient shivering during Wada test provides insight into human thermoregulation. *Epilepsia, 51*(5), 745–751. https://doi.org/10.1111/j.1528-1167.2009.02398.x

Snyder, P. J., Novelly, R. A., & Harris, L. J. (1990). Mixed speech dominance in the intracarotid Amytal procedure: Validity and criteria issues. *Journal of Clinical & Experimental Neuropsychology, 12*, 629–643. https://doi.org/10.1080/01688639008401007

Strauss, E., Satz, P., & Wada, J. (1990). An examination of the crowding hypothesis in epileptic patients who have undergone the carotid amytal test. *Neuropsychologia, 28*(11), 1221–1227. https://doi.org/10.1016/0028-3932(90)90057-u

Szabó, C. A., & Wyllie, E. (1993). Intracarotid amobarbital testing for language and memory dominance in children. *Epilepsy Research, 15*(3), 239–246. https://doi.org/10.1016/0920-1211(93)90061-B

Szantroch, M., Bala, A., Rysz, A., Żyłkowski, J., & Marchel, A. (2019). Experience of adverse events with cerebral propofol testing in patients with drug resistant epilepsy. *Scientific Reports, 9*(1), 1–7. https://doi.org/10.1038/s41598-018-36031-w

Telfeian, A. E., Berqvist, C., Danielak, C., Simon, S. L., & Duhaime, A. C. (2002). Recovery of language after left hemispherectomy in a sixteen-year-old girl with late-onset seizures. *Pediatric Neurosurgery, 37*(1), 19–21. https://doi.org/10.1159/000065096

Wada, J. (1949). A new method for the determination of the side of cerebral speech dominance: A preliminary report on the intracarotid injection of sodium amytal in man. *Igaku Seibutsugaku, 14*, 221–222.

Wada, J. A. (2008). A fateful encounter: Sixty years later—reflections on the Wada test. *Epilepsia, 49*(4), 726–727. https://doi.org/10.1111/j.1528-1167.2008.01515_6.x

Wada, J. A., & Kosaka, B. (2009). The Wada test—60th year anniversary update in epilepsy surgery. In H. Coenen, F. B. H. So, & T. A. M. A. Henriquez (Eds.), *Textbook of stereotactic and functional neurosurgery* (pp. 2587–2616). Springer. https://doi.org/10.1007/978-3-540-69960-6_154.

Wada, J., & Rasmussen, T. (1960). Intracarotid injection of sodium amytal for the lateralization of cerebral speech dominance: Experimental and clinical observations. *Journal of Neurosurgery, 17*(2), 266–282. https://doi.org/10.3171/jns.1960.17.2.0266

Waseem, H., Osborn, K. E., Schoenberg, M. R., Kelley, V., Bozorg, A. M., Benbadis, S. R., & Vale, F. L. (2015). Predictors of surgical outcome in medically-resistant temporal lobe epilepsy with bilateral features on pre-operative evaluation. *Clinical Neurology and Neurosurgery, 139*, 199–205. https://doi.org/10.1016/j.clineuro.2015.10.016

Whitman, L. A., Morrison, C. E., Becske, T., Barr, W. B., & Carlson, C. (2012). The intracarotid amobarbital procedure: When is it worth repeating? *Epilepsia, 53*(4), 721–727. https://doi.org/10.1111/j.1528-1167.2011.03399.x

Williams, J., & Rausch, R. (1992). Factors in children that predict performance on the intracarotid amobarbital procedure. *Epilepsia, 33*(6), 1036–1041. https://doi.org/10.1111/j.1528-1157.1992.tb01755.x

8
Polysomnography

Sarah E. Cook, Marjorie Soltis, and Paula-Ann E. Chin

Polysomnography (PSG) is a term derived from both Greek and Latin roots consisting of "polys" (many), "somnus" (sleep), and "graphein" (to write) (Polysomnography, 2008). Also known as a sleep study, this diagnostic procedure allows clinicians to examine various components of sleep and respiratory information through the collection of electrographic and electrophysiological data. The resulting data, in conjunction with a patient's medical history and presenting symptoms, allows for the diagnosis of various sleep disorders (e.g., obstructive sleep apnea, narcolepsy, and various parasomnias) (Gottlieb & Punjabi, 2020).

Laboratory-based PSG is a standard of care ordered for individuals presenting with certain sleep-related concerns. Set within a sleep lab, which can be located in a number of different places (e.g., lab within a hotel, a freestanding clinic, or lab within a hospital), patients are seen by a certified sleep technologist who aids in the physiological monitoring to complete the sleep study. During a sleep study, patients are monitored and connected to different electrodes, sensors, and belts including electroencephalogram and electrooculogram electrodes (on the head), electromyogram electrodes (on the extremities), electrocardiogram electrodes (primarily on the chest), sensors around the nose and mouth to measure airflow, thorax and abdomen belts to measure chest/abdomen rise correlating with respiratory effort, and a pulse oximeter. While all of this equipment can be intimidating, after the patient is connected to all monitoring devices, they are encouraged to sleep (generally from around 10 p.m. to 6 a.m.) in order to complete the study. Customarily, the sleep technologist monitors the study overnight via video from a nearby room and evaluates and scores the data while the patient is sleeping. Evaluation and scoring typically entails assessing sleep stage, body position and movements, and respiratory status/events. The initial results amassed by the sleep technologist are then reviewed by a sleep medicine physician who verifies the data accuracy, formulates a diagnosis, and provides treatment and/or management recommendations (Berry et al., 2020).

Sarah E. Cook, Marjorie Soltis, and Paula-Ann E. Chin, *Polysomnography*. In: *Medical and Laboratory Examinations for Neuropsychologists*. Edited by: Thomas J. Farrer and R. Brock Frost, Oxford University Press.
 DOI: 10.1093/9780197619537.003.0009

While laboratory-based polysomnography is the conventional diagnostic tool for sleep-related disorders and concerns, a home sleep study is an alternative tool sometimes used by clinicians. Generally retrieved from a sleep lab or through the mail, a home sleep study kit allows the patient to attach included sensors and belts themselves, but electrodes are rarely included. Through the home sleep study kit, patients are able to perform the study from the comfort of their home and are responsible for returning the equipment the following day for analysis. Home sleep studies allow for more convenient, and cheaper, screening for sleep apnea but are not without some drawbacks. This diagnostic tool tends to underestimate apnea severity and may also miss the diagnosis of other sleep-related conditions due to the limited data received in comparison to in-lab polysomnography (Bianchi & Goparaju, 2017).

Data Collected in Polysomnography

The *AASM Manual for the Scoring of Sleep and Associated Events*, by the American Academy of Sleep Medicine, is the main reference guide to understanding what parameters should be evaluated and reported during polysomnography (Berry et al., 2020). The manual breaks the data down into subcategories including general parameters, sleep scoring data, arousal events, cardiac events, movement events, and respiratory events, which will each be covered next.

General Parameters

It is recommended that general parameters include electroencephalogram (EEG) derivations (measuring brainwave activity), electrooculogram (EOG) derivations (measuring eye movement), chin electromyogram (measuring chin muscle tone), leg electromyogram (measuring leg movements), airflow signals, respiratory effort signals, oxygen saturation, body position, electrocardiogram (ECG; measuring cardiac activity), and synchronized video (in order to visualize the patient during the study).

Sleep Scoring Data

Generally entered by a sleep technologist, data derived from each of the sleep study's "epochs," or 30-second intervals, is scored by the sleep technologist to delineate if the patient was awake, in rapid eye movement (REM) sleep, or in a stage of non-REM sleep (N1, N2, or N3). Other recommended sleep scoring data includes the following (most measured in minutes):

- "Lights out" clock time
- "Lights on" clock time
- Total sleep time (TST; time spent in sleep stages N1, N2, N3, and REM sleep combined)
- Total recording time ("lights out" to "lights on" time)
- Sleep latency ("lights out" to first epoch of any sleep stage)
- Stage R (REM) latency (sleep onset to first epoch of stage REM sleep),
- Wake-after-sleep onset, or WASO (total recording time minus sleep latency, minus total sleep time)
- Percent sleep efficiency (TST divided by total recording time multiplied by 100)
- Time in each sleep stage
- Percent of total sleep time in each stage

Arousal Events

Reporting the number of arousals during the study, as well as the arousal index (i.e., the number of arousals × 60/ total sleep time), is recommended to provide a better sense of the level of disruption in a patient's sleep period. Of note, this can be influenced by the fact that the patient is in a new environment and is connected to several wires. This novel environment often leads to more arousals than typically experienced at home under normal sleeping conditions.

Cardiac Events

The polysomnography should include information on cardiac data including the average heart rate during sleep, highest heart rate during sleep, and highest heart rate during recording as well as occurrences of any abnormal heart rhythms with descriptions of observational behavior or events.

Movement Events

Movements in sleep can be very important and may act as contributory factors to a variety of sleep disorders. Given this potential contribution, it is recommended to provide commentary on a number of movement events should they occur during the study. Sleep technologists are tasked to include information on the number of periodic limb movements of sleep (PLMS), number of PLMS with arousals, the PLMS index (PLMS × 60/TST), and PLMS arousal index (PLMS with arousals × 60/TST). As it is normal (and expected) to have loss of muscle tone (i.e., atonia) during REM sleep, it is optional to include comments on REM without atonia. In certain conditions, such as REM sleep behavior disorder, patients may maintain muscle tone and therefore additional commentary on the observation may help aid in this diagnosis.

Respiratory Events

While there are numerous reportable respiratory events, only some are recommended to be reported while many others remain optional. The recommended reportable respiratory events include:

- Number of obstructive apneas (a drop in measurable respiratory effort by ≥90% for ≥10 seconds)
- Number of mixed apneas (apnea associated with absent inspiratory effort in the initial portion of the event followed by resumption of inspiratory effort in the second portion of the event)
- Number of central apneas (apnea associated with absent inspiratory effort throughout the entire period of absent airflow)
- Number of hypopneas (a drop in measurable respiratory effort by ≥30% for ≥10 seconds **with** a ≥3% oxygen desaturation **or** the event is associated with an arousal)
- Number of apneas + hypopneas
- Apnea index (# obstructive apneas + # central apneas + mixed apneas) × 60/TST)
- Hypopnea index (# hypopneas × 60/TST)
- Apnea-hypopnea index (# apneas + # hypopneas) × 60/TST)
- Arterial oxygen saturation (mean value)
- Minimum oxygen saturation during sleep

Lastly, the occurrence of Cheyne-Stokes breathing in adults is recorded. Cheyne-stokes breathing is scored whenever there are episodes of ≥3 consecutive central apneas (and/or hypopneas) separated by a crescendo and decrescendo change in breathing amplitude with a cycle length of ≥40 seconds, **and** there are ≥5 central apneas/hypopneas per hour of sleep associated with the crescendo/decrescendo breathing pattern recorded over ≥2 hours of monitoring

(Berry et al., 2020). Cheyne-Stokes breathing is associated with arrival at a high altitude, brain injury or brain tumor, high intracranial pressure, or acute cardiogenic pulmonary edema, and can be a clinical sign observed during end-of-life (Pacheco & Wright, 2022).

Summary Statements

The sleep medicine physician responsible for reading the study is tasked with summarizing the findings, with specific focus on sleep diagnoses, EEG abnormalities, ECG abnormalities, and behavioral observations.

Multiple Sleep Latency Test

When evaluating specifically for central disorders of hypersomnolence (e.g., narcolepsy and idiopathic hypersomnia), sleep medicine specialists will complete a multiple sleep latency test (MSLT) which is sometimes called a daytime nap study and has similar measurement procedures and parameters. Ideally, an MSLT should be completed when the patient is clinically stable and when treatments for existing sleep disorders are effective and well-established. The MSLT should be performed when the patient has been having consistent sleep patterns demonstrated by completion of a 2-week sleep diary prior to the MSLT, ideally with approximately 7 hours of sleep per night. Some medications may affect the MSLT results, especially those with stimulating or sedating properties. Antidepressants may also affect MSLT results by causing suppression/modulation of rapid eye movement (REM) sleep. Given the potential adverse effects on MSLT results, a brief discontinuation of certain medication may be required of patients for at least 2 weeks prior to testing.

The MSLT should be performed after an in-lab PSG, as outlined above, to verify the quality and quantity of sleep obtained (minimum 7 hours in bed with 6 hours of sleep). The MSLT should consist of five nap trials which begin 1.5–3 hours after completion of the PSG with each subsequent trial beginning 2 hours after the start of the prior trial. When evaluating the MSLT results, clinicians evaluate both the time it takes the patient to fall asleep during each nap as well as the onset to REM sleep. A patient is considered to have pathologic hypersomnolence if their mean sleep latency is 8 minutes or less; the results may be suggestive of narcolepsy if the patient has a sleep-onset REM period (SOREMP) in two or more of the five naps with a mean sleep latency of 8 minutes or less. If the patient has comorbid sleep apnea, their apnea should be appropriately treated during the polysomnogram and the nap trials (typically through the use of a continuous positive airway pressure device; CPAP).

Neuropsychology and Common Sleep Disorders Diagnosed with Polysomnography

There are three main classes of sleep disorders diagnosable through the use of polysomnography. This next section of the chapter will focus on each type of disorder and its relevance to the practice of clinical neuropsychology. First, sleep-disordered breathing conditions will be discussed, with a particular focus on obstructive sleep apnea, which is one of the most common disorders discovered through PSG. Then, we will briefly review hypersomnias and parasomnias.

Sleep-Disordered Breathing

Sleep Apnea

Sleep apnea is characterized as recurrent complete, or partial, cessations in breathing for at least 10 seconds (called apneas or hypopneas, respectively). These cessations lead to arousal from sleep and/or drop in blood oxygen saturation. Sleep apnea is further classified as either obstructive sleep apnea or central sleep apnea based on the mechanism responsible for disruption in airflow. Obstructive sleep apnea (OSA) occurs when muscles relaxing in the throat narrow the upper airway, whereas central sleep apnea (CSA) is caused by dysfunction of the area of the brain controlling breathing and carbon dioxide monitoring (a brainstem function) (Gottlieb & Punjabi, 2020; Strohl, 2022).

Obstructive Sleep Apnea

Obstructive sleep apnea is defined by the presence of at least five episodes of apneas and/or hypopneas per hour resulting from narrowing of airway despite normal respiratory muscle effort. A downstream effect of the recurrent cessations in breathing from OSA is excessive daytime sleepiness. OSA severity, subcategorized as mild, moderate, or severe, is determined by the apnea hypopnea index (AHI) as indicated below (Gottlieb & Punjabi, 2020):

- Mild OSA: 5–14 events per hour
- Moderate OSA: 15–29 events per hour
- Severe OSA: 30 or more events per hour

Risk factors for OSA include age (with increased age correlating with greater risk), sex (men are at greater risk), head and neck anatomy, body weight, and family history (Patel, 2019). OSA is rather common, with a cohort study finding that 17.4% of women and 33.9% of men in the United States, aged 30 to 70 years, had at least mild OSA (Peppard et al., 2013) although figures vary greatly from study to study. The prevalence of OSA has been rising in recent decades, likely attributable to increases in obesity, the aging population, greater recognition of the effect of sleep on health, and the increasing availability of diagnostic testing (i.e., PSG) (Patel, 2019). However, many people with symptoms never bring up nor are they asked about their sleep by their health provider for proper evaluation which is of great consequence since OSA has become more widely recognized as being associated with serious and adverse health and social consequences. These include daytime fatigue, changes in mood and cognition, hypertension and risk of cardiac events, metabolic syndrome, and increased risk of stroke. Several studies have also found relationships between OSA and risk of motor vehicle accidents, occupational injuries and/or absenteeism, reduced quality of life, and marital strain (Olaithe et al., 2018).

The clinical presentation of OSA varies. Common symptoms include snoring, not feeling refreshed in the morning and/or awakening with headache that resolves without intervention, excessive daytime sleepiness with greater propensity to fall asleep during passive activity, frequent awakenings at night with or without recognized gasping or choking, nocturia, nighttime gastroesophageal reflux, and resistant hypertension. Gottleib and Punjabi (2020) offer a more detailed clinical presentation of OSA including the prevalence of each symptom in patients presenting for PSG.

Central Sleep Apnea

Central sleep apnea (CSA) is caused by a problem in the areas of the brainstem that control breathing (i.e., pons and medullary tracts). Normally certain chemoreceptors in the brainstem continuously monitor the level of carbon dioxide in circulating blood and when carbon dioxide levels are elevated, the brainstem signals respiratory muscles to breathe deeper and faster to remove the excess carbon dioxide from the body. However, with CSA, the brain stem does not register changes in carbon dioxide level correctly. Consequently, moments of paused, shallow, or slowed breathing from reduced respiratory effort occur during sleep. CSA may occur as the result of medical conditions or insults such as stroke, traumatic brain injury, heart failure, or brain tumor. CSA can also occur in individuals with chronic opioid medication use and is also more prevalent in those living at higher altitudes. Overall, CSA is much rarer than OSA with a prevalence rate of less than 1% of the population (Strohl, 2022).

Treatment of Sleep Apnea

There are multiple treatment options available for sleep apnea. The primary treatment for OSA and CSA is a positive airway pressure (PAP) device that acts as a splint to keep the airway open and prevent collapse. It involves wearing a mask over the nose or over both the mouth and nose that is attached to an air pump. Traditionally there have been two types of PAP devices: the CPAP, which offers a continuous pressure, and the BiPAP, which offers one pressure during inhalation and another for exhalation. More recent technology is an AutoPAP or auto-adjusting PAP device, which automatically adjusts pressure levels in real time to meet breathing needs, which often change throughout the night depending on the stage of sleep. PAP treatment, regardless of device, is highly effective but dependent on adherence, which can be challenging for some individuals with sleep apnea (Gottlieb & Punjabi, 2020).

For those with mild OSA, behavioral treatments are sometimes offered first to attempt to reduce symptoms, such as positional changes to enforce side-sleeping or sleeping on an incline (Srijithesh et al., 2019). Patients may also be counseled to increase exercise (Kline et al., 2011), reduce alcohol consumption (Simou et al., 2018), and/or attempt lowering body weight (Hudgel et al., 2018), as these lifestyle changes can sometimes reduce OSA severity. Additionally, it is important for medical providers to complete a medication review to ensure that the individual is not taking medications that can impact sleep disordered breathing, such as sedatives, hypnotics, muscle relaxers, opioids, as well as tricyclic antidepressants. For individuals who cannot tolerate PAP treatment and for whom these alternatives are not sufficient for symptom reduction and/or improved sleep quality, some other options have been developed, including mandibular repositioning devices (oral device worn over teeth to reposition jaw), surgical modifications to the airway, and most recently a hypoglossal nerve stimulation device (Gottlieb & Punjabi, 2020; see Chapter 1 for discussion of cranial nerves).

Mechanism of Cognitive Change in Obstructive Sleep Apnea

Evidence suggests that the cognitive and emotional changes commonly seen in OSA may relate to structural brain changes, oxidative stress, neuroinflammation, and alterations in normal sleep architecture. Such findings include altered white matter integrity and functional network connections, changes in brain metabolites (e.g., increased carbon dioxide, reduced blood oxygen saturation), and alterations in regional gray matter volume (including the hippocampus, basal ganglia, thalamus, cingulate, insula, medulla, cerebellar, and frontal regions) (Gosselin et al., 2019). In addition, recurrent hypoxemia can directly affect the health of neuronal cell bodies and axons. The hypoxia that accompanies OSA may also cause endothelial cell dysfunction, thereby weakening the blood-brain barrier responsible for protecting the brain from potential circulating toxins and pathogens (Song et al., 2018).

Given these physiological changes, one proposed mechanism for cognitive change secondary to sleep-disordered breathing is the creation of vasculopathy due to vasoconstriction, and increased nocturnal blood pressure resulting in microvascular brain changes that induce cognitive changes (Zimmerman & Aloia, 2012). For a detailed review on potential mechanisms that link OSA and cognitive decline, see Gosselin (2019).

Research Findings on Cognition with Sleep-Disordered Breathing and/or OSA

The following section will detail the results of some individual studies as well as results from meta-analytic work examining cognitive effects in individuals with sleep-disordered breathing and/or obstructive sleep apnea. Of note, there has been variability in results to date, possibly because some study control groups were composed of patients referred for sleep studies without a consequent diagnosis of sleep apnea. Nonetheless, these patients likely had sleep symptoms requiring evaluation that could potentially be a cause of cognitive change. Additionally, there are far fewer studies on individuals with severe sleep apnea compared to mild or moderate, possibly due to the higher prevalence of less severe apnea (Fulda & Schulz, 2001). Further criticisms include concern that research to date has failed to identify a consistent relationship between severity of cognitive impairment and severity of sleep disturbance. Also, many study designs have been unable to differentiate the contribution of sleep fragmentation from blood gas changes, since they both occur in OSA (Olaithe, 2015).

Hoth et al. (2013) sought to examine the association between hypoxemia and cognitive performance. The investigators separated participants into lower and higher hypoxemia groups and then matched the groups on age, gender, and severity of respiratory disturbance (AHI). There were no group differences in education, BMI, daytime sleepiness, or depressive symptoms between these matched groups. Rather counterintuitively, they found that those with higher hypoxemia performed better on learning and memory than those with lower hypoxemia. It was opined that those with higher hypoxemia had greater adaptation to the condition and thus benefited from some protection.

Similarly, another study found that younger individuals displayed a greater impact of hypoxemia on their cognition compared to older adults. These findings were again thought to be due to the same mechanism: habituation, or adaptation, to hypoxemia over time (Mathieu et al., 2008). Interestingly, the converse result has also been found. In a study by Alchanatis and colleagues (2007), younger participants seemed able to compensate whereas older adults had more impairment thought to be due to sleep fragmentation and hypoxemia. In one study that attempted to separate the effects of sleep fragmentation and hypoxemia, researchers found that it was the fragmentation, and not hypoxemia, that was associated with changes in cognition. However, the study also suggests that there is a strong relationship between the two variables of fragmentation and blood gas abnormalities (r = 70%) (Olaithe, 2015).

A comprehensive review of the literature by Fulda and Schulz (2001) found that there was no difference between OSA (mild or severe) and controls on perceptual functioning. Those with mild OSA were no different from controls on a broad range of attention and vigilance tasks, however, there were variable findings for those with moderate OSA compared to controls on these tasks. Across studies, those with severe OSA demonstrated impaired sustained attention but not for other types of attention reviewed. Furthermore, there was inconclusive evidence on a difference between those with OSA and controls on reasoning, motor, or construction tasks. Driving simulation deficit was noted for moderate-to-severe OSA compared to controls. There was limited evidence of memory difficulty in those with mild OSA, but moderate OSA has shown reduced learning. On verbal fluency measures, those with severe OSA appeared deficient.

In a meta-review by Bucks and colleagues (2013) on the cognitive impact of OSA, they reported that the majority of the reviews included in their investigation reported deficits in attention and/or vigilance, executive function, and delayed memory recall regardless of the severity of OSA. In a more recent meta-analysis of nineteen studies by Stranks and Crowe (2016), psychomotor speed and executive functions were found to have the largest difference, whereas the cognitive domains of memory, motor control, construction, attention, and speed of cognitive processing were more mildly affected. The authors noted that the results were similar to an earlier meta-analysis by Beebe and colleagues (2003) suggesting that effects have not been changing over time.

Research on the Effectiveness of Treatment

Other research has sought to investigate the associations between OSA severity and the effectiveness of CPAP treatment on cognition. A study conducted by Kushida et al. (2012) remains to date the largest clinical trial examining the effectiveness of treatment with CPAP (active treatment versus sham) in those with newly diagnosed OSA. Through this clinical trial, investigators found that while subjective and objective measures of daytime sleepiness improved for those with moderate-to-severe OSA, there was only a significant difference on one measure of executive functioning between treatment groups at the 2-month marker which was not evident at the later 6-month follow-up. In a study comparing mild-to-moderate OSA in three treatment arms (CPAP, oral appliance, or placebo pill) with age-matched community controls, findings showed that those with untreated OSA had significantly lower working memory and psychomotor function. Following treatment with CPAP for 3 months, improvements were seen in some neuropsychological domains compared to baseline (i.e., in verbal fluency and psychomotor function). Investigators subsequently examined the importance of dose-response and found that individuals most compliant with CPAP had incremental improvements in psychomotor speed and vigilance although the differences between compliance groups were not statistically significant (Jackson et al., 2018). Similarly, prior research by Munoz et al. (2000) also explored the effects of CPAP in those with OSA. For this study, conducted after 1 year of treatment, results indicated significant improvement in daytime sleepiness reports as well as performance on a vigilance task when compared to pretreatment performance (Muñoz et al., 2000). These findings suggest it is possible that longer treatment is needed for neurocognitive benefits to be appreciated.

Literature review also reveals that various imaging studies have been conducted to gain further insight into the neuroanatomical implications of OSA. One such study, conducted by Xiong and colleagues (2017), examined MRI outcomes after one year of compliant CPAP use. These investigators found that individuals who reported residual sleepiness showed significantly different mean diffusivity and radial diffusivity in analysis of whole brain white matter integrity. Furthermore, they also displayed altered fractional anisotropy and mean diffusivity values (markers of white matter integrity) in specific white matter regions when compared to the participants who reported significantly improved fatigue. These neuroanatomical findings may account for differences in treatment response among compliant individuals (Xiong et al., 2017). CPAP treatment has shown increased hippocampal volume and gray matter thickness in frontal regions, and was associated with improvements in short-term memory and executive functions. A study conducted by Castronovo and colleagues examined white matter integrity and cognition in OSA using treatment-naive male patients with severe OSA compared to age and education-matched healthy male control participants (Castronovo et al., 2014). Participants, re-evaluated after 3 and 12 months of treatment, were found to have not only reduced daytime sleepiness, but also improvements in nearly all cognitive measures. There were limited changes to diffusion tensor imaging (DTI) metrics after 3 months of treatment; however changes were more notable after 12 months.

Dementia Risk and Benefits of Treatment

Thus far, the research presented in this chapter has primarily related to risks and outcomes for middle-aged adults. However, when discussing risk factors in preceding sections, it was briefly mentioned that the risk of OSA increases with age. Since sleep is important for clearing the brain of metabolic waste, including the waste products implicated in the pathogenesis for several dementia syndromes (e.g., Aβ42, tau) (Fultz et al., 2019), it is not surprising that sleep disorders may increase risk of cognitive decline in older adults. This next section will cover age-related issues with regard to OSA as well as the benefit of treatment in old age.

In a prospective study of almost 300 women, investigators found that sleep-disordered breathing, particularly a measure of hypoxia on home PSG, was associated with increased risk of developing dementia over the 5 years of follow-up (Yaffe et al., 2011). A meta-analysis examining dementia risk which included over 200,000 middle-to-late age adults found that adults with OSA were 26% more likely to develop significant cognitive decline or dementia over the follow-up interval (ranging from 3–15 years) (Leng et al., 2017).

A retrospective study using data from the Alzheimer's Disease Neuroimaging Initiative found that a diagnosis of OSA was associated with an earlier age of progression from mild cognitive impairment to dementia. This study also found that use of CPAP was associated with delayed progression (Osorio et al., 2015). In a randomized clinical trial of patients with Alzheimer's disease with newly diagnosed OSA, investigators found modest, but statistically significant, improvements in composite cognitive scores associated with only 3 weeks of therapeutic CPAP (Ancoli-Israel et al., 2008). Given these findings, early identification and treatment of OSA is an important dementia prevention strategy.

Disorders of Hypersomnolence

Although this chapter has primarily focused on sleep-disordered breathing (the most prevalent type of condition diagnosed with PSG), the following section will further delineate another type of sleep disorder for which polysomnography is used to aid diagnosis: hypersomnolence. Hypersomnolence refers to excessive sleepiness either during the day or leading to prolonged nighttime sleeping. This differs from lack of energy or tiredness in that individuals with hypersomnia often fall asleep even during inappropriate times (e.g., during conversation or while eating or driving), and there is limited relief from fatigue regardless of time spent asleep. A comprehensive review of the two most common disorders of hypersomnolence: narcolepsy and idiopathic hypersomnia, is provided below.

Narcolepsy

Narcolepsy is a disorder of the sleep-wake cycle characterized by excessive daytime sleepiness, cataplexy (total or partial sudden loss of muscle tone while awake in response to emotion), and sometimes hypnagogic and/or hypnopompic hallucinations (false visual perceptions while falling asleep or immediately upon awakening, respectively), and/or sleep paralysis (regaining awareness on awakening without the physical ability to move momentarily) (Anderson, 2021). There is noted heterogeneity in the presentation such that about 30%–50% of patients with narcolepsy do not have all of the above symptoms (Kushida et al., 2005). Symptoms often start in adolescence or early adulthood. While the exact cause of narcolepsy is not fully understood, multiple neurotransmitter system abnormalities have been implicated, including noradrenaline, serotonin, dopamine, and acetylcholine systems, as well as evidence of dysfunction to the hypocretin (or orexin) system (Naumann et al., 2006). There has been some discrepancy regarding structural abnormality correlates of narcolepsy, with earlier research finding no neuroimaging differences (Cluydts & Verstraeten, 2002). However, a more recent systematic review of narcolepsy indicated several abnormalities have been replicated, albeit with heterogeneity of samples and small sample sizes, including reduced gray matter volumes in several structures (including frontal and temporal lobes as well as hypothalamus, cingulate cortex, amygdala, insular cortex, hippocampus, and thalamus), reduced fractional anisotropy on diffusion tensor imaging in the white matter of cingulate and orbitofrontal regions, reduced cerebral blood flow and brain metabolism in cingulate cortex and frontal lobes, and increased activity of insular cortex, amygdala, and inferior frontal gyri (see Wada, 2019, for review).

As noted earlier in the chapter, a PSG with a REM latency of less than 20 minutes may signify narcolepsy, and therefore a multiple sleep latency test (MSLT), also discussed earlier in the chapter, is subsequently used to confirm diagnosis. Other biomarkers are diagnostically useful, including human leukocyte antigen (HLA) testing since more than 90% of patients with narcolepsy carry the DQB1-0602 marker, and CSF analysis, which often shows abnormally low hypocretin-1/orexin-A (Kushida et al., 2005). Unfortunately, there is no known cure for narcolepsy, and treatment includes both behavioral (e.g., rigid sleep schedule, daily exercise, limiting consumption of caffeine) as well as pharmacologic methods. Individuals with narcolepsy should avoid taking medications that can cause drowsiness (e.g., antihistamines, opioids, sedatives) and may benefit from medications that promote wakefulness including stimulants (e.g., modafinil, armodafinil, methylphenidate), sodium oxybate, as well as newer agents (e.g., pitolisant, solriamfetol) (Anderson, 2021).

Due to sudden sleep attacks, some individuals with narcolepsy can carry out automatic, familiar tasks while in a sleep state and then have retrograde amnesia for completing the task (Anderson, 2021). Given the role of acetylcholine in memory functioning, early studies examined memory functioning in individuals with narcolepsy on neuropsychological tests, and to date evidence of impairment remains inconclusive. However, this may potentially be due to small sample sizes and inconsistency in measurement across studies. Studies evaluating aspects of attention and vigilance, especially long, repetitive tasks, suggest that those with narcolepsy are negatively impacted compared to controls. Thus, researchers speculate that the resultant cognitive deficits of narcolepsy may be due to reduced cognitive processing resources available as a consequence of the constant need for monitoring and maintaining alertness (Naumann et al., 2006).

Idiopathic Hypersomnia

Another disorder of hypersomnolence is called idiopathic hypersomnia (IH). It is characterized by excessive daytime sleepiness coupled with long sleep times, but without the cataplexy, hallucinations, and sleep paralysis seen in narcolepsy. Individuals with idiopathic hypersomnia are difficult to awaken and report continued drowsiness and inertia during the daytime. They also often report feeling unrefreshed even after long naps. PSG results are usually normal, but MSLT demonstrates short sleep latencies (<8 minutes) and 2 or fewer REM periods. The condition seems to first present in adolescence or young adulthood, and is more common in women and those with a family history of IH (Dauvilliers et al., 2022). It is often treated similarly to narcolepsy.

IH appears to be less common than narcolepsy, but without clear consensus about biological mechanisms or biomarkers, as well as the fact that symptoms can be common in general populations, there is less known about this condition. In the few studies that have been reviewed by Filardi and colleagues (2021), individuals with IH were noted to have difficulties with vigilance with increased reaction times over prolonged tasks compared to controls. No significant differences have been noted for memory or executive functions when comparing those with IH to controls.

Parasomnias

Parasomnias, which are disruptive sleep disorders involving unusual movements or actions during sleep, may serve as important symptoms of various neurological conditions. In this section, two common parasomnias significant to the practice of clinical neuropsychology will be reviewed. While parasomnias include nightmares and night terrors, sleepwalking, sleep-related eating, and sleep paralysis, we will only focus briefly on REM sleep behavior disorder and periodic limb movement disorder.

REM Sleep Behavior Disorder

REM sleep behavior disorder (RSBD) is characterized by the enacting of dreams during rapid eye movement (REM) sleep. During this time, movements mimicking dream content occur rather than atonia (lack of muscle tone), which is typical during REM sleep. Patients may hit or kick, thrash, sit up, or stand up, or fall out of bed as part of the enactment. These movements may cause serious physical injury to the individual with RSBD and/or their bed partner, and often lead to separate sleep locations for the safety of all parties. The key PSG finding suggestive of RSBD is electromyographic (EMG) abnormality during REM sleep. However, such enactment is typically episodic and therefore may not be observed during one night of PSG monitoring (Zhang et al., 2008). Research has shown that up to 93% of individuals with RSBD eventually develop dementia with Lewy bodies or another neurodegenerative condition associated with deposition of alpha-synuclein, if they are followed and live long enough (Donaghy & McKeith, 2014). Thus, this disorder is of particular relevance to neuropsychological practice as it can be an early indicator of a neurodegenerative condition.

Periodic Limb Movement Disorder

As the name implies, periodic limb movement disorder (PLMD) occurs when there are intermittent, but repetitive, involuntary movements of the lower extremities while sleeping that may cause brief arousals from sleep. Similar to RDBD, the main finding on polysomnography for PLMD is an electromyographic (EMG) abnormality. However, if sleep arousals are caused by the movement, that would also be noted by electroencephalogram (EEG) data. Limb movements during sleep are relatively common in the general population and most individuals with restless leg syndrome experience limb movement in their sleep. PLMD also often accompanies narcolepsy and RSBD. Clinically, true PLMD not accounted for by another sleep disorder remains uncommon. Although onset can vary, it is more common with age. The etiology of PLMD is not entirely clear; however, there are implications of dopamine contribution. Furthermore, some medications and caffeine have shown to worsen symptoms (Drakatos et al., 2021).

Unsurprisingly, given the possibility of multiple arousals during sleep in PLMD and the cognitive effects of sleep fragmentation, studies have shown some impact of PLMD on cognition. In a longitudinal study of over 2,500 older men without dementia who had an in-home PSG, the men with a high score of periodic limb movements in sleep showed greater decline in Trails B score over 3–4 years (those with periodic limb movement index [PLMI] ≥30 were 48% more likely to experience >1 SD change). This was independent of sleep efficiency, nocturnal hypoxemia, or use of dopaminergic medications (e.g., levodopa or dopamine agonists). Further analysis once excluding men with Parkinson-related conditions resulted in similar findings (Leng et al., 2016).

Assessing for Possible Sleep Disturbance

Given the potential impact of sleep disorders on cognition, a thorough clinical interview as part of a neuropsychological evaluation should cover some basic questions about sleep. Such questions may be open-ended such as, "how is your sleep?" to elicit preliminary information. Covering topics such as hours slept, difficulty with sleep onset or maintenance, degree of feeling refreshed in the morning, awakening with headaches, presence of snoring and/or witnessed apneas, dream enactment behavior (e.g., kicking, thrashing, falling out of bed), and nightmares serve as a good basis for uncovering some common issues with sleep. These interviews may lead to recommendations for further assessment and/or treatment by their physician.

Some commonly used measures have been developed to assess sleep quality or problems and may be helpful to include in clinical evaluations. A common self-report measure is the Epworth Sleepiness Scale (score of ≥11 warrants further evaluation for a sleep disorder; Johns, 1991). Another, called STOP-BANG, was originally designed for surgical screening but is useful in screening for OSA (Chung et al., 2008). Other sleep-related issues may be better assessed with more broad questionnaires, such as the Pittsburgh Sleep Quality Index (Buysse et al., 1989) or Insomnia Severity Index (Morin et al., 2011). These three instruments are available in the public domain, although there are several other options.

It is important to note that while snoring and subjectively reported daytime sleepiness are symptoms of, and highly correlated with diagnosis of, sleep-related breathing disorders, not everyone experiencing these symptoms has a sleep disorder. Thus, objective measurement of sleep is important for proper identification (Kushida et al., 2005).

Case Example

The patient "Mr. D," age 73 at time of evaluation, is a right-handed white male referred for neuropsychological re-evaluation by his neurologist due to progressive cognitive concerns. He is married and presented with his wife. He is high-school educated and reported being an average student overall. Early educational history includes retention for one grade in elementary school, but average educational performance in high school with no known neurodevelopmental concerns. He worked in sales at a telecommunications company for the entirety of his career and had retired over a decade before his first evaluation.

Mr. D initially presented to neurology at age 68 with concerns about memory loss, such as forgetting conversations and relying more on GPS for navigation. He was fully independent for IADLs. Medical history was significant for hypertension. Both patient and spouse reported that he was sleeping about 14 hours a day yet remained fatigued. His wife further reported that he snored and was observed to have intermittent "choking" episodes in his sleep. When asked about mood and psychiatric history, he endorsed increased irritability and anhedonia and had a prior diagnosis of depression. Medications included aspirin, citalopram, triamterene-hydrochlorothiazide, vitamin D3, and vitamin B12. He scored 27 out of 30 on the MMSE (Folstein et al., 1975) upon clinical screening in neurology. Given subjective concerns and medical history, initial differential diagnoses from his neurologic evaluation included mild cognitive impairment versus obstructive sleep apnea. The neurologist referred him for polysomnography (PSG), brain MRI, and neuropsychological evaluation.

A split-night PSG was completed and revealed an overall apnea-hypopnea index (AHI) of 27/hr. The REM AHI was 12/hr, non-REM AHI was 32/hr, supine AHI was 40/hr, and lateral AHI was 12/hr. A CPAP pressure of 7 CWP with heated humidification was recommended. At his initial follow-up a few months later in sleep medicine, he presented reporting compliance with CPAP. This was verified by his machine; recorded usage was 93% of days with an average usage time more than 5 hours per night. Brain MRI with NeuroQuant was also completed and revealed age-appropriate volume loss and periventricular white matter increased T2-FLAIR consistent with chronic ischemic small vessel disease. There was no disproportionate lobar atrophy. The NeuroQuant software estimated the bilateral hippocampal volume at the 40th percentile for age.

As seen in Table 8.1, at his initial neuropsychological evaluation in 2018, Mr. D demonstrated deficits in some executive functioning tasks (e.g., difficulty maintaining set and complex sequencing), working memory, phonemic fluency, visual-spatial planning, and reduced acquisition and retrieval of new information without amnesia. Behaviorally, he seemed concrete in thinking and exhibited some utilization behavior during testing. The etiology was believed to be mixed, including small-vessel ischemia as well as history of sleep apnea. While he had mildly elevated mood symptoms, it was concluded to not fully account for the degree of deficit given acceptable performance on administered and embedded validity measures. With IADLs minimally impacted by his cognitive difficulties, the patient was diagnosed with non-amnestic mild cognitive impairment.

At his 1-year neurology follow-up, Mr. D reported initially observing improvement in cognitive symptoms with continued CPAP therapy, however believed that improvements had tapered off as he again started to appreciate cognitive decline. He was again referred for neuropsychological evaluation and seen approximately 18 months after the initial evaluation (2019). Upon interview by the neuropsychologist, he indicated that he had tapered off use of CPAP in recent months due to perceived lack of benefit and inconvenience. Two new medications, levothyroxine and quetiapine (as a sleep aid), were started in the interim in conjunction with his medication list of aspirin, citalopram, triamterene-hydrochlorothiazide, vitamin D3, and vitamin B12. His thyroid values had returned to within normal limits by the time of testing. Results of the re-evaluation, as noted in the middle column of Table 8.1, revealed a few areas of decline, including simple and complex sequencing as well as semantic fluency. Conversely, there were multiple areas of testing that improved from baseline, including working memory, processing speed, aspects of executive function, visual and verbal memory, response inhibition, and phonemic fluency. Other measures were relatively equivalent. Behaviorally, he seemed impetuous at times and utilization behavior was again noted. Diagnosis was not changed.

Mr. D was lost to neurology follow-up for almost 3 years (possibly related to the pandemic) and returned with continued complaints about worsening cognition. He had ceased all use of CPAP during the intervening years. A bedside MOCA (Nasreddine et al., 2005) of 12/30 prompted a referral for neuropsychological re-evaluation. At the most recent neuropsychological evaluation (2022), he reported continued poor sleep and daytime fatigue. There was no report of or evidence of REM sleep behavior or hallucinations. Mood was again reported as depressed and irritable. One new medication, tamsulosin, was added to his medication regimen of aspirin, citalopram, triamterene-hydrochlorothiazide, vitamin D3, vitamin B12, levothyroxine, and quetiapine since his last evaluation. The patient and spouse also had concerns about declining memory and difficulties with some IADLs. Examples of IADL difficulties included having forgotten to turn off the stove on numerous occasions, and issues monitoring food while cooking, resulting in a call to 911 for intense smoke in the home. His wife reported that he was still managing his own medications without apparent difficulty. Given observed cognitive decline, he had ceased driving per the recommendation of the referring neurologist.

Repeat testing (see last column of Table 8.1) revealed deficits in cognitive flexibility, basic processing speed (including word reading speed), generation

Table 8.1 Neuropsychological evaluation results for Mr. D

	April 2018	November 2019	December 2022
Age at evaluation	68	70	73
Intellectual Functioning			
Wechsler Adult Intelligence Test–Fourth Ed. Index Score(%)			
WAIS-IV VCI	83(13)	83(13)	83(13)
WAIS-IV PRI	71(3)	79(8)	77(6)
WAIS-IV WMI	71(3)	77(6)	69(2)
WAIS-IV PSI	81(10)	89(23)	81(10)
WAIS-IV FSIQ	72(3)	78(7)	74(4)
WAIS-GAI	75(5)	79(8)	78(7)
Subtest Raw Score(ACSS)			
Similarities	20(8)	16(7)	16(7)
Vocabulary	—	16(5)	24(7)
Information	7(6)	13(9)	—
Block Design	12(4)	12(4)	16(6)
Matrix Reasoning	6(5)	8(8)	6(6)
Visual Puzzles	6(6)	7(7)	—
Digit Span	15(4)	19(7)	15(4)
Arithmetic	9(6)	7(5)	7(5)
Symbol Search	18(7)	19(8)	14(6)
Coding	35(6)	40(8)	33(7)
Executive Functions			
Trail-Making Test			
Trial A Raw[T]	48[40]	59[37]	43[46]
Trial A errors	0	1	3
Trial B Raw[T]		>300[—]	>300[—]
Trial B errors	2	6	2
Stroop			
Word Raw[T]	79[37]	78[37]	60[25]
Color Raw[T]	49[32]	48[31]	38[23]
Color-Word Raw[T]	17[35]	25[43]	15[34]
Interference[T]	−12[38]	−4[46]	−8[42]
Wisconsin Card Sorting			
Correct/Administered	83/128	62/128	—
Categories(%)	1(11–16)	2(>16)	—
Failure to Maintain Set	5	1	—
Errors[T]	45[45]	66[39]	—
Perseverative Responses[T]	41[41]	42[42]	—
Perseverative Errors[T]	36[40]	37[41]	—
Non-Perseverative Errors[T]	9[57]	29[37]	—

continued

Table 8.1 *continued*

	April 2018	November 2019	December 2022
Memory			
Wechsler Memory Scale-IV			
Logical Memory I(Age SS)	26(8)	31(10)	18(5)
Logical Memory II(Age SS)	17(9)	20(11)	8(6)
Percent Retained	81	91	62
Recognition(Cumulative %)	19(51–75)	20(>75)	17(26–50)
Hopkins Verbal Learning Test			
Trial 1	3	3	6
Trial 2	5	7	4
Trial 3	6	7	5
Total Recall[T]	14[31]	17(37)	15[34]
Delayed Recall[T]	0[≤20]	7(44)	5[38]
Percent Retained[T]	0[≤20]	100[56]	100[56]
Recognition	10/12	11/12	11/12
False Positive	4/12	2/12	4/12
Discrimination[T]	6[25]	9[42]	7[32]
Benton Visual Memory Test-Revised			
TI[T]	2[35]	5[50]	2[36]
TII[T]	1[<20]	7[48]	5[39]
TIII[T]	3[23]	8[47]	6[39]
Total Recall[T]	6[23]	20[48]	13[36]
Delayed Recall[T]	5[36]	8[50]	3[29]
Recognition (Cum. %)	4(6–10)	5(>16)	5(>16)
Discrim. Index (Cum. %)	4(11–16)	3(3–5)	4(11–16)
Language			
Boston Naming Test-II Raw[T]	53[45]	53[48]	49[44]
COWAT[T]	13[26]	17[31]	9[20]
Animals[T]	23[62]	15[47]	16[47]
Visual Spatial/Construction			
JOLO Raw + Correction(SS)	20+2(22)	21+2(40)	20+2(22)
Rey-O Complex Figure(%)	21.5(<1)	16.0(<1)	—
BVMT-R Copy Raw	11	9	10
Mood			
Geriatric Depression Scale	10/30	13/30	7/30

Note: Normative values derived from test manuals with the exception of Trail-Making, Boston Naming Test, COWAT, and Animal Fluency, in which Heaton norms correcting for age, education, sex, and race were utilized.

of words beginning with certain sounds, and milder reductions in basic attention and working memory. He had reduced acquisition of new information across memory tasks, but without rapid forgetting or amnesia on recognition testing. Naming, semantic word knowledge, and semantic fluency were intact. Visual construction, visual reasoning, and inhibition also appear intact. Much of his performance across cognitive domains remained relatively stable in comparison to his last evaluation in 2019, however, declined performances were noted for basic attention and working memory capacity, word reading speed, and phonemic fluency as well as some aspects of learning and retention (contextual verbal learning and visual learning and retention). Behaviorally, he was noted to be impulsive in responding. The overall pattern continued to be one of deficits impacting executive skills, including executive aspects of memory, and language suggestive of primarily frontal subcortical involvement. Findings were considered secondary to vascular risks, including untreated sleep apnea and neuroimaging findings.

Conclusion

Sleep is vital for optimal daytime physical and cognitive functioning in part because it allows metabolic waste to be cleared from the brain. This chapter reviewed the most common sleep disorder, obstructive sleep apnea, which can be easily detected and treated to reduce the cognitive and cardiac consequences of untreated apnea that have been documented. Less-common sleep disorders, including hypersomnias (narcolepsy and idiopathic hypersomnolence) and parasomnias (REM sleep behavior disorder and periodic limb movement disorder) were also reviewed. Because of the negative cognitive consequences of some of these conditions as already outlined, it is important for clinical neuropsychologists to assess for sleep quality and concerns as a potential treatable cause of presenting cognitive symptoms.

References

Alchanatis, M., Zias, N., Deligiorgis, N., Liappas, I., Chroneou, A., Soldatos, C., & Roussos, C. (2007). Comparison of cognitive performance among different age groups in patients with obstructive sleep apnea. *Sleep and Breathing, 12*(1), 17–24. https://doi.org/10.1007/s11325-007-0133-y

Ancoli-Israel, S., Palmer, B. W., Cooke, J. R., Corey-Bloom, J., Fiorentino, L., Natarajan, L., Liu, L., Ayalon, L., He, F., & Loredo, J. S. (2008). Cognitive effects of treating obstructive sleep apnea in Alzheimer's disease: A randomized controlled study. *Journal of the American Geriatrics Society, 56*(11), 2076–2081. https://doi.org/10.1111/j.1532-5415.2008.01934.x

Anderson D. (2021). Narcolepsy. *Journal of the American Academy of Physician Assistants, 34*(6), 20–25. https://doi.org/10.1097/01.JAA.0000750944.46705.36

Beebe, D. W., Groesz, L., Wells, C., Nichols, A., & McGee, K. (2003). The neuropsychological effects of obstructive sleep apnea: A meta-analysis of norm-referenced and case-controlled data. *Sleep*, *26*(3), 298–307. https://doi.org/10.1093/sleep/26.3.298

Benedict, R. (1997a). *Brief visuospatial memory test–Revised.* Psychological Assessment Resources.

Benton, A. L., Varney, N. R., & Hamsher, K. D. (1978). Visuospatial judgment. *Archives of Neurology*, *35*(6), 364–367. https://doi.org/10.1001/archneur.1978.00500300038006

Berry, R. B., Quan, S. F., Abreu, A. R., et al., for the American Academy of Sleep Medicine. (2020). *The AASM manual for the scoring of sleep and associated events: Rules, terminology and technical specifications*, Version 2.6, www.aasmnet.org, American Academy of Sleep Medicine.

Bianchi, M. T., & Goparaju, B. (2017). Potential underestimation of sleep apnea severity by at home kits: Rescoring in-laboratory polysomnography without sleep staging. *Journal of Clinical Sleep Medicine*, *13*(4), 551–555. https://doi.org/10.5664/jcsm.6540

Brandt, J., & Benedict, R. (1997). *Hopkins verbal learning test–Revised.* Psychological Assessment Resources.

Bucks, R. S., Olaithe, M., & Eastwood, P. (2013). Neurocognitive function in obstructive sleep apnoea: A meta-review. *Respirology (Carlton, Vic.)*, *18*(1), 61–70. https://doi.org/10.1111/j.1440-1843.2012.02255.x

Buysse, D. J., Reynolds, C. F., Monk, T. H., Berman, S. R., & Kupfer, D. J. (1989). The Pittsburgh sleep quality index: A new instrument for psychiatric practice and research *Psychiatry Research*, *28*(2), 193–213. https://doi.org/10.1016/0165-1781(89)90047-4

Castronovo, V., Scifo, P., Castellano, A., Aloia, M. S., Iadanza, A., Marelli, S., Cappa, S. F., Strambi, L. F., & Falini, A. (2014). White matter integrity in obstructive sleep apnea before and after treatment. *Sleep*, *37*(9), 1465–1475. https://doi.org/10.5665/sleep.3994

Chung, F., Yegneswaran, B., Liao, P., Chung, S. A., Vairavanathan, S., Islam, S., Khajehdehi, A., & Shapiro, C. M. (2008). STOP questionnaire: A tool to screen patients for obstructive sleep apnea. *Anesthesiology*, *108*(5), 812–821. https://doi.org/10.1097/ALN.0b013e31816d83e4

Cluydts, R., & Verstraeten, E. (2002). The neuropsychology of sleep disorders. In D'Haenen, J. A. den Boer, & P. Willner (Eds.), *Biological Psychiatry* (pp. 1275–1284). Wiley. https://doi.org/10.1002/0470854871.CHXXIV6

Daulatzai, M. A. (2015). Evidence of neurodegeneration in obstructive sleep apnea: Relationship between obstructive sleep apnea and cognitive dysfunction in the elderly. *Journal of Neuroscience Research*, *93*(12), 1778–1794. https://doi.org/10.1002/jnr.23634

Dauvilliers, Y., Bogan, R. K., Arnulf, I., Scammell, T. E., St Louis, E. K., & Thorpy, M. J. (2022). Clinical considerations for the diagnosis of idiopathic hypersomnia. *Sleep Medicine Reviews*, *66*, 101709. https://doi.org/10.1016/j.smrv.2022.101709

Donaghy, P. C., & McKeith, I. G. (2014). The clinical characteristics of dementia with Lewy bodies and a consideration of prodromal diagnosis. *Alzheimer's Research & Therapy*, *6*(4), 46. https://doi.org/10.1186/alzrt274

Drakatos, P., Olaithe, M., Verma, D., Ilic, K., Cash, D., Fatima, Y., Higgins, S., Young, A. H., Chaudhuri, K. R., Steier, J., Skinner, T., Bucks, R., & Rosenzweig, I. (2021). Periodic limb movements during sleep: A narrative review. *Journal of Thoracic Disease*, *13*(11), 6476–6494. https://doi.org/10.21037/jtd-21-1353

Filardi, M., D'Anselmo, A., Agnoli, S., Rubaltelli, E., Mastria, S., Mangiaruga, A., Franceschini, C., Pizza, F., Corazza, G. E., & Plazzi, G. (2021). Cognitive dysfunction in central disorders of hypersomnolence: A systematic review. *Sleep Medicine Reviews, 59*, 101510. https://doi.org/10.1016/j.smrv.2021.101510

Folstein, M. F., Folstein, S. E., & McHugh, P. R. (1975). Mini-mental state. *Journal of Psychiatric Research, 12*(3), 189–198. https://doi.org/10.1016/0022-3956(75)90026-6

Fulda, S., & Schulz, H. (2001). Cognitive dysfunction in sleep disorders. *Sleep Medicine Reviews, 5*(6), 423–445. https://doi.org/10.1053/smrv.2001.0157

Fultz, N. E., Bonmassar, G., Setsompop, K., Stickgold, R. A., Rosen, B. R., Polimeni, J. R., & Lewis, L. D. (2019). Coupled electrophysiological, hemodynamic, and cerebrospinal fluid oscillations in human sleep. *Science, 366*(6465), 628–631. https://doi.org/10.1126/science.aax5440

Gagnon, K., Baril, A. A., Gagnon, J. F., Fortin, M., Decary, A., Lafond, C., Desautels, A., Montplaisir, J., & Gosselin, N. (2014). Cognitive impairment in obstructive sleep apnea. *Pathologie Biologie, 62*(5), 233–240. https://doi.org/10.1016/j.patbio.2014.05.015

Golden, C. J. (1978). *Stroop color and word test: A manual for clinical and experimental uses.* Stoetling.

Gosselin, N., Baril, A. A., Osorio, R. S., Kaminska, M., & Carrier, J. (2019). Obstructive sleep apnea and the risk of cognitive decline in older adults. *American Journal of Respiratory and Critical Care Medicine, 199*(2), 142–148. https://doi.org/10.1164/rccm.201801-0204pp

Gottlieb, D. J., & Punjabi, N. M. (2020). Diagnosis and management of obstructive sleep apnea: A review. *JAMA, 323*(14), 1389–1400. https://doi.org/10.1001/jama.2020.3514

Heaton, R. K., Chelune, G. J., Talley, J. L., Kay, G. G., & Curtiss, G. (1993). *Wisconsin card sorting test manual: Revised and expanded.* Psychological Assessment Resources.

Heaton, R. K., Miller, S. W., Taylor, M. J., & Grant, I. (2004). *Revised comprehensive norms for an expanded Halstead-Reitan battery: Demographically adjusted neuropsychological norms for African American and caucasian adults.* Psychological Assessment Resources.

Hilsendager, C. A., Zhang, D., McRae, C., & Aloia, M. (2016). Assessing the influence of obesity on longitudinal executive functioning performance in patients with obstructive sleep apnea syndrome. *Obesity Research & Clinical Practice, 10*(1), 33–40. https://doi.org/10.1016/j.orcp.2015.04.010

Hoth, K. F., Zimmerman, M. E., Meschede, K. A., Arnedt, J. T., & Aloia, M. S. (2013). Obstructive sleep apnea: Impact of hypoxemia on memory. *Sleep and Breathing 17*(2), 811–817. https://doi.org/10.1007/s11325-012-0769-0

Hudgel, D. W., Patel, S. R., Ahasic, A. M., Bartlett, S. J., Bessesen, D. H., Coaker, M. A., Fiander, P. M., Grunstein, R. R., Gurubhagavatula, I., Kapur, V. K., Lettieri, C. J., Naughton, M. T., Owens, R. L., Pepin, J. L., Tuomilehto, H., Wilson, K. C., & American Thoracic Society Assembly on Sleep and Respiratory Neurobiology (2018). The role of weight management in the treatment of adult obstructive sleep apnea. An official American Thoracic Society clinical practice guideline. *American Journal of Respiratory and Critical Care Medicine, 198*(6), e70–e87. https://doi.org/10.1164/rccm.201807-1326ST

Jackson, M. L., McEvoy, R. D., Banks, S., & Barnes, M. (2018). Neurobehavioral impairment and CPAP treatment response in mild-moderate obstructive sleep apnea. *Journal of Clinical Sleep Medicine, 14*(1), 47–56. https://doi.org/10.5664/jcsm.6878

Johns M. W. (1991). A new method for measuring daytime sleepiness: The Epworth sleepiness scale. *Sleep, 14*(6), 540–545. https://doi.org/10.1093/sleep/14.6.540

Kerner, N. A., & Roose, S. P. (2016). Obstructive sleep apnea is linked to depression and cognitive impairment: Evidence and potential mechanisms. *American Journal of Geriatric Psychiatry, 24*(6), 496–508. https://doi.org/10.1016/j.jagp.2016.01.134

Kielb, S. A., Ancoli-Israel, S., Rebok, G. W., & Spira, A. P. (2012). Cognition in obstructive sleep apnea-hypopnea syndrome (OSAS): Current clinical knowledge and the impact of treatment. *NeuroMolecular Medicine, 14*(3), 180–193. https://doi.org/10.1007/s12017-012-8182-1

Kline, C. E., Crowley, E. P., Ewing, G. B., Burch, J. B., Blair, S. N., Durstine, J. L., Davis, J. M., & Youngstedt, S. D. (2011). The effect of exercise training on obstructive sleep apnea and sleep quality: A randomized controlled trial. *Sleep, 34*(12), 1631–1640. https://doi.org/10.5665/sleep.1422

Kushida, C. A., Littner, M. R., Morgenthaler, T., Alessi, C. A., Bailey, D., Coleman, J., Jr., Friedman, L., Hirshkowitz, M., Kapen, S., Kramer, M., Lee-Chiong, T., Loube, D. L., Owens, J., Pancer, J. P., & Wise, M. (2005). Practice parameters for the indications for polysomnography and related procedures: An update for 2005. *Sleep, 28*(4), 499–523. https://doi.org/10.1093/sleep/28.4.499

Kushida, C. A., Nichols, D. A., Holmes, T. H., Quan, S. F., Walsh, J. K., Gottlieb, D. J., Simon, R. D., Jr., Guilleminault, C., White, D. P., Goodwin, J. L., Schweitzer, P. K., Leary, E. B., Hyde, P. R., Hirshkowitz, M., Green, S., McEvoy, L. K., Chan, C., Gevins, A., Kay, G. G., Bloch, D. A., . . . Dement, W. C. (2012). Effects of continuous positive airway pressure on neurocognitive function in obstructive sleep apnea patients: The Apnea Positive Pressure Long-term Efficacy Study (APPLES). *Sleep, 35*(12), 1593–1602. https://doi.org/10.5665/sleep.2226

Leng, Y., Blackwell, T., Stone, K. L., Hoang, T. D., Redline, S., & Yaffe, K. (2016). Periodic limb movements in sleep are associated with greater cognitive decline in older men without dementia. *Sleep, 39*(10), 1807–1810. https://doi.org/10.5665/sleep.6158

Leng, Y., McEvoy, C. T., Allen, I. E., & Yaffe, K. (2017). Association of sleep-disordered breathing with cognitive function and risk of cognitive impairment: A systematic review and meta-analysis. *JAMA Neurology, 74*(10), 1237–1245. https://doi.org/10.1001/jamaneurol.2017.2180

Mathieu, A., Mazza, S., Décary, A., Massicotte-Marquez, J., Petit, D., Gosselin, N., Malo, J., & Montplaisir, J. (2008). Effects of obstructive sleep apnea on cognitive function: A comparison between younger and older OSAS patients. *Sleep Medicine, 9*(2), 112–120. https://doi.org/10.1016/j.sleep.2007.03.014

Morin, C. M., Belleville, G., Bélanger, L., & Ivers, H. (2011). The Insomnia Severity Index: Psychometric indicators to detect insomnia cases and evaluate treatment response. *Sleep, 34*(5), 601–608. https://doi.org/10.1093/sleep/34.5.601

Muñoz, A., Mayoralas, L. R., Barbé, F., Pericás, J., & Agusti, A. G. (2000). Long-term effects of CPAP on daytime functioning in patients with sleep apnoea syndrome. *European Respiratory Journal, 15*(4), 676–681. https://doi.org/10.1034/j.1399-3003.2000.15d09.x

Nasreddine, Z. S., Phillips, N. A., Bédirian, V., Charbonneau, S., Whitehead, V., Collin, I., Cummings, J. L., & Chertkow, H. (2005). The Montreal Cognitive Assessment, MoCA: A brief screening tool for mild cognitive impairment. *Journal of the American Geriatrics Society, 53*(4), 695–699. https://doi.org/10.1111/j.1532-5415.2005.53221.x

Naumann, A., Bellebaum, C., & Daum, I. (2006). Cognitive deficits in narcolepsy. *Journal of Sleep Research, 15*(3), 329–338. https://doi.org/10.1111/j.1365-2869.2006.00533.x

Olaithe, M., Bucks, R. S., Hillman, D. R., & Eastwood, P. R. (2018). Cognitive deficits in obstructive sleep apnea: Insights from a meta-review and comparison with deficits observed in COPD, insomnia, and sleep deprivation. *Sleep Medicine Reviews, 38*, 39–49. https://doi.org/10.1016/j.smrv.2017.03.005

Osorio, R. S., Gumb, T., Pirraglia, E., Varga, A. W., Lu, S. E., Lim, J., Wohlleber, M. E., Ducca, E. L., Koushyk, V., Glodzik, L., Mosconi, L., Ayappa, I., Rapoport, D. M., de Leon, M. J., & Alzheimer's Disease Neuroimaging Initiative. (2015). Sleep-disordered breathing advances cognitive decline in the elderly. *Neurology, 84*(19), 1964–1971. https://doi.org/10.1212/WNL.0000000000001566

Osterrieth, P. A. (1944). Le test de copie d'une figure complexe. *Arch Psychol, 30*, 206–356.

Pacheco, D., & Wright, H. (2022). Cheyne-Stokes respirations. *Sleep Foundation.*

Patel, S. R. (2019). Obstructive sleep apnea. *Annals of Internal Medicine, 171*(11), ITC81–ITC96. https://doi.org/10.7326/AITC201912030

Peppard, P. E., Young, T., Barnet, J. H., Palta, M., Hagen, E. W., & Hla, K. M. (2013). Increased prevalence of sleep-disordered breathing in adults. *American Journal of Epidemiology, 177*(9), 1006–1014. https://doi.org/10.1093/aje/kws342

Polysomnography. (2008). *Gale encyclopedia of medicine.* (2008). Retrieved November 1, 2022, from https://medical-dictionary.thefreedictionary.com/polysomnography

Reitan, R. M. (1956). *Trail making test: Manual for administration, scoring, and interpretation.* Indiana University Press.

Seelye, A., Mattek, N., Howieson, D., Riley, T., Wild, K., & Kaye, J. (2015). The impact of sleep on neuropsychological performance in cognitively intact older adults using a novel in-home sensor-based sleep assessment approach. *Clinical Neuropsychologist, 29*(1), 53–66. https://doi.org/10.1080/13854046.2015.1005139

Simou, E., Britton, J., & Leonardi-Bee, J. (2018). Alcohol and the risk of sleep apnoea: A systematic review and meta-analysis. *Sleep Medicine, 42*, 38–46. https://doi.org/10.1016/j.sleep.2017.12.005

Song, X., Roy, B., Kang, D. W., Aysola, R. S., Macey, P. M., Woo, M. A., Yan-Go, F. L., Harper, R. M., & Kumar, R. (2018). Altered resting-state hippocampal and caudate functional networks in patients with obstructive sleep apnea. *Brain and Behavior, 8*(6), e00994. https://doi.org/10.1002/brb3.994

Srijithesh, P. R., Aghoram, R., Goel, A., & Dhanya, J. (2019). Positional therapy for obstructive sleep apnoea. *Cochrane Database of Systematic Reviews, 2019*(11). https://doi.org/10.1002/14651858.CD010990.pub2

Stranks, E. K., & Crowe, S. F. (2016). The cognitive effects of obstructive sleep apnea: An updated meta-analysis. *Archives of Clinical Neuropsychology, 31*(2), 186–193. https://doi.org/10.1093/arclin/acv087

Strohl, K. P. (2022). *Central sleep apnea.* Merck & Co. https://www.merckmanuals.com/professional/pulmonary-disorders/sleep-apnea/central-sleep-apnea

Vaessen, T. J., Overeem, S., & Sitskoorn, M. M. (2015). Cognitive complaints in obstructive sleep apnea. *Sleep Medicine Reviews, 19*, 51–58. https://doi.org/10.1016/j.smrv.2014.03.008

Wada, M., Mimura, M., Noda, Y., Takasu, S., Plitman, E., Honda, M., Natsubori, A., Ogyu, K., Tarumi R., Graff-Guerrero, A., & Nakajima, S. (2019). Neuroimaging correlates of narcolepsy with cataplexy: A systematic review. *Neuroscience Research, 142*, 16–29. https://doi.org/10.1016/j.neures.2018.03.005

Wechsler, D. (2008). *WAIS-IV administration and scoring manual.* Pearson Assessment.

Wechsler, D. (2009). *Wechsler memory scale–Fourth edition (WMS–IV).* Pearson Assessment.

Xiong, Y., Zhou, X. J., Nisi, R. A., Martin, K. R., Karaman, M. M., Cai, K., & Weaver, T. E. (2017). Brain white matter changes in CPAP-treated obstructive sleep apnea patients with residual sleepiness. *Journal of Magnetic Resonance Imaging, 45*(5), 1371–1378. https://doi.org/10.1002/jmri.25463

Yaffe, K., Laffan, A. M., Harrison, S. L., Redline, S., Spira, A. P., Ensrud, K. E., Ancoli-Israel, S., & Stone, K. L. (2011). Sleep-disordered breathing, hypoxia, and risk of mild cognitive impairment and dementia in older women. *JAMA, 306*(6), 613–619. https://doi.org/10.1001/jama.2011.1115

Yesavage, J. A., Brink, T., Rose, T. L., Lum, O., Huang, V., Adey, M., & Leirer, V. O. (1982). Development and validation of a geriatric depression screening scale: A preliminary report. *Journal of Psychiatric Research, 17*(1), 37–49. https://doi.org/10.1016/0022-3956(82)90033-4

Zhang, J., Lam, S. P., Ho, C. K., Li, A. M., Tsoh, J., Mok, V., & Wing, Y. K. (2008). Diagnosis of REM sleep behavior disorder by video-polysomnographic study: Is one night enough? *Sleep*, 31(8), 1179–1185.

Zimmerman, M. E., & Aloia, M. S. (2012). Sleep-disordered breathing and cognition in older adults. *Current Neurology and Neuroscience Reports, 12*(5), 537–546. https://doi.org/10.1007/s11910-012-0298-z

9

Lab Work Relevant for Cognitive Functioning

David P. Sheppard, Kathleen F. Pagulayan, Rebecca Hendrickson, and Brett Parmenter

Introduction

Neuropsychologists working in healthcare settings often integrate multiple sources of information in their comprehensive assessments of cognition and behavior. While lab values might be available, many neuropsychologists have not received formal training in the relationship between these metrics and a patient's clinical or behavioral presentation. This chapter provides a review of adult lab values most relevant to clinical practice in an outpatient setting, with select references to younger adults as it relates to typical age-related lab variability; demographic (e.g., age, gender) variables that have known influence on normative ranges of lab values are included when possible. Historically, normative ranges for some laboratory values have been set differently depending on the identified race/ethnic background of the patient; more recently, such practices have undergone a re-evaluation, and professional societies have moved toward recommendations to remove such race- or ethnicity-based criteria from the interpretation of laboratory values. In general, it is important for neuropsychologists to understand that the normal range of lab values can vary in some demographic groups, so interpretation can be context-dependent. Also, like the variability associated with our neuropsychological testing methods, the normative ranges of lab values can be laboratory-dependent due to variation in factors such as equipment, calibration, and methods.

The purpose of this chapter is to provide an overview of the types of lab values that are most relevant to clinical neuropsychologists working in an outpatient setting, their typical ranges, and factors that can affect their level. It is also important to start with a cautionary note regarding the scope of practice of a neuropsychologist; the information in this chapter is intended to assist in the comprehensive evaluation of factors that can affect cognition, including possibly reversible causes of cognitive decline. However, this review is not

David P. Sheppard et al., *Lab Work Relevant for Cognitive Functioning*. In: *Medical and Laboratory Examinations for Neuropsychologists*. Edited by: Thomas J. Farrer and R. Brock Frost, Oxford University Press.
 DOI: 10.1093/9780197619537.003.0010

comprehensive, is not medical training, and does not reflect the full complexity of reading lab values such as when there are multiple medical conditions present. To this end, this chapter is not intended to provide the medical knowledge that is needed to recommend or otherwise order specific lab panels. Instead, neuropsychologists are encouraged to collaborate with medical providers when there is concern about an underlying condition that could be identified via lab work, or when there are questions about the contribution of medical conditions evidenced by abnormal lab values.

How to Consider Lab Values as a Neuropsychologist

Lab values are viewed as part of the larger clinical picture much in the same way that a single neuropsychological test is a part of the broader neuropsychological evaluation. In this way, neuropsychologists may consider lab values much in the same way they interpret results from a single list-learning task or even a single battery of tests, which would not necessarily be used as standalone evidence for real-world cognitive functioning, much less a definitive diagnosis of a brain disorder. There are multiple reasons for cautiously interpreting lab values. First, a single timepoint of a lab value may not provide reliable or sufficient information needed for clinical decision-making. This is because a specific lab value can often be most meaningfully evaluated when compared to that individual's prior lab values (e.g., a change from previous levels indicating a notable decline or increase in values), much in the same way serial neuropsychological testing can be used to investigate intra-individual changes over time. Second, like neuropsychological tests, lab values have error due to measurement variability. Relatedly, as a marker of biological function, lab values can demonstrate error related to individual differences or multiple ongoing medical processes that may not necessarily be attributed to the specific physiological process being assessed by that lab, such that labs may be falsely abnormal due to interacting findings of other lab values. For example, M protein is elevated in myeloma and can cause artificial inflation of the values measured for bilirubin and phosphate. These inflated values do not reflect actual increases in bilirubin and phosphate in the body, but nevertheless have significance because the measurement error itself may be an indicator of the myeloma diagnosis. This example highlights the challenges of using labs to inform an understanding of someone's overall health, particularly without formal medical training, but labs can nevertheless be useful as pieces of evidence to support or refute neuropsychological findings. Together, neuropsychologists should consider lab values to be a part of the larger clinical picture, and as a supportive but not definitive feature of the neuropsychological evaluation.

Lab values can be informative to a neuropsychological evaluation in different ways depending on the clinical question, patient characteristics, and available labs. Lab work can be useful for ruling out certain acute processes that may overlap behaviorally with the potential diagnosis being assessed as part of the neuropsychological evaluation. For example, if the referral question is aimed at assessing dementia, laboratory screening for a specific nutritional deficiency could help rule out those conditions that can mimic dementia. Although individual lab values vary with regard to associations with cognition, it is critical that neuropsychologists appreciate that lab values may return to normal before cognitive functioning, and it is also possible that lab values returning to normal levels may predict a subsequent return to cognitive baseline. Thus, clinical information collected via clinical interview can be just as important for establishing timelines of cognition related to physiological processes that may complicate clinical diagnosis. Clinical information is also needed surrounding the health history of the individual patient. For example, patients with acute abnormalities in lab values resulting in encephalopathy may demonstrate more dramatic cognitive changes when experiencing their first instance of encephalopathy, compared to those with a prior encephalopathy history (Bajaj et al., 2010). Thus, it is recommended that neuropsychologists use extreme caution when using lab values to inform clinical decisions surrounding specific diagnoses and during capacity evaluations. For medical providers, lab values can be utilized to provide supportive evidence for a given diagnosis, such as specific viral markers present in cerebrospinal fluid (CSF) to diagnosis syphilis. Finally, lab values are critical for identifying ongoing pathological process in the body. In the example above where bilirubin labs can indicate that bilirubin values are indeed high, *or* could indicate that there is something else going on in the body (e.g., multiple myeloma) making that test invalid, another lab value (in this case M protein) might be used to identify the presence of high levels of monoclonal antibody protein in the blood indicating a likelihood of a multiple myeloma diagnosis. For the purposes of neuropsychological practice, these examples serve as evidence that lab values: (1) can at times be unreliable indicators of body function, (2) are ideally interpreted in the context of other clinical information, and (3) are best used for clinical decision-making when this is performed in concert with medical providers who have training in potential factors that are likely to give rise to unreliable testing results.

Basic Laboratory Blood Work

Neuropsychologists might encounter lab values in the context of preexisting routine or emergently ordered labs in the medical record. In general, outside of specific diagnostic workups, routine lab work often involves blood and sometimes urinalysis. Routine blood tests and panels can vary by provider and by specific laboratory availability, but can often include metabolic markers (e.g., hemoglobin A1c) and complete blood count (CBC). One common standardized panel is the comprehensive metabolic panel (CMP), which is often given as part of a standard checkup to assess for metabolic abnormalities. Of note, the term "metabolic functioning" is an umbrella term for overall cellular functioning that can be assessed in part by blood labs, but most often is used to describe tests of hemoglobin A1c, lipid panels, and glucose. The CMP includes labs of glucose, calcium, sodium, potassium, carbon dioxide, chloride, albumin, total protein, alkaline phosphatase (ALP), alanine transaminase (ALT), aspartate aminotransferase (AST), bilirubin, and blood urea nitrogen (BUN). A basic metabolic panel (BMP) is equivalent to the CMP but includes only 8 tests (excludes liver and protein tests). Other tests that may be more regularly ordered include thyroid panels and essential nutrients (e.g., vitamin D, vitamin B12). Urinalysis can be used to assess for markers for diagnosing or monitoring chronic disorders affecting vital organs (e.g., kidney or liver disease). Beyond more routinely given labs, many lab tests are used in a context-specific manner to supplement common panels. These are highly specific to a given patient and their clinical presentation and may involve specific additional tests from blood or urine, but also may include lumbar puncture (to measure substances in CSF) or hair testing. However, results from lumbar puncture or hair testing and their relationship to neuropsychological functioning are outside the scope of this chapter. In general, these specific lab tests and panels are used to assess for various processes as shown in Table 9.1.

Although preexisting records are helpful during chart review, there may be instances where neuropsychologists may be interested in additional lab work that could help inform diagnoses and clinical decision-making for the healthcare team. Clinical neuropsychologists in the United States follow the Guidelines for Psychological Practice in Health Care Delivery Systems put forth by the American Psychological Association (APA, 2013), which specifies that psychologists operate within their scope of practice while also applying expertise to the full spectrum of health issues via timely communication with other healthcare professionals. As part of our clinical expertise in the assessment of cognition and brain-behavior relationships, this can include the recognition that certain neuropsychological assessment questions cannot be fully answered without additional information, and consultation with referring or consulting medical providers about potential additional evaluations (in this case lab test) that could help clarify a clinical picture or diagnostic question can be a critical component of our comprehensive assessment role. Having a depth of knowledge about specific tests is critical for neuropsychologists during these conversations and allows

Table 9.1 Medical processes that have known effects on cognitive functioning and their associated laboratory tests

Process	Labs That May Be Used	Indication(s)
Infection	CBC Glucose Urinalysis for UTI Specific virus/bacteria test LP for specific infections (e.g., meningitis)	• Assess whether infectious process is present • Diagnosis acute or new infections (e.g., HIV)
Liver function	Albumin, globulin, total protein, ALT, ALP, AST, GGT	• Assess for liver disease or damage
Kidney function	BUN, CR	• Assess for anemia and kidney function
Thyroid function	TSH, T4, T3, parathyroid hormone	• Hypothyroidism, hyperthyroidism
Acute toxicity	Substance use Therapeutic drug toxicity Tests for heavy metals and toxins	• Medication toxicity (e.g., examining levels compared to prescribed dose, or looking at family member medication lists in case of misuse) • Toxic levels of therapeutic drugs • Toxic levels of environmental toxins
Nutritional deficiencies	Vitamin B12, folate, niacin, vitamin D, calcium, phosphorus, sodium, potassium chloride, glucose, magnesium	• Assess for diet nutritional deficits, and impacts of known diagnoses (e.g., Crohn's disease, Celiac disease) • Assess for anemia
Medication monitoring	Specific medication compounds or their metabolites	• Assess medication levels during titration or weaning • Assess response to medication
Disease monitoring or diagnosis	Hemoglobin A1c Specific antibodies tests	• Monitor diabetes management (A1c) • Monitoring biomarkers related to disease (e.g., CD4 count in HIV) • Diagnosing new diseases (e.g., prion disease)

continued

Table 9.1 *continued*

Process	Labs That May Be Used	Indication(s)
Delirium	CBC (and in particular, WBC) Urinalysis for UTI Sodium BUN, CR LP for specific etiologies Substance use tests	• Assess for UTI, other infections, or other medical contributors • Assess for hyponatremia • Assess for severe renal dysfunction • Assess for acute intoxication
Inflammatory processes	LP for antinuclear antibodies Erythrocyte sedimentation rate blood test	• Assess for inflammatory processes • Assess for anemia
Autoimmune disorders	LP for antinuclear antibodies Vitamin B12 (especially for identifying pernicious anemia)	• Identify specific autoimmune disorders
Acute intoxication or toxicity	Individual substance tests CMP (in particular liver function tests)	• Alcohol or other substance use (e.g., amphetamines, barbiturates, benzodiazepines, cocaine, opioids, THC) • Medication toxicity (e.g., examining levels compared to prescribed dose, or examining family member medication lists in case of misuse) • Liver function tests are commonly evaluated because toxins so readily affect liver function

Note: CBC = complete blood count; LP = lumbar puncture; UTI = urinary tract infection; BUN = blood urea nitrogen; CR = creatinine; ALT = alanine transaminase; ALP = alkaline phosphatase; AST = aspartate aminotransferase; GGT = gamma-glutamyl transferase; TSH = thyroid stimulating hormone; T4 = thyroxine; T3 = triidothyronine; HIV =; CD4 = cluster of differentiation 4 T-cells; THC = tetrahydrocannabinol; CMP = comprehensive metabolic panel.

neuropsychologists to discuss these recommendations without stepping outside of our scope of practice. In summary, it is recommended that neuropsychologists consider lab values as one element of the larger clinical picture in their neuropsychological evaluation. In certain settings, it may be useful to include recent relevant lab values into neuropsychological reports either to provide evidence of diagnostic features, or to provide evidence against alternative etiologies to cognitive dysfunction.

We now offer a review of common labs and how these relate to neuropsychological functioning. We offer this as a starting point for the neuropsychologist and note that it should not be considered an exhaustive or comprehensive list of the labs that are relevant to cognitive functioning. It is also important to note that all of the ranges presented below can vary according to lab, and as such, this should be used as a general guide and not a definitive guideline for interpretation.

Metabolic Functioning Labs

Electrolytes

Electrolytes are chemical substances that become ions in a solution and acquire the capacity to conduct electricity (Table 9.2). In the human body, electrolyte balance is essential for normal function of cells and organs. Common electrolytes for multiple organ systems are sodium and potassium, while other electrolytes can be indicators of function for specific organ systems (e.g., blood urea nitrogen for kidney function; Gounden et al., 2022; Hosten, 1990).

Table 9.2 Common electrolytes and their associated reference ranges

Electrolyte or Mineral	Normal	High	Low
Calcium	8.5–10 mg/mL	Hypercalcemia > 10 mg/mL	Hypocalcemia < 8.5 mg/mL
Chloride	96 to 106 mmol/L	Hyperchloremia	Hypochloremia < 95 mmol/L
Magnesium	1.7 to 2.2 mg/dL	Hypermagnesemia > 2.2 mg/dL	Hypomagnesemia < 1.7 mg/dL
Phosphorous	2.5 to 4.5 mg/dL (0.81 to 1.45 mmol/L)	Hyperphosphatemia > 4.5 mg/dL	Hypophosphatemia < 2.5 mg/dL
Potassium	3.5 to 5.0 mmol/L	Hyperkalemia 5.1–6 mmol/L (mild to moderate) > 6 mmol/L (severe)	Hypokalemia 3–3.4 mmol/L (mild) 2.5–3 mmol/L (moderate) < 2.5 mmol/L severe
Sodium	135 and 145 milliequivalents per liter (mEq/L)	Hypernatremia (> 145 Eq/L)	Hyponatremia (< 135 Eq/L)

mmol/L= millimoles per liter; mg/mL= milligrams per milliliter; mg/dL= milligrams per deciliter

- **Sodium** is critical for cellular function, with mildly abnormal levels having the potential to have subtle effects of physical and mental functioning, while more extreme abnormalities can result in significant symptom burden or even death. Sodium levels in the body reflect a balance between dietary sodium intake and renal secretion. Medications also affect sodium levels, with anabolic steroids, corticosteroids, antibiotics, cough medicines, laxatives, and oral contraceptives increasing sodium, and ACE inhibitors, diuretics, heparin, NSAIDS, and tricyclic antidepressants decreasing sodium. **Hyponatremia** is the most common electrolyte disorder, with possible causes including congestive heart failure, certain antidepressant medications, and water intoxication. Older adults are at increased risk for hyponatremia (Filippatos et al., 2017). Acute symptoms of hyponatremia can include delirium, while chronic hyponatremia can include lethargy, disorientation, and other cognitive changes. As it relates to cognition, normal sodium is generally defined as 136 or greater. However, sodium levels can be associated with cognitive impairment and decline in older men (age 65+) with normal serum sodium levels or mild (< 136) hyponatremia (Nowak et al., 2018). Sodium serum levels < 125 generally result in weakness, and levels < 115 can result in acute cognitive changes including confusion, lethargy, stupor, or coma. Chronic symptoms associated with hyponatremia include lethargy, disorientation, and other cognitive changes. **Hypernatremia** is associated with dehydration and kidney disease. Chronic hypernatremia is better tolerated compared to chronic hyponatremia, but it can still result in cognitive difficulties. In general, stabilization of sodium levels leads to normalization of cognitive function. However, if too rapid, sodium stabilization can result in central pontine myelinolysis, which can include spastic quadriparesis, pseudobulbar palsy, respiratory paralysis, locked-in syndrome, and even death.
- **Potassium** serves the role of regulating fluid balance and cardiac functioning, as well as muscle and nerve functioning. Diet significantly impacts levels of potassium, which can be found in legumes, soybeans, potatoes, fruits, vegetables, poultry, fish, milk, and yogurt. Excessive potassium is removed by the kidneys, and kidney disease, eating disorders, and alcohol use disorders all have impacts on potassium levels. Other contributors to potassium changes include diarrhea, vomiting, and excessive sweating. Medications that increase potassium include antibiotics, histamines, and lithium, while medications that decrease potassium levels include laxatives, insulin, and certain diuretics. Potassium also has important interactions with sodium and can contribute to reductions in blood pressure. The normal range of serum potassium levels varies among laboratories, but is approximately is 3.6–5.2 mEq/L. Too little serum potassium is termed **hypokalemia**

and is a common electrolyte disturbance for which hospitalization is a risk factor; hypokalemia symptoms tend to be associated with level and duration of potassium change. Mild hypokalemia (3.0–3.5 mEq/L) can include symptoms of constipation, heart palpitations, fatigue, muscle weakness or spasms, and tingling numbness. Severe hypokalemia (< 2.5–3.0 mEq/L) can result in muscle twitches and cramps, severe muscle weakness possibly leading to paralysis, low blood pressure, lightheadedness, and arrhythmias, and hypokalemia <2.5 mEq/L can be life-threatening. **Hyperkalemia** (>5.5 mmol/L) is more common among individuals with kidney disease and those who have diabetes. Both severe hypokalemia and severe hyperkalemia can be associated with life-threatening cardiac arrhythmias.

- **Chloride** is an important electrolyte for body fluids, and is typically examined in conjunction with sodium or potassium because it is typically abnormal when these ion levels are abnormal. The typical reference interval for adults is 96 to 106 milliequivalents per liter (mEq/L).

Blood Sugar Levels

Glucose can vary in the body by time of day, recently eating, caffeine, pregnancy, and stress (including trauma, anesthesia, infections, burns). Normal levels of glucose during fasting are <100 mg/dL, and levels between 100 and 125 mg/dL are in the range of prediabetes. Medications also can affect glucose levels, with glucose increasing due to tricyclic antidepressants, corticosteroids, estrogens, and lithium. Glucose decreases with acetaminophen, alcohol, MAOIs, and propranolol. There also are medications (e.g., statins that permeate the blood-brain barrier) with significant effects on glucose, although the specific mechanism of these is unclear. Individuals with Type I or Type II diabetes mellitus can experience cognitive and motor changes during hypoglycemia; cognitive impairments can be more consistently observed during episodes of moderate to severe hypoglycemia. In contrast, there is less clear evidence for clinically significant effects of hypoglycemia among individuals living without diabetes.

Hemoglobin A1c provides a measure of average glucose over the past 2 to 3 months, such that higher hemoglobin A1c reflects higher average glucose. Hemoglobin is found in red blood cells and becomes glycated with sugar when blood sugar is consistently high. The normal range for hemoglobin A1c is between 4% and 5.6%, with levels between 5.7% and 6.4% reflecting an increased risk for diabetes. Levels of 6.5% or higher indicate diabetes. Higher hemoglobin A1c is associated with increased risk of diabetes-related health complications (Mohr et al., 2022), with the goal of people living with diabetes

Table 9.3 Fasting glucose and hemoglobin A1c ranges in the normal range and ranges according to diabetes risk status

Lab	Normal	Pre-Diabetes	Diabetes
Glucose (Fasting)	< 100 mg/dL	100–125 mg/dL (pre-diabetes)	≥ 126
Hemoglobin A1c	4%–5.6%	5.7%–6.4%	≥ 6.5%

to keep hemoglobin A1c below 7%. One important note is that other disorders, such as anemia, kidney disease, and liver disease, and dietary supplements (e.g., vitamins C and E), can result in abnormal results of A1c labs (Radin, 2014) (Table 9.3).

Renal Function

Kidney functioning is critical for cleaning toxins from the bloodstream and for transforming waste into urine. **Uremia**, or increased nitrogenous waste compounds in the blood, can develop due to deteriorating kidney functioning, often due to disease or injury. Uremia can cause fluid accumulation, hormone and electrolyte imbalance, and other conditions such as hyponatremia, hyperkalemia, hypermagnesemia, hypocalcemia, and metabolic acidosis. Primary renal labs include **blood urea nitrogen** (BUN) and **creatinine**. BUN is typically increased when kidney function declines, and in adults the normal range is 6–24 mg/dL. BUN reference levels can vary by sex assigned at birth, with men having typical levels between 8–24 mg/dL and women having a reference range of 6–21 mg/dL. BUN can be affected by level of protein intake dehydration, and severe liver disease (Gounden et al., 2022). Creatinine is an indicator of waste products in the blood due to muscle activity that are normally removed from the blood by the kidneys. Thus, creatinine levels increase with poorer kidney functioning. Normal levels of creatinine can vary by age and sex, with normal levels in adults from 0.6 to 1.2 mg/dL. Creatinine can be affected by muscle mass and protein intake, so much so that creatinine can be significantly lower in someone with very low muscle mass and limited protein intake, or conversely, significant increases can be seen after consumption of red meat (Gounden et al., 2022).

Both acute renal dysfunction and chronic kidney disease (CKD) can affect both central and peripheral nervous systems. Acute renal dysfunction can result in symptoms including neuropathies, altered mental status, and motor disturbances, and cognitive changes most often include impaired concentration/attention (Arnold et al., 2016). In more severe renal dysfunction, there can be symptoms of delirium, hallucinations, coma, and seizures. In general,

Table 9.4 Typical renal function labs and associated reference ranges

Renal Function Lab	Typical Range
Blood Urea Nitrogen (BUN)	8–24 mg/dL (men) 6–21 mg/dL (women)
Creatinine (CR)	0.6–1.2 mg/dL (men) 0.5–1.1 mg/dL (women)

cognitive difficulties are more pronounced with rapid development of uremia to uremic encephalopathy, which can result in delirium. CKD may cause more subtle changes that can be observed on neuropsychological testing, with domains most commonly affected being attention/orientation, language, abstract reasoning, executive functions, and memory (Berger et al., 2016) (Table 9.4).

Thyroid Function

Thyroid function is important for neuropsychologists to consider specifically because of its effects on cognition and mood. Hypothyroidism has associated symptoms of psychomotor slowing, apathy, and depression. Severe hypothyroidism can result in "myxedema madness," (Asher, 1949) which is rare but can include psychotic symptoms including paranoia and hallucinations along with attention and memory deficits. Hyperthyroidism is associated with symptoms of anxiety, irritability, distractibility, restlessness, emotional lability, attention difficulties, impaired memory, and difficulties with calculations (Ritchie & Yeap, 2015). Thyroid function labs typically involve **thyroid stimulating hormone (TSH)**, **thyroxine (T4)**, and **triidothyronine (T3)**. Normal/euthyroid TSH range is 0.4–4.0 mIU/L. Atypical thyroid function is defined using thyroid labs and indicated as subclinical hypothyroidism (TSH > 4.0 with normal T4), hypothyroidism (TSH > 4.0 with low T4), subclinical hyperthyroidism (TSH < 0.4 with normal T4/T3), and hyperthyroidism (TSH < 0.4 with high T4/T3). Note that TSH and thyroid function have an inverse relationship, such that a higher TSH reflects lower thyroid function because TSH is released by the pituitary gland to direct thyroid T4 and T3 production. Critically, what is considered subclinical versus an indication for treatment depends on specific symptoms and specific clinical presentation; in individuals with normal TSH but convincing symptoms of hypothyroidism assessment of free T4 may still be indicated, and in individuals with elevated TSH, normal free T4, and symptoms considered likely to be due to hypothyroidism treatment may still be indicated.

Additionally, hyperthyroidism can also be caused by other factors that can be observed in other lab values (e.g., extremely low calcium, lithium use), while other therapeutic drugs (e.g., carbamazepine) can contribute to altered thyroid function even if not resulting in hypothyroidism (Welsh et al., 2016). Another related lab is **parathyroid hormone**, which can be altered due to hypercalcemia or hypocalcemia, which can be related to hyperparathyroidism. Although hyperparathyroid levels may be associated with poorer cognition and possibly dementia risk, there is generally limited quality research establishing this relationship (Lourida et al., 2015). Hypocalcemia is itself associated with fatigue and confusion, and among individuals with hyperparathyroidism, lower serum calcium levels can be associated with cognitive dysfunction (Sardella et al., 2021). One important note for the neuropsychological evaluation is that TSH and other thyroid hormones can be altered during times of physiological stress, which can include behavioral presentation of acute mania (Zhao et al., 2021). For this reason, a TSH lab collected during a stress episode (e.g., during an ICU admission) may not reflect the same thyroid functioning as a lab that was collected during a routine clinic visit, and "abnormal" levels observed during times of stress may in fact reflect a temporary, normal stress response (Walter et al., 2012).

While abnormal thyroid function can have immediate effects on neuropsychological functioning, epidemiological studies suggest that both hypothyroidism and hyperthyroidism may increase risk for future neuropsychological changes and dementia. Tan and colleagues (2008) utilized the Framingham Study dataset and observed in a sample with a mean age of 71 that incident Alzheimer's disease (AD) at an average follow-up interval of 12.7 years was associated with high and low TSH in women, but not in men. The Thyroid Epidemiology, Audit, and Research Study (Vadiveloo et al., 2011) found that among participants from a large sample with a mean age of 66 years and at an average 5.6-year follow-up interval, those with TSH levels < 0.4 mlU/L at baseline had a 2-fold increased risk of dementia compared to those with euthyroid and subclinical hyperthyroidism. The Danish Nationwide Register study (Thvilum et al., 2021) found that hypothyroidism was associated with increased risk of dementia in a sample with an average age of 56, with an observed 12% increased risk of dementia for each 6 months of TSH above 4.0 IU/L (Table 9.5).

Other tests measuring general endocrinological functioning and relevant to cognition include plasma cortisol level, which can be elevated in Cushing's disease and abnormally low in Addison's disease. Plasma cortisol is measured dynamically using the dexamethasone suppression test, in which dexamethasone (a steroid that also binds to the same receptor as cortisol) is administered to the patient typically in the evening and then cortisol levels are measured the next morning. Both Cushing's syndrome and Addison's disease are associated with mental status changes. Cushing's syndrome is associated with

Table 9.5 Thyroid functioning lab values and associated reference ranges

Clinical Status	Lab Findings
Normal/Euthyroid	Normal **TSH** (0.4–4.0 mIU/L)
Subclinical hypothyroidism	High **TSH** (> 4.0 mIU/L) Normal **Total T4** (5.0 to 12.0 μg/dL)
Hypothyroidism	High **TSH** (> 4.0 mIU/L) Low **Total T4** (< 5.0 μg/dL)
Subclinical hyperthyroidism	Low **TSH** (< 0.4 mIU/L) Normal **Total T4** (5.0 to 12.0 μg/dL) and **Total T3** (80–220 ng/dL)
Hyperthyroidism	Low **TSH** (< 0.4 mIU/L) High **Total T4** (> 12.0 μg/dL) and **Total T3** (> 220 ng/dL)

neuropsychological difficulties compared to healthy controls in the areas of attention, speed of information processing, visual perception and construction, and executive functions (Frimondt-Møller et al., 2019). Addison's disease can be associated with difficulties compared to healthy control participants in the areas of attention and verbal learning (Schultebraucks et al., 2015; Blacha et al., 2021).

Complete Blood Count

The complete blood count (CBC) provides data on white blood cells (WBC), platelets (Plt), and red blood cells (RBC). In general, CBC is useful for the detection of anemia, polycythemia, infection, and inflammation factors that can contribute to altered mental status (George-Gay and Parker, 2003). However, generalized increases in CBC values can be seen due to dehydration or even increased stress. WBC can be particularly helpful for identifying infectious diseases and leukemia or other cancers, with normal ranges for adults ranging from 4.30–10.00 K/μL, with levels > 11.00 K/μL being very elevated and indicative of infection or other processes (e.g., trauma). However, WBC counts greater than 10.00 K/μL can also be due to chronic low-grade inflammation, acute stress, and even those who present with acute mania. A rapid rise over a short period of time is particularly concerning for an acute infection. Platelet counts can be reduced in the case of certain substance use (e.g., certain antipsychotic medications). Platelet count reference ranges for adults are from 140,000–450,000 cells/microliter (cells/mcL).

Normal RBC counts for men are 5–6 million cells/mcL, and for women 4–5 million cells/mcL. Of particular utility are labs related to RBC function: hematocrit (Hct) describes the proportion of RBC to plasma in blood, while hemoglobin (Hb) is the oxygen-carrying protein in RBC that transports oxygen and carbon dioxide. Normal Hct varies with regard to sex, with the reference interval in adult women being from 36% to 44% and 41% to 50% in adult men. Normal Hb levels can vary according to sex and age. Specifically, normal Hb levels in men are 14.0 to 17.0 g/dL and in women are 12.0 to 15.0 g/dL. Newborns and infants also have lower Hb, and Hb decreases in older age. In addition, Hb can be higher due to other factors such as time of day (e.g., morning tends to have higher Hb), dehydration, living at altitude, polycythemia vera disorder, smoking, and lung disease (e.g., COPD, emphysema, pulmonary fibrosis).

Low Hb is referred to as anemia and reflects having too few RBC. There are a number of potential causes of anemia, and this is in part due to disruptions in one of the multiple components that make up Hb (e.g., iron, vitamin B12, folate). In general, anemia is caused by the body making too few red blood cells, the body destroying too many red blood cells, or the body losing too many red blood cells. Microcytic anemia refers to anemia with smaller than average RBC and is often seen in iron-deficient anemia. Serum iron is a specific lab that can be helpful for identifying iron-deficient anemia. Normocytic anemia is having too few (but normal-sized) RBC, which could be due to low RBC production or significant blood loss. Macrocytic anemia means having larger than average RBC size, but not enough Hb overall, which can be due to changes in DNA synthesis during RBC production that can be seen in certain conditions, including low vitamin B12 or folate, or alcohol use. As it relates to cognition, anemia generally is associated with poorer global cognitive function, but this association may be specific to measures of processing speed and executive functions, with less impact of anemia on memory. Moreover, cognitive dysfunction due to anemia occurs regardless of type (e.g., macrocytic vs. microcytic anemia). There is no strong evidence that anemia contributes to neurodegeneration or dementia risk. However, high concentrations of iron can lead to risk factors of neurodegeneration, such as oxidative stress and inflammation. Anemia also can lead to cardiac arrhythmias and may be related to inadequate cerebral oxygenation, which may in turn lead to impaired cerebral perfusion and impact cerebral function. The reticulocyte count provides information about the body's rate of production of new RBC and is often increased in anemia, particularly when it is due to blood loss or RBC destruction (Table 9.6).

Liver Function

Liver function tests (LFTs) include laboratory values that can help to indicate both whether liver damage has occurred, and, if so, where. **Albumin**, **globulin**, and **total protein** (most commonly reported as albumin and total protein, with

Table 9.6 Typical CBC labs and associated reference ranges

CBC Element	Typical Range
Red Blood Count	5–6 million cells/mcL (men) 4–5 million cells/mcL (women)
Hemoglobin	14–17 g/dL (men) 12–15 g/dL (women)
Hematocrit	41%–50% (men) 36%–44% (women)
White blood cell count	4.30–10.00 K/μL
Platelet count	140,000–450,000 cells/mcL

Note: Like most lab value ranges, CBC reference ranges can vary depending on laboratory range.

globulin inferred) measure major sources of protein in the blood (Lala et al., 2023). Albumin is the most abundant protein produced by the liver, and is lower in individuals with liver disease or damage. The reference interval for serum albumin is 3.4 to 5.4 g/dL. Globulins are made by both the liver and the immune system, and have a reference interval of 2.3–3.5 gm/dL; the ratio of albumin and globulin can help to interpret abnormalities in either value. Total protein is a sum of albumin and globulin and has a reference range of 6.0–8.4 gm/dL. Another important consideration are levels of enzymes that tend to increase following liver damage. **Alkaline phosphatase (ALP)** is found in the liver and the bones, and is important in the breaking down of proteins. Elevated ALP may indicate liver damage/disease or bone disease. **Alanine aminotransferase (ALT)** helps convert proteins into energy for liver cells and has a reference interval for adults of 1–21 units/L. **Aspartate aminotransferase (AST)** is an enzyme that helps metabolize amino acids. The AST reference interval for adults is 7–27 units/L. Elevations in ALT and AST are the most commonly observed and followed markers of both acute and chronic liver damage or disease. Interestingly, AST and ALT also can return to normal in the context of chronic disease (Johnston, 1999). In hepatitis, which refers to inflammation of the liver and can be due to multiple causes, there is a typical pattern whereby albumin, globulin, and total protein are normal, but albumin/globulin ratio > 1, ALP is elevated 1–2 times what is normal, and both ALT and AST are increased into the thousands with ALT being greater than AST. In contrast, the pattern observed in liver cirrhosis, which is scarring of the liver due to various potential liver diseases, involves below-normal total protein and albumin, above-normal globulin, albumin/globulin ratio < 1, ALP elevated 1–2 times normal, and elevated ALT and AST. **Gamma-glutamyl transferase (GGT)** is a less commonly

Table 9.7 Typical liver function tests (LFTs) and associated reference ranges

Liver Function Test Lab	**Typical Range**
Serum albumin	3.4–5.4 g/dL
Globulin	2.3–3.5 gm/dL
Total protein	6.0–8.4 gm/dL
Alkaline phosphatase (ALP)	44–147 IU/L
Alanine aminotransferase (ALT)	1–21 U/L
Aspartate aminotransferase (AST)	7–27 U/L
Bilirubin	0.3–1.0 mg/dL

used test that at higher levels indicates a sign of liver disease or damage; some evidence also suggests that higher GGT may be a risk factor for dementia and cognitive decline (Praetorius & Johansson, 2018). **Bilirubin** can be higher when the liver is not excreting bile effectively, and it has an adult reference range of 0.3–1.0 mg/dL.

Liver disease can result in toxic levels of **ammonia** in the blood, which can result in brain dysfunction defined as hepatic encephalopathy (HE). Over 40% of patients with cirrhosis will develop HE. The West Haven Criteria for grading the extent of HE are: trivial lack of awareness, euphoria or anxiety, shortened attention span, impairment of addition or subtraction (Grade 1); lethargy or apathy, minimal disorientation of time or place, subtle personality changes, inappropriate behavior, lethargy or apathy (Grade 2); somnolence to semi-stupor but responsiveness to verbal stimuli, confusion, gross disorientation (Grade 3); and coma (Grade 4). Ammonia levels are typically less than 30 micromol/L in healthy adults, and ammonia > 50 can indicate toxic levels. However, causes of abnormal ammonia levels are not always clear (Table 9.7).

Vitamins Associated with Cognitive Function

Vitamin B12 is found mainly in meat, eggs, and milk. Normal serum B12 levels are generally above 160 picograms per milliliter (pg/mL). Values less than 160 pg/mL can result in cognitive symptoms in adults, but older adults may experience cognitive symptoms even at normal or borderline levels. Individuals with abnormal plasma vitamin B12 binding antibodies may also experience cognitive symptoms at normal B12 levels. Substances that can decrease B12 include alcohol, anticonvulsants, and oral contraceptives. Possible causes of vitamin B12 deficiency include pernicious anemia (an autoimmune disorder impacting B12 absorption in the stomach), digestive diseases affecting absorption of vitamin B12 (e.g., Crohn's disease or celiac disease), low levels of stomach acid or intrinsic factor, medications for gastroesophageal reflux disease, low vitamin B12 in diet, and gastric surgeries. Neurologic symptoms associated with low vitamin B12 include psychomotor slowness, confusion, memory difficulties, depression, peripheral neuropathy, superficial sensory impairment, burning paresthesia, and vision changes. Low vitamin B12 also can cause demyelinating myelopathy of the posterior and lateral columns, which can contribute to impaired vibratory sense, limb weakness, spasticity, increased muscle stretch reflexes, and extensory plantar responses. Low vitamin B12 also is associated with increased risk of hyperhomocystinemia, renal disease, blood clots, migraine headaches, epilepsy, and decreased integrity of the blood-brain barrier.

Folic acid or **folate**, the natural form of vitamin B9, is found in many foods. Serum folate of 3 ng/ml or less is abnormally low and may result in cognitive symptoms. Neurological symptoms due to low folate are similar to those seen in vitamin B12 deficiencies.

Niacin (vitamin B3) is important for the metabolism of macronutrients such as carbohydrates, protein, and fat. Optimal levels of niacin are 0.5 to 8.45 μg/mL. Deficits in niacin are often due to chronic dietary neglect (e.g., chronic and severe alcohol use) or malabsorption conditions. Niacin deficit results in pellagra, which can include symptoms associated with dementia (e.g., memory impairment, psychomotor slowing, poor concentration, anxiety, depression, irritability, lability, headache, insomnia), dermatitis (sunburn-like changes involving brown discoloration or lesions), and diarrhea.

Vitamin D is a fat-soluble vitamin that promotes calcium absorption, bone growth, immune function, and reduction of inflammation. It is obtained through skin sun exposure, food, and nutritional supplements. Levels < 20 ng/mL are considered deficient. Symptoms associated with deficient vitamin D include mood changes, bone loss, muscle cramps or weakness, bone and join pain, and fatigue. Chronic or severe vitamin D deficiency can lead to hypocalcemia (low calcium levels in the blood), which can lead to secondary hyperparathyroidism (overactive parathyroid glands attempting to regulate blood calcium levels). Vitamin D deficiencies are common, with approximately 35% of adults in the United States having vitamin D deficiency. Vitamin D deficiency affects individuals across the age spectrum, but the prevalence is highest in the elderly and among patients who are either hospitalized or living in nursing homes (Sizar et al., 2022), with over 60% of the elderly population in the United States estimated to have vitamin D deficiency (Palocios & Gonzalez, 2014).

Calcium also can be measured directly and has a reference interval in adults of 8.5–10.5 mg/mL.

Phosphorus also can be measured and is associated with bone metabolism; it has a reference interval of 3.0–4.5 µg/dL.

There has been significant interest in determining whether vitamin D deficiency is associated with risk for dementia, but findings are limited because dementia can itself be associated with decreased vitamin D making it difficult to establish causality. A longitudinal study over 20 years examined a large sample of adults with an average baseline age of 57 and found that individual cognitive performance changes were not associated with baseline vitamin D levels. However, a meta-analysis has suggested that severe vitamin D deficiency (< 10 ng/mL) is associated with increased risk for dementia (Sommer et al., 2017), with dementia risk higher for those with more extreme deficiencies (Chai et al., 2019). Of note, vitamin D supplementation over 1 year in patients previously diagnosed with AD (Jia et al., 2019) and mild cognitive impairment (MCI; Yang et al., 2020) has shown some improvement in cognitive function compared to placebo groups, yet this effect was not previously demonstrated in larger randomized clinical trials (Rossom et al., 2012). Given these remaining questions surrounding whether risk of cognitive decline is mitigated by vitamin D supplementation, there are no current practice recommendations for treatment of dementia vis-à-vis vitamin supplementation (Table 9.8).

Table 9.8 Common vitamins and their associated reference ranges

Vitamin	Typical Range
Vitamin B12	> 160 pg/mL
Vitamin B9 (folate)	3.0–17.0 ng/mL
Niacin	0.5–8.45 µg/mL
Vitamin D	20–50 ng/mL
Calcium	8.5–10 mg/mL
Phosphorous	3.0–4.5 µg/dL

Conditions Associated with Altered Vitamin Absorption

Celiac disease (CD) is an autoimmune disorder with intestinal inflammation in response to gluten, which is the protein found in wheat, rye, and barley. Common symptoms include diarrhea and weight loss, and it is associated with increased total cholesterol and high-density lipoproteins. Treatment requires a gluten-free diet, which can be associated with nutritional deficiencies in minerals (calcium, iron, magnesium, zinc) and vitamins (particularly vitamin B12, vitamin D, and folate). Patients with CD also are at 3- to 4-fold increased risk for hypothyroidism, and it is unclear if a gluten-free diet normalizes thyroid function. Suspicion for CD should be raised in patients with hypothyroidism who are adherent to thyroid medications but not responding to treatment, as CD can cause malabsorption of levothyroxine. Neurologic symptoms associated with CD can appear at any time during the course of the disease and may even be the first or only symptom of CD. Neurologic symptoms can be present in as many as 35% of adult patients. Headaches are common, occurring in nearly half of people living with CD; approximately 75% of patients with headaches associated with CD experience improvement with a gluten-free diet. Seizures also are associated with CD, with people living with epilepsy being 2.3 times more likely to have CD. For individuals living with both epilepsy and CD, gluten-free diets may help but do not typically lead to complete resolution of seizures. Other symptoms and disorders associated with CD include depression, anxiety, irritability, ADHD, and eating disorders.

Inflammatory bowel disease includes ulcerative colitis (UC) affecting the large intestine, and Crohn's disease affecting any part of the digestive track (but the small intestine is involved in most cases). Crohn's disease activity is associated with vitamin B12 malabsorption, which can be a risk factor for iron-deficiency anemia. GI hemorrhage also increases risk for iron-deficiency anemia. Crohn's disease is often treated with corticosteroids. Crohn's disease can affect the frontal cortex and the hippocampus, and cognitive domains most often affected in Crohn's disease include memory, attention, and executive functions (Hopkins et al., 2021). However, there is some mixed evidence surrounding what contributes to cognitive difficulties in this group, with some studies providing conflicting evidence for disease activity and possible vitamin malabsorption contributing to cognitive difficulties (Golan et al., 2016) and others not showing this association (Clarke et al., 2020). Neuropsychologists also should examine other factors (e.g., anemia, corticosteroid use) that may also contribute to cognitive changes, particularly because acute effects of corticosteroids also involve difficulties with executive functions, recent memory, and long-term memory. Chronic corticosteroid use is associated with difficulties in recent memory. Similar nutritional deficits can also be seen in individuals who undergo gastric bypass and sleeve gastrectomy.

Therapeutic Drug Monitoring

Lab values can be useful for determining the extent to which therapeutic drugs or medications may be affecting behavior and cognition. For example, labs may be used during early stages of taking a new medication as dosing is titrated, or for monitoring for long-term toxicity. Lab values also can be helpful for diagnosing undertreatment, monitoring adherence, monitoring/detecting drug interactions, or guiding withdrawal of therapy (Kang & Lee, 2009). In another example, labs may be used to assess for abrupt level changes in one medication (e.g., lithium) due to another medication (e.g., starting an NSAID). Common medication classes that are monitored include steroids, antibiotics, cardiovascular medications, anti-epileptic drugs (AEDs), medications prescribed for autoimmune diseases, and psychiatric medications (e.g., lithium, valproic acid). AED medications are particularly important to monitor due to their use in patients with other risk factors for cognitive difficulties (e.g., due epileptiform activity) and because these can have neurotoxic effects. Neuropsychologists may often also need to consider the effects of antipsychotic or other neurotropic medications.

Substance Use Monitoring

Blood labs or urinalysis are also commonly used to assess for acute intoxication in the context of substance use disorder treatment. Clinically, this is often undertaken to rule out acute intoxication when there is suspected delirium or other acute behavioral symptoms, but urinalysis also is helpful for monitoring of drug dependency or during emergency situations (e.g., to determine potential causes of drug toxicity). Although not exhaustive, Table 9.9 lists the common substances and their respective time periods in which they can be detected

Table 9.9 Common illicit substances and associated durations of detection of substance or metabolites via urinalysis

Substance	Duration of Detection
Alcohol	3 to 10 hours
Amphetamines	24 to 48 hours
Barbiturates	3–14 days, but sometimes weeks
Benzodiazepines	Up to 6 weeks with high level use
Cocaine	2 to 4 days; up to 10 to 22 days with heavy use
Opioids	1 to 2 days
Tetrahydrocannabinol (THC)	6 to 11 weeks with heavy use

(or their metabolites can be detected) via urinalysis. Like lab values generally, substance monitoring is dependent on the lab location and test, and may be more or less sensitive or specific at different testing locations. Again, while neuropsychologists are not interpreting these results, it is helpful for them to be familiar with the half-life of a particular substance and how this might be contributing to a person's neuropsychological presentation. Also, urinalysis drug screens may not always consist of the same series of substance tests. There also are important caveats to the list below, such as clonazepam not typically being identified in standard urinalysis for benzodiazepines; this is problematic because patients may be incorrectly identified as diverting medications or being non-adherent, which could lead to acute withdrawal.

Environmental Toxins and Heavy Metals

Lab tests can be useful for identifying individual toxins/chemicals due to environmental exposure. In most cases of acute toxicity, patients will experience physical symptoms such as GI distress, tremor, or muscle weakness, and/or will have identifiable abnormalities on neurologic exam. Signs and symptoms of chronic, low-dose exposure, however, may be more subtle. In these cases, laboratory evaluation for environmental or heavy metal toxicity is usually undertaken based on potential personal or occupational exposures identified in the patient's history. Common toxins that have neurologic effects include lead, mercury, manganese (particularly among those with a history of welding), copper, aluminum, arsenic, and other various heavy metals. Poisoning due to heavy metal exposure can result in nervous system symptoms that may mimic delirium, but also can have various other behavioral effects. Patients who present with behavioral or psychiatric changes that are unexplained could benefit from further interview surrounding possible exposure and follow-up lab tests.

Cerebrospinal Fluid

In clinical settings, lumbar puncture (LP) for assessing makeup of cerebrospinal fluid (CSF), rather than for therapeutic target of alleviating intracranial pressure, is most commonly used when there is suspicion for an infection but also can assess for cancers, multiple sclerosis, trauma, or other disorders affecting the nervous system. In these instances, LP can be useful for identifying specific antibodies and infectious processes, or for identifying RBC or WBC, proteins, and glucose. Other uses of LP include identification of paraneoplastic syndromes (via identification of oligoclonal bands and specific antibodies), meningitis, amyloid angiopathy, neurosarcoidosis, and in some instances to assess for CSF effects of intracranial, and specifically subarachnoid, hemorrhage due to trauma or stroke. CSF can be assessed for 14-3-3 proteins as a sensitive test for prion disorders and Creutzfeldt-Jakob disease (CJD). Antinuclear antibodies can be examined to investigate whether there might be autoimmune conditions.

Case Examples

In each of the following examples, we consider what specific lab values or series of values might be most useful for assisting in supporting or refuting suspected etiologies of cognitive symptoms.

- **Case A:** A 67-year-old Black woman is referred by Neurology for an outpatient neuropsychological evaluation to assess for dementia. She is reporting cognitive difficulties with an insidious, non-acute, onset. She has no significant medical history but scored a 20 on the Montreal Cognitive Assessment (MoCA) completed in the neurologist's office. Her assessment results showed reduced memory and processing speed with some mild executive dysfunction. The etiology of her difficulties is unclear at this time, as she has no known medical risk factors or psychiatric comorbidities, and is sleeping well. You want to rule out reversible causes of cognitive decline and recommend as such to the referring provider, including evaluating for vitamin deficiencies (B12, folate, vitamin D), blood glucose abnormalities, thyroid abnormalities, kidney and liver function, a toxicology screen and infectious processes that can affect cognitive function.
- **Case B:** Patient B is an 80-year-old White woman with a history of recent urinary tract infections (UTIs) who presents to your outpatient clinic with recent disorientation and fluctuating mental status, according to family reporting. Her family indicated that she was hospitalized 3 months prior, at which time she developed delirium. Her family said the delirium had been improving until approximately 2 weeks ago when she seemed to have increased periods of confusion and disorientation. This case would likely benefit from similar lab values as described for Patient A, but with additional labs being helpful to rule-out acute etiologies, such as new infections or other conditions that can cause encephalopathy, such as acute kidney or liver function changes. Therefore, additional labs for consideration would be urinalysis with urine culture, LFTs, CMP, and CBC.

- **Case C:** You are seeing Patient C with hypothyroidism and Celiac disease (CD) for evaluation of cognitive difficulties, including memory changes and reduced attention. You review the medical records and do not see any lab work for the past 6 months (pre-dating the reported cognitive changes). You are aware that CD has been associated with malabsorption of levothyroxine, and so consult with the patient's primary physician about checking thyroid function, as this could explain the new onset cognitive symptoms. In addition, an assessment of vitamins (especially vitamin B12 and folate) would be important because of possible nutritional deficiencies if the CD is poorly managed. Creatinine also would be important to further assess cognitive changes potentially related to kidney disease.

Summary

Lab values can provide critical information for neuropsychologists who are trying to understand factors associated with cognitive changes. We want to emphasize the importance of neuropsychologists being mindful of their scope of practice and not try to diagnose medical conditions or vitamin deficiencies based on lab values. Nonetheless, it is important for neuropsychologists to have knowledge about how certain lab values and associated medical conditions can impact neuropsychological functioning and the patients we see. This includes an appreciation of which labs can be useful for consideration of certain etiologies, as well as an understanding of what body systems are reflected by certain labs. Given the various contributions of lab values to cognition and risk for cognitive decline, it is critical that neuropsychologists familiarize themselves with available labs to support diagnostic impressions and treatment recommendations. This chapter aimed to provide an overview of the labs most commonly encountered by neuropsychologists in clinical settings.

References

American Psychological Association. (2013). Guidelines for psychological practice in health care delivery systems. *American Psychologist*, *68*(1), 1–6. https://doi.org/10.1037/a0029890

Arnold, R., Issar, T., Krishnan, A. V., & Pussell, B. A. (2016). Neurological complications in chronic kidney disease. *JRSM Cardiovascular Disease*, *5*, 2048004016677687. https://doi.org/10.1177/2048004016677687

Asher, R. (1949). Myxoedematous madness. *British Medical Journal, 2*(4627), 555–562. https://doi.org/10.1136/bmj.2.4627.555. ISSN 0007-1447. JSTOR 25373396. PMC 2051123. PMID 18148089.

Bajaj, J. S., Schubert, C. M., Heuman, D. M., Wade, J. B., Gibson, D. P., Topaz, A., Saeian, K., Hafeezullah, M., Bell, D. E., Sterling, R. K., Stravitz, R. T., Luketic, V., White, M. B., & Sanyal, A. J. (2010). Persistence of cognitive impairment after resolution of overt hepatic encephalopathy. *Gastroenterology, 138*(7), 2332–2340. https://doi.org/10.1053/j.gastro.2010.02.015

Berger, I., Wu, S., Masson, P., Kelly, P. J., Duthie, F. A., Whiteley, W., Parker, D., Gillespie, D., & Webster, A. C. (2016). Cognition in chronic kidney disease: A systematic review and meta-analysis. *BMC Medicine, 14*(1), 206. https://doi.org/10.1186/s12916-016-0745-9

Blacha, A. K., Rahvar, A. H., Flitsch, J., van de Loo, I., Kropp, P., & Harbeck, B. (2021). Impaired attention in patients with adrenal insufficiency—Impact of unphysiological therapy. *Steroids, 167*, 108788. https://doi.org/10.1016/j.steroids.2020.108788

Chai, B., Gao, F., Wu, R., Dong, T., Gu, C., Lin, Q., & Zhang, Y. (2019). Vitamin D deficiency as a risk factor for dementia and Alzheimer's disease: An updated meta-analysis. *BMC Neurology, 19*(1), 1–11.

Clarke, G., Kennedy, P. J., Groeger, J. A., Quigley, E. M., Shanahan, F., Cryan, J. F., & Dinan, T. G. (2020). Impaired cognitive function in Crohn's disease: Relationship to disease activity. *Brain, Behavior, & Immunity—Health, 5*, 100093. https://doi.org/10.1016/j.bbih.2020.100093

Filippatos, T. D., Makri, A., Elisaf, M. S., & Liamis, G. (2017). Hyponatremia in the elderly: Challenges and solutions. *Clinical Interventions in Aging, 12*, 1957–1965. https://doi.org/10.2147/CIA.S138535

Frimodt-Møller, K. E., Møllegaard Jepsen, J. R., Feldt-Rasmussen, U., & Krogh, J. (2019). Hippocampal volume, cognitive functions, depression, anxiety, and quality of life in patients with Cushing syndrome. *Journal of Clinical Endocrinology & Metabolism, 104*(10), 4563–4577. https://doi.org/10.1210/jc.2019-00749

George-Gay, B., & Parker K. (2003). Understanding the complete blood count with differential. *Journal of Perianesthesia Nursing: Official Journal of the American Society of PeriAnesthesia Nurses, 18*(2), 96–114. https://doi.org/10.1053/jpan.2003.50013

Golan, D., Gross, B., Miller, A., Klil-Drori, S., Lavi, I., Shiller, M., Honigman, S., Almog, R., & Segol, O. (2016). Cognitive function of patients with Crohn's disease is associated with intestinal disease activity. *Inflammatory Bowel Diseases, 22*(2), 364–371. https://doi.org/10.1097/MIB.0000000000000594

Gounden, V., Bhatt, H., & Jialal, I. (2022). Renal function tests. *StatPearls*. https://www.ncbi.nlm.nih.gov/books/NBK507821/

Graveling, A. J., Deary, I. J., & Frier, B. M. (2013). Acute hypoglycemia impairs executive cognitive function in adults with and without type 1 diabetes. *Diabetes Care, 36*(10), 3240–3246. https://doi.org/10.2337/dc13-0194

Hopkins, C. W. P., Powell, N., Norton, C., Dumbrill, J. L., Hayee, B., & Moulton, C. D. (2021). Cognitive impairment in adult inflammatory bowel disease: A systematic review and meta-analysis. *Journal of the Academy of Consultation-Liaison Psychiatry, 62*(4), 387–403. https://doi.org/10.1016/j.psym.2020.10.002

Hosten, A. O. (1990). BUN and creatinine. In H. K. Walker, W. D. Hall, & J. W. Hurst (Eds.), *Clinical methods: The history, physical, and laboratory examinations* (3rd ed.). Butterworths. http://www.ncbi.nlm.nih.gov/books/NBK305/

Jia, J., Hu, J., Huo, X., Miao, R., Zhang, Y., & Ma, F. (2019). Effects of vitamin D supplementation on cognitive function and blood Aβ-related biomarkers in older adults with Alzheimer's disease: A randomised, double-blind, placebo-controlled trial. *Journal of Neurology, Neurosurgery, and Psychiatry*, *90*(12), 1347–1352. https://doi.org/10.1136/jnnp-2018-320199

Johnston, D. E. (1999). Special considerations in interpreting liver function tests. *American Family Physician*, *59*(8), 2223–2230.

Kang, J. S., & Lee, M. H. (2009). Overview of therapeutic drug monitoring. *Korean Journal of Internal Medicine*, *24*(1), 1–10. https://doi.org/10.3904/kjim.2009.24.1.1

Lala, V., Zubair, M., & Minter, D. A. (2023). Liver function tests. In *StatPearls*. StatPearls Publishing. http://www.ncbi.nlm.nih.gov/books/NBK482489/

Lourida, I., Thompson-Coon, J., Dickens, C. M., Soni, M., Kuźma, E., Kos, K., & Llewellyn, D. J. (2015). Parathyroid hormone, cognitive function and dementia: A systematic review. *PloS One*, *10*(5), e0127574. https://doi.org/10.1371/journal.pone.0127574

Mohr, D. C., Zhang, L., Prentice, J. C., et al. (2022). Association of hemoglobin A1c time in range with risk for diabetes complications. *BMJ Open Diabetes Research and Care*, *10*, e002738. https://doi.org/10.1136/bmjdrc-2021-002738

Nowak, K. L., Yaffe, K., Orwoll, E. S., Ix, J. H., You, Z., Barrett-Connor, E., Hoffman, A. R., & Chonchol, M. (2018). Serum sodium and cognition in older community-dwelling men. *Clinical Journal of the American Society of Nephrology: CJASN*, *13*(3), 366–374. https://doi.org/10.2215/CJN.07400717

Palacios, C., & Gonzalez, L. (2014). Is vitamin D deficiency a major global public health problem? *Journal of Steroid Biochemistry and Molecular Biology*, *144*(Pt A), 138–145. https://doi.org/10.1016/j.jsbmb.2013.11.003

Praetorius Björk, M., & Johansson, B. (2018). Gamma-glutamyltransferase (GGT) as a biomarker of cognitive decline at the end of life: Contrasting age and time to death trajectories. *International Psychogeriatrics*, *30*(7), 981–990. https://doi.org/10.1017/S1041610217002393

Radin, M. S. (2014). Pitfalls in hemoglobin A1c measurement: When results may be misleading. *Journal of General Internal Medicine*, *29*(2), 388–394. https://doi.org/10.1007/s11606-013-2595-x. Epub 2013 September 4. PMID: 24002631; PMCID: PMC3912281.

Ritchie, M., & Yeap, B. B. (2015). Thyroid hormone: Influences on mood and cognition in adults. *Maturitas*, *81*(2), 266–275. https://doi.org/10.1016/j.maturitas.2015.03.016

Rossom, R. C., Espeland, M. A., Manson, J. E., Dysken, M. W., Johnson, K. C., Lane, D. S., LeBlanc, E. S., Lederle, F. A., Masaki, K. H., & Margolis, K. L. (2012). Calcium and vitamin D supplementation and cognitive impairment in the women's health initiative. *Journal of the American Geriatrics Society*, *60*(12), 2197–2205. https://doi.org/10.1111/jgs.12032

Sardella, A., Bellone, F., Morabito, N., Minisola, S., Basile, G., Corica, F., & Catalano, A. (2021). The association between hypoparathyroidism and cognitive impairment: A systematic review. *Journal of Endocrinological Investigation*, *44*(5), 905–919. https://doi.org/10.1007/s40618-020-01423-1

Schultebraucks, K., Wingenfeld, K., Heimes, J., Quinkler, M., & Otte, C. (2015). Cognitive function in patients with primary adrenal insufficiency (Addison's disease). *Psychoneuroendocrinology*, *55*, 1–7. https://doi.org/10.1016/j.psyneuen.2015.01.025

Sizar, O., Khare, S., Goyal, A., et al. (2022). Vitamin D deficiency. *StatPearls*. https://www.ncbi.nlm.nih.gov/books/NBK532266/

Sommer, I., Griebler, U., Kien, C., Auer, S., Klerings, I., Hammer, R., . . . Gartlehner, G. (2017). Vitamin D deficiency as a risk factor for dementia: A systematic review and meta-analysis. *BMC Geriatrics*, *17*(1), 1–13.

Tan, Z. S., Beiser, A., Vasan, R. S., Au, R., Auerbach, S., Kiel, D. P., Wolf, P. A., & Seshadri, S. (2008). Thyroid function and the risk of Alzheimer disease: The Framingham Study. *Archives of Internal Medicine*, *168*(14), 1514–1520. https://doi.org/10.1001/archinte.168.14.1514

Thvilum, M., Brandt, F., Lillevang-Johansen, M., Folkestad, L., Brix, T. H., & Hegedüs, L. (2021). Increased risk of dementia in hypothyroidism: A Danish nationwide register-based study. *Clinical Endocrinology*, *94*(6), 1017–1024. https://doi.org/10.1111/cen.14424

Umapathy, S., Dhiman, R. K., Grover, S., Duseja, A., & Chawla, Y. K. (2014). Persistence of cognitive impairment after resolution of overt hepatic encephalopathy. *American Journal of Gastroenterology*, *109*(7), 1011–1019. https://doi.org/10.1038/ajg.2014.107

Vadiveloo, T., Donnan, P. T., Cochrane, L., & Leese, G. P. (2011). The Thyroid Epidemiology, Audit, and Research Study (TEARS): Morbidity in patients with endogenous subclinical hyperthyroidism. *Journal of Clinical Endocrinology and Metabolism*, *96*(5), 1344–1351. https://doi.org/10.1210/jc.2010-2693

Walter, K. N., Corwin, E. J., Ulbrecht, J., Demers, L. M., Bennett, J. M., Whetzel, C. A., & Klein, L. C. (2012). Elevated thyroid stimulating hormone is associated with elevated cortisol in healthy young men and women. *Thyroid Research*, *5*(1), 13. https://doi.org/10.1186/1756-6614-5-13

Welsh, K. J., & Soldin, S. J. (2016). Diagnosis of endocrine disease: How reliable are free thyroid and total T3 hormone assays? *European Journal of Endocrinology*, *175*(6), R255–R263. https://doi.org/10.1530/EJE-16-0193.

Yang, T., Wang, H., Xiong, Y., Chen, C., Duan, K., Jia, J., & Ma, F. (2020). Vitamin D supplementation improves cognitive function through reducing oxidative stress regulated by telomere length in older adults with mild cognitive impairment: A 12-month randomized controlled trial. *Journal of Alzheimer's Disease: JAD*, *78*(4), 1509–1518. https://doi.org/10.3233/JAD-200926

Zhao, S., Zhang, X., Zhou, Y., et al. (2021). Comparison of thyroid function in different emotional states of drug-naïve patients with bipolar disorder. *BMC Endocrine Disorders*, *21*(1), 210. https://doi.org/10.1186/s12902-021-00869-5

10

Visual Diagnostics for Neuropsychologists

Lynn A. Schaefer, Howard D. Pomeranz, and Carol Kave Chamoff

Abbreviations

cerebrovascular accident (CVA)
computerized tomography (CT)
decibels (dB)
electroencephalogram (EEG)
intraocular pressure (IOP)
motor vehicle accident (MVA)
occupational therapy (OT)
optical coherence tomography (OCT)
progressive supranuclear palsy (PSP)
retinal nerve fiber layer (RNFL)
traumatic brain injury (TBI)
Wechsler Adult Intelligence Scale (WAIS)

Neuropsychologists are likely to encounter patients with visual impairment, from visual field and oculomotor deficits to higher-order visual dysfunction, such as disorders of visual perception or visuospatial functioning that involve visual cortical networks. Visual disturbances can be present in patients following a cerebrovascular accident (CVA) or brain injury, or neurodegenerative conditions. Visual impairments can impact the patient's ability to perform near point neuropsychological evaluations and participate in intervention techniques utilized as part of rehabilitation. These patients may have already received an optometric or ophthalmic examination for their visual complaints, but consideration should also be given to referring them for a neuro-ophthalmic assessment.

Lynn A. Schaefer, Howard D. Pomeranz, and Carol Kave Chamoff, *Visual Diagnostics for Neuropsychologists*.
In: *Medical and Laboratory Examinations for Neuropsychologists*. Edited by: Thomas J. Farrer and R. Brock Frost, Oxford University Press. © Oxford University Press (2026). DOI: 10.1093/9780197619537.003.0011

Tests Used to Assess Visual Dysfunction

Neuro-Ophthalmic Exam

Neuro-ophthalmology is a subspecialty of ophthalmology and neurology that deals with neurological disorders that affect vision. Neuro-ophthalmology can be thought of as encompassing two aspects of visual function: afferent and efferent. Afferent visual function deals with issues involving the visual (sensory) processing pathway, starting in the retina and progressing through the optic nerves, optic chiasm, and the post-chiasmal visual pathway, including the optic tracts, lateral geniculate nuclei, optic radiations, occipital cortex, and the visual association areas in the temporal and parietal lobes. Efferent visual function broadly covers the processes involved in ocular movement, ocular alignment, issues of accuracy in eye movement involving ocular saccades and pursuit movements, nystagmus, and abnormalities involving the eyelids, face, and orbit. Afferent functions include acuity, contrast, color vision, visual fields, and higher cortical visual functioning. Efferent functions include pupil response and eye movements. Abnormalities involving either afferent or efferent ophthalmic function can result in vision deficits that can range from subtle to profound.

As is true for neurology in general, the history obtained as part of the neuro-ophthalmologic exam is critical in helping to determine the cause of a patient's visual symptoms. Specific details as to the timing of onset of symptoms, whether the visual symptoms are intermittent or constant, and other details of the patient's medical history together with the ophthalmic exam findings, are important in determining whether the patient has a vision issue of afferent, efferent, or combined origin, as well as for determining the necessary workup and management for the patient's visual issues. Typically, management may include corrective lenses, referral to an appropriate professional (e.g., occupational therapy), medication, and/or surgery.

Elements of the neuro-ophthalmic exam include the following: (1) visual acuity, (2) pupil exam, (3) ocular motility (testing cranial nerves III, IV, and VI—see discussion of cranial nerve exam in Chapter 1), (4) visual fields (Figure 10.1), (5) external exam, (6) slit lamp/penlight exam, (7) intraocular pressure, and (8) fundus exam. Testing that may also be performed as part of the neuro-ophthalmic exam includes (1) color vision testing, (2) Hertel exophthalmometry, (3) Maddox rod testing, (4) automated perimetry, (5) optical coherence tomography, and (6) refraction. Where indicated, normal values are included.

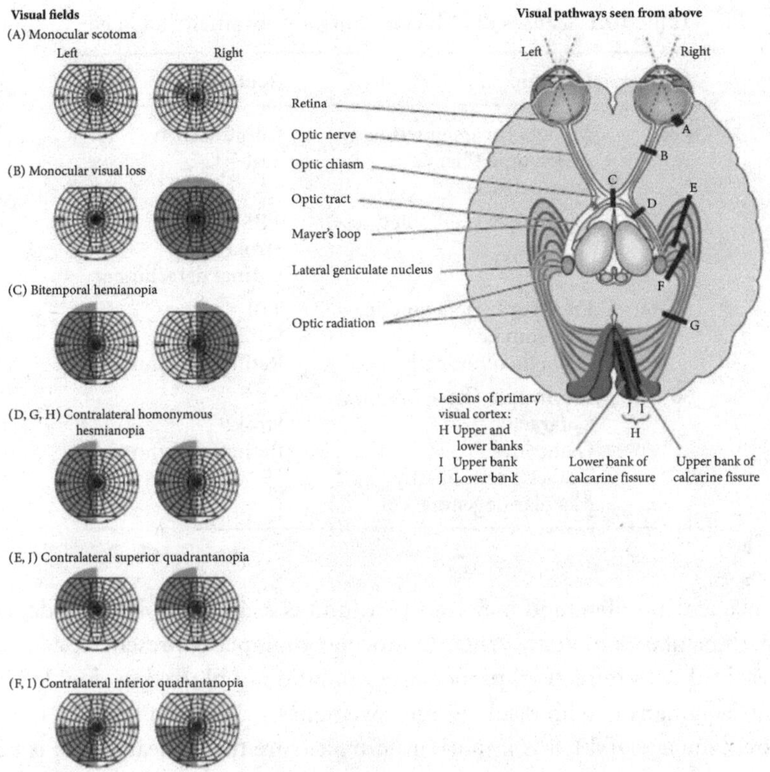

Figure 10.1 Basic structures and pathways involved in vision, with corresponding field deficits. From Hal Blumenfeld (2021), *Neuroanatomy through clinical cases*, 3rd ed. Oxford University Press. Reproduced with permission.

Visual Acuity

Determination of best corrected visual acuity (a process termed "refraction") is an important aspect of the neuro-ophthalmic exam. Visual acuity is defined as sharpness of vision, measured by the ability to discern letters or numbers at a given distance (typically 20 feet) according to a fixed standard (Table 10.1). Ocular structures involved in determining visual acuity include those involved in the transmission of light from the front to the back of the eye, including the cornea, lens, and retina. Common concerns with loss of acuity include any pathology affecting the visual pathway from the front to the back of the eye, such as a corneal opacity, cataract, vitreous hemorrhage, macular degeneration, or a problem affecting the optic nerve. Loss of visual acuity and the presence of an afferent pupillary defect in an eye suggests that an optic neuropathy may be present. If an afferent pupillary defect is not present, the loss of visual acuity may be due

Table 10.1 Causes of reduced acuity, by chronicity and age

Age	Chronic	Acute
<30	Myopia (nearsightedness) Amblyopia ("lazy eye") astigmatism	Conjunctivitis TBI
40–50	Presbyopia (farsightedness)	TBI Stroke Retinal detachment
50–60	Presbyopia (farsightedness) Glaucoma Diabetic retinopathy	TBI Stroke Retinal detachment
60–70+	Presbyopia (farsightedness) Cataract Glaucoma Diabetic retinopathy Macular degeneration	TBI Stroke Retinal detachment

to a macular problem and may lead to a fundus exam to look for evidence of retinal/macular pathology. When monocular diplopia is present, it can often be resolved with refraction, particularly resolution of blurred or double vision due to astigmatism, with resultant improvement in visual acuity. In an individual over the age of 40, it is important to make sure that if near vision is being tested, corrective lenses for presbyopia should be utilized to obtain an accurate assessment of near visual acuity.

- Normal near point distance by age is: at 10 years of age—7 cm; at 30 years of age—14 cm; at 40 years of age—25 cm; at 45 years of age and older 33 cm.

Pupils

Pupils should be assessed for reactivity (brisk, sluggish, or non-reactive) to light and/or accommodation, shape (round, oval, or irregular), difference in size between the two eyes (anisocoria) and whether that difference is the same in bright versus dim lighting conditions, and presence of a relative afferent pupillary defect (dilated in only one eye; also known as Marcus Gunn pupil), which is usually indicative of optic nerve dysfunction (Table 10.2). A non-briskly reacting pupil could be due to an afferent condition, such as loss of vision due to a retinal or optic nerve problem or alternatively due to an efferent problem affecting the parasympathetic pupillary pathways, such as a third nerve palsy, Adie's tonic pupil (pupil remains dilated in bright light) or light-near dissociation in the setting of a dorsal midbrain syndrome. Anisocoria can be due to problems affecting either the sympathetic or parasympathetic pupil pathways that control pupil dilation and constriction respectively.

Table 10.2 Pupillary abnormalities, afferent versus efferent

Afferent	Efferent
Optic neuropathy	3rd cranial nerve palsy
Marcus-Gunn pupil (can be seen in MS, retinal detachment, and severe glaucoma)	Adie's tonic pupil Dorsal midbrain syndrome Internal ophthalmoplegia

Ocular Motility

The ocular motility exam assesses if there is a deficit in the movement or alignment of the eyes in the horizontal, vertical, and/or torsional planes (Table 10.3). Inability to fully move the eye may be due to problems affecting the cranial nerves (third, fourth, or sixth nerve palsy; see also Chapter 1 for cranial nerve exam), the ocular muscles (such as eye muscle enlargement due to thyroid eye disease, muscle inflammation due to orbital myositis or a space-occupying lesion in the orbit that mechanically restricts movement), or the neuromuscular junction (such as myasthenia gravis). In addition to the extent of movement, accuracy in making ocular saccades and ocular pursuit movements is also assessed and can be abnormal in the setting of neurological conditions such as CVA or Parkinson's disease or in the setting of a concussion/post-concussion syndrome. Diagnostic and clinical implications of oculomotor disorders will be discussed below. Nystagmus (involuntary eye movements) or other disorders of involuntary binocular eye movement, such as opsoclonus (rapid, non-rhythmic eye movements), may also be present.

Visual Fields

Assessment of visual fields (central versus peripheral vision) can be made either by confrontational visual field testing using the examiner's hand or an object manually presented within different aspects of the patients expected visual field, or by formal visual field perimetry using a kinetic/moving stimulus or a static/non-moving stimulus. The temporal retina projects onto the nasal visual field and the nasal retina projects onto the temporal visual field. The superior retina projects onto the inferior visual field and vice versa. The optic radiations passing through Meyer's loop, in the temporal lobe, carry visual information from the inferior retina and superior visual fields, while the optic radiations passing through the parietal lobe carry visual information from the superior retina and inferior visual fields. The central most aspect of the visual field is located at the posterior most aspect of the visual cortex in the occipital lobe, while more peripheral aspects of the visual field map more anteriorly within the visual cortex. Determining the pattern of visual field loss is important for diagnostic purposes in localizing a visual pathway abnormality to the retina, optic nerve,

Table 10.3 Examining ocular motility signs

Ocular Motility Sign	Examination Technique	Some Possible Causes of Dysfunction
Extraocular movements	• Assess ability to move eyes in six cardinal positions of gaze (left, right, and four corners of a square) • Cover/uncover • Alternate cover	• 3rd, 4th, or 6th cranial nerve dysfunction • Thyroid eye disease • Orbital myositis • Myasthenia gravis
Tracking saccades Pursuit movement	• Follow small circles in both directions • Follow between fingers, left to right and back and up and down • Follow a pen's movement	• CVA • Parkinson's disease • Post-concussive syndrome
Binocular eye movements	Near point of convergence	• Parkinson's disease • Post-concussive syndrome/TBI • PSP

chiasm or retro-chiasmal pathway. For example, a homonymous visual field deficit would be a sign of pathology in the post-chiasmal visual pathway.

- The normal visual field extends 100 degrees temporally, 60 degrees nasally, 60 degrees superiorly and 70 degrees inferiorly. The physiological scotoma (blind spot) is located 15 degrees temporally to central fixation, where the optic nerve exits the globe. From a practical standpoint, manual visual field testing occurs with the examiner sitting far enough away from the patient to be able to compare one's own visual field to that of the patient.

External Exam

The patient is assessed for abnormalities such as facial weakness or asymmetry, proptosis, excessive or absent blinking of the eyelids, ptosis, or periorbital edema or erythema. The globes are gently palpated to assess for the presence of an orbital mass or resistance to retropulsion.

Slit Lamp/Penlight Exam

The anterior structures of the eye including the eyelids, conjunctiva, sclera, cornea, anterior chamber, iris, and lens are examined either with an external light source or at higher magnification using an office or portable slit lamp. Conditions affecting these structures such as blepharitis, trichiasis, chemosis, corneal opacity, keratoconus, iritis, and cataract can be assessed with the slit lamp.

Intraocular Pressure (IOP) Measurement

IOP can be measured with a hand-held or office tonometer. IOP elevation can be an indicator of glaucoma, an orbital mass, or impedance of outflow of venous blood from the orbit due to a vascular lesion such as a carotid-cavernous fistula.

- The normal range of intraocular pressure is 10–21 mm Hg.

Fundus Exam

The fundus is the inside, back surface of the eye. It consists of the retina, macula, optic disc, fovea, and blood vessels. The fundus can be observed with direct or indirect ophthalmoscopy or by fundus photography. The optic disk is evaluated for elevation, swelling, pallor/atrophy, changes in appearance of the vasculature overlying and around the optic disk, and for the presence of retinal fluid, hemorrhages, cotton wool spots (opaque white patches on retina, often secondary to ischemia), or other anomalies in structure. The retina is similarly evaluated for any abnormalities affecting the macula and/or peripheral retina such as a retinal artery or vein occlusion, macular edema, macular degeneration, a retinal detachment, or retinal inflammation.

- Results noted as positive or negative; there are no specific values.

Additional Neuro-Ophthalmic Testing

Color

Color vision testing can be performed by asking a patient to identify objects of different colors such as colored pencils. Other means of testing color vision include using books of color plates such as Ishihara color plates or Farnsworth D15 or D100 color testing to more formally evaluate for specific axes of color vision loss (such as red-green or blue-yellow). Red-green is most commonly affected and more so in males. Color vision loss can be hereditary or an indicator of retinal or optic nerve dysfunction. Although less common, cortical lesions can also impact color vision (e.g., achromatopsia).

Hertel

Proptosis or exophthalmos (protrusion) of the globe can be objectively assessed with exophthalmometry. The most common causes of proptosis are thyroid eye disease and orbital masses such as a cavernous hemangioma.

- The normal amount of ocular protrusion from the lateral orbital rim to the corneal apex is on average 14–21 mm in adults. While there is limited data on normal values across all ethnicities, the upper limits of normal protrusion of globe are as follows: Caucasian men: < 22 mm; Caucasian women: < 20 mm; Black men: < 25 mm; Black women: <23 mm (Migliori & Gladstone, 1984).

Maddox Rod

The Maddox rod consists of a red-colored disk in which a series of prisms have been embedded to project a red colored line to the retina when the Maddox rod is held in front of the eye and a light is projected through it. Assessment of the horizontal and vertical alignment of the eyes in primary gaze and other directions of gaze can be performed by projecting a light to both eyes while the Maddox rod is held in front of one eye. This can assist the examiner in determining whether the ocular alignment is normal. An abnormal pattern of ocular alignment may indicate cranial neuropathy, eye muscle pathology, or neurological cause.

Perimetry

Automated or manual static or kinetic perimetry are ancillary procedures, which can be used to determine whether visual field loss is present. The visual field defect can involve central fixation (a central, cecocentral, or paracentral scotoma), or may affect the peripheral visual field while sparing central fixation (such as an arcuate, altitudinal, or homonymous visual field defect). Visual field deficits may occur due to CVA or brain injury, as mentioned below.

- Normal mean deviation values on automated perimetry are typically between 0 dB and −2 dB.

Optical Coherence Tomography (OCT)

OCT of the retinal nerve fiber layer (RNFL) and macula has revolutionized the evaluation of the optic disk and macula by providing a noninvasive and quick image of these structures through the non-dilated eye that is often more accurate and provides more information than direct or indirect fundoscopy. For example, OCT of the RNFL can provide visualization of buried optic disk drusen, objectively measure the degree of elevation of the optic disk, and help to differentiate between different causes of optic disk swelling such as papilledema. OCT of the macula provides images of the cross-sectional structure of the retina which is not observable with fundoscopy. Another advantage to this process is that the images can be archived to monitor the health of the fundus serially.

Measurement of the circumpapillary retinal nerve fiber layer (RNFL) thickness is given as a global average by quadrant and by clock hour for each eye on the OCT. Summary measurements are given in a table form. Thickness maps of the RNFL show RNFL measurements around the optic disk with warmer colors corresponding to thicker values in a classic butterfly-shaped distribution around the optic disk. Deviation from normal thickness is also illustrated in a deviation map to highlight potential areas of RNFL loss. RNFL thickness is also graphed in a "TSNIT" plot in which the circumferential RNFL thickness measurements are "unrolled" into linear form and presented from the temporal to inferior quadrants. In normal eyes, this plot should demonstrate a characteristic double hump with the thickest RNFL measurements peaking in the inferior and superior regions.

- Patient measurements are plotted over a normative measurement database. Interpretation of the probability of abnormal values is aided by color coding with red, green, yellow, and white shading. Red color highlights values in the lowest 1% of normative values and is considered outside of normal limits. Yellow flags values from 1st to 5th percentile and suggests borderline abnormality. Green indicates the middle 90% of normal from the 5th to 95th percentile, and white flags values greater than the 95th percentile of normal subjects. Plots over normative values compare patient measurements with those from subjects of similar age.

Diagnostic and Clinical Implications

Deficits in visual functioning can involve sensory (afferent) and/or motor functioning (efferent) in patients. Among cortical afferent functions, visual information is processed in both ventral and dorsal cortical pathways. The ventral cortical pathway is involved in object recognition and analyzing form ("what"), while the dorsal cortical pathway is involved in spatial localization ("where"). Ventral stream deficits include visual agnosias such as apperceptive and associative agnosia, or the inability to perceive or recognize familiar objects, respectively; prosopagnosia (inability to recognize faces); and color agnosia. Dorsal stream deficits frequently involve apraxia, or difficulty locating objects in space and with motor planning; examples include constructional apraxia and dressing apraxia. A third system, known as the superior temporal sulcus system, has been proposed and is thought to involve motion perception (Scott & Schoenberg, 2011). Thus, while the apparatus of vision (i.e., the components of the eye and related structures) may be intact, patients with illness or injury to the brain can exhibit visual abnormalities that are due to damage to higher-order visual cortical areas, which can affect aspects of cognition.

Neurodegenerative conditions such as Alzheimer's disease, dementia with Lewy bodies, and parkinsonian disorders (such as Parkinson's disease and progressive supranuclear palsy [PSP]) can all exhibit visual impairment (Holden et al., 2019; Pelak, 2022). Alzheimer's disease can impair higher cortical visual functioning of both streams, but particularly of the dorsal stream as reflected by commonly observed difficulty with motion detection (Mapstone et al., 2003). Alzheimer's patients can also exhibit difficulty with saccadic eye movements on tests of ocular motility, which has been explored as an early biomarker (Wu et al., 2020). Additionally, visual search has also been shown to be frequently impaired, particularly when set size is large or items are too physically similar to discriminate (Foldi et al., 2005). Posterior cortical atrophy, often considered a subtype of Alzheimer's disease and a visual variant, can present with visual field defects (particularly a left homonymous hemianopsia), poor color vision as measured by not being able to identify numbers on the Ishihara color plates, and decreased stereopsis (Olds et al., 2020). Interestingly, in describing their clinical sample, Olds et al. reported that difficulty reading, such as missing words and skipping lines on the page, was the main symptom reported by these patients on presentation to the neuro-ophthalmologist. Kaeser and others (2015) additionally found elements of Balint's syndrome, consisting of optic ataxia, oculomotor apraxia, and, particularly, simultanagnosia, in patients with posterior cortical atrophy. Dementia with Lewy bodies can result in constructional apraxia and, differentiating it from Alzheimer's disease, visual illusions called pareidolia (Uchiyama, 2012), in which common images are envisioned superimposed on unrelated objects (i.e., seeing a man on the moon). Parkinson's disease can present with convergence deficits on tests of ocular motility as well as dry eye (Pelak, 2022). Weil and others (2016) reported that visual disturbances appear to increase as the Parkinson's disease pathology progresses, from decreased color vision and contrast sensitivity to difficulties with object and motion processing in more severe disease. Similarly, PSP can result in convergence insufficiency and diplopia (Holden et al., 2019).

In a review, Sand et al. (2013) found that approximately 30% of post-stroke patients suffered from some type of visual impairment, affecting overall functioning and quality of life. Depending on the location of the CVA, visual deficits can include eye movement disorders (i.e., nystagmus), visual field cuts (especially homonymous hemianopia), neglect, and decreased visual acuity (Sand et al., 2013). A recent study of thalamic infarction indicated oculomotor abnormalities, including vertical gaze palsy, cranial nerve palsies (i.e., cranial nerves III and VI), and visual field defects (Moon et al., 2021).

As in stroke, brain injuries may cause visual impairments depending on the location and severity of the injury. More severe traumatic brain injury (TBI) may lead to ocular injury and optic neuropathies. However, even mild TBI and concussion has been associated with oculomotor difficulties such as with smooth pursuit, saccades, convergence, accommodation, and the vestibulo-ocular reflex (Ventura et al, 2014). For example, Gallaway and others (2017) found oculomotor deficits, such as accommodation insufficiency, convergence insufficiency and saccadic dysfunction in individuals with concussion and Suleiman et al. (2020) noted vestibular abnormalities in a population of individuals suffering from post-concussion symptoms. Deficits in smooth pursuit and saccadic eye movements were also noted after sports concussion and may serve as a biomarker for concussion (Murray et al., 2020). Thus, oculomotor and visual impairments can be used diagnostically as well as clinically.

Importance to Neuropsychological Assessment

Vision plays an important role in our ability to interact with and adapt to our environment and function effectively. It is a primary receptor for cognitive, motor, communication, and emotional functions (Morton, 2004). Changes in a person's ability to process visual information can affect all aspects of functional performance. Visual disorders can impact a wide variety of daily functional tasks including reading and writing, safe functional mobility, dressing, cooking, independent management of financial and medical matters, filling out forms, and even facial recognition and processing of nonverbal communication (Scheiman, 2002). The neuropsychological assessment of a patient's cognitive and behavioral functioning often requires the patient to sustain attention to visual information at near point. It is, therefore, essential that the neuropsychologist consider the patient's visual abilities prior to testing, or at least consider the patient's visual abilities when interpreting the results of neuropsychological assessment.

In a rehabilitation setting, a multidisciplinary approach often involves occupational therapists (OTs) experienced in visual assessment and rehabilitation. A thorough evaluation of visual skills is a critical part of a comprehensive OT evaluation of functioning. It should be noted that unless there has been direct trauma to the eye or evidence of significant ocular pathology, most acute care hospitals and rehabilitation centers do not routinely provide ophthalmologic or optometric evaluations that can correct visual problems. Therefore, the following are visual skills often evaluated by the OT; while some overlap with functions assessed by the neuro-ophthalmologist, the method of assessment is different and focuses on a patient's daily functioning. According to the American Occupational Therapy Association (1999), OT interventions may include the remediation or restoration of performance limitations that are due to impairment in biological, physiological, psychological or neurological processes. Interventions may include adaptations of tasks, processes, or environments and the teaching of compensatory strategies in addition to disability prevention and health promotion education.

Occupational Therapy Visual Evaluation and Treatment

Near Visual Acuity

Deficits in near acuity can result from a variety of conditions. Hyperopia, or farsightedness, is caused by a shortened shape of the eye resulting in blurring at near point. Presbyopia is an age-related decrease in the accommodative, or focusing, power of the lens. Accommodative insufficiency, which will also cause blurring at near point, is often seen in post concussive injuries and other neurologic conditions. Macular degeneration, more commonly seen in older patients, will also impact near visual acuity, as will diabetic retinopathy (Scheiman, 2002). Without the appropriate plus (biconvex) lenses, the patient with hyperopia, presbyopia, or accommodative insufficiency will experience near point blurring, which can greatly impact their ability to process near point information on neuropsychological tests. The OT can assess near point acuity utilizing a variety of near vision cards including the Snellen chart, Allen Figures, Tumbling "E" chart, and the near point Broken Wheel Test (Scheiman, 2002). The alternative cards to the Snellen chart are useful for evaluating patients who are unable to process letters and numbers accurately due to perceptual impairments, unfamiliarity with English letters, or language deficits. The OT will always assess the patient with any appropriate corrective lenses they currently use to determine their best functional acuity.

Correcting near acuity lies outside the practice of occupational therapy; lenses and prisms can be prescribed by neuro-ophthalmologists or neuro-optometrists (or behavioral optometrists). However, every effort should be made to ensure that patients have access to their most recently prescribed lenses to enable them to fully participate in evaluations relying on visual function. For more on working with the patient's visual preferences and on accommodations for assessment, see Schaefer and Pick (2022).

Visual Field Deficits

Peripheral field deficits such as hemianopsias and quadrantanopsias are frequently seen in patients with brain injuries from trauma, stroke, and neoplasms. Chronic central field deficits, or scotomas, may result from macular degeneration and diabetic retinopathy. Central scotomas, especially those impacting the macula, will result in significant near point blurring. Glaucoma and retinitis pigmentosa may also impair the visual field (Scheiman, 2002). While the OT cannot accurately plot visual field loss through computerized testing, hemianopsias and quadrantopias can be easily identified using confrontation testing (Zoltan, 2007). Use of the Amsler grid can help identify central scotomas. Visual field deficits will impact the patient's ability to accurately see all of the visual information presented during neuropsychological testing. Loss of the right visual field has a particularly significant impact on reading comprehension, as the right visual field is the field of anticipation when reading modern languages that are scanned from left to right. Reading with a right homonymous hemianopsia is typically slow and halting which can have a significant impact on comprehension.

The OT can help patients learn how to compensate more effectively for their visual field loss through increasing their awareness of the loss and training effective eye and head movements in the direction of the impaired field (Zoltan, 2007). The use of partial occlusion of the intact visual field during training activities can facilitate the patient's ability to make increased saccadic eye movements toward their impaired field. It is important that the OT educate the patient and their family about the impact of their field loss on safe mobility and provide them with suggestions for environmental modifications that can increase safety and independence.

Convergence

Convergence is the efferent ability to medially rotate the eyes in a coordinated fashion as visual information is brought closer to the individual. The normal convergence near point is 3–4 inches from the bridge of the nose. Convergence insufficiency is a very common sequalae of neurological injury, as mentioned above (Scheiman, 2002). The individual with a convergence insufficiency will have difficulty sustaining attention to near point tasks and may even experience diplopia at near point. The longer the duration of the near point task, the more difficulty the patient will experience. The patient may complain of headache or ocular discomfort at near point, but their only symptom may be impaired performance on near point tasks as a result of the inability to sustain attention. For the patient experiencing a convergence insufficiency, it may be necessary for the neuropsychologist to break testing into several shorter sessions to prevent near point fatigue from interfering with performance.

The OT will assess the patient's convergence near point by having them follow a target presented in midline as it is slowly moved closer to the patient's face. The convergence near point is where the patient either reports diplopia or the therapist notes that one of the patient's eyes ceases medial rotation.

There are a variety of training activities that the OT can integrate into treatment to improve convergence skills and therefore the ability to sustain attention at near point. Brock string exercises, pencil push-ups, and pointer/straw activities can all be used during OT treatment and as part of a home exercise program to improve convergence skills.

Ocular Alignment—Tropia (Strabismus)

Tropia is a condition in which one eye deviates from its normal alignment. Typically, a single eye is affected; however, alternating tropia, variations in which eye is the fixating eye depending on the nature of the task, has been observed. Individuals who suddenly acquire a tropia as a result of brain injury may have sustained damage to cranial nerves III, IV, or VI, which can result in misalignment of the eyes. If the patient has visual capability in both eyes and there was not a preexisting strabismus from childhood, this will result in diplopia (Scheiman, 2002).

The OT will assess the presence of a tropia utilizing the Cover Test. The patient will not be able to function efficiently with diplopia and may complain of headache, dizziness, and nausea. They may attempt to eliminate the diplopia through closing or covering one eye. The OT can temporarily eliminate the diplopia by utilizing opaque or translucent occlusion of one eye. While this will enable the patient to participate in near point testing without diplopia, the neuropsychologist must be aware that occluding one eye, while eliminating diplopia, will also eliminate a portion of the visual field on the occluded side (Scheiman, 2002).

Ocular Alignment—Phoria

Phoria is the natural posture of the eyes at rest. Some people's eyes tend to posture medially (esophoria) and some people's eyes tend to posture laterally (exophoria). Within a certain degree this is normal and will not create visual symptoms. Excessive exophoria can be a sequalae of brain injury (Scheiman, 2002). The symptoms for the patient are similar to those caused by a convergence insufficiency—difficulty sustaining attention to near point activities, headache, ocular discomfort, and even diplopia at near point. The OT can assess the presence and degree of phoria utilizing the Alternate Cover Test and the Bernell phoria measure card (Scheiman, 2002). The presence of a significant degree of exophoria should alert the neuropsychologist that the patient's performance on near point testing may be impaired, particularly as the duration of a task increases. The OT can utilize many of the same treatments used to address convergence insufficiency to reduce excessive exophoria and the neuropsychologist may want to break testing into several shorter sessions to prevent near point fatigue from interfering with performance.

Oculomotor Skills

The patient's ability to follow a moving target (visual pursuit), quickly move the eyes from one area of the visual field to another (saccadic eye movements) and visual scanning (visual search, reading) will be assessed by the OT through careful observation of eye movements as well as through standardized tests such as the King-Devick Test (Galleta et al., 2016) and the Developmental Eye Movement Test (Zoltan, 2007). Impaired oculomotor skills will cause the patient to miss visual details and can have a significant impact on reading comprehension.

The OT can integrate a wide variety of oculomotor training activities to improve these skills, such Marsden Ball activities, flashlight pursuits and saccades, and scanning activities such as word searches and reading tasks. During treatment, the therapist will stress the elimination of compensatory head movements during activities. The OT may also teach the patient effective compensatory strategies such as using a straight edge and finger pointing when reading to help guide eye movements as well as advising a switch to larger print reading material or using an e-reader that enables the patient to enlarge the type font, as a larger font decreases the effort needed to scan effectively. The neuropsychologist might consider having the patient use these compensatory strategies to facilitate the patient's best performance during testing.

Neuropsychological Correlates of Test Findings

Neuropsychological tests for visual impairment tend to assess higher-order visual dysfunction, such as disorders of visual perception or visuospatial functioning, rather than visual field and oculomotor deficits per se. However, assessment of visual scanning and visual search, which require oculomotor demands, can be achieved by the use of cancellation tests (whether of letters, numbers, or figures) and tools such as the Trail Making Test, whereas hemispatial attention and neglect can also be assessed with these or via line bisection (Lezak et al, 2012). In fact, a neglect (typically right hemisphere mediated attentional deficit) should be ruled out before assuming a visual field cut or difficulty with eye movements. Some authors have subsumed cancellation tasks and line bisection under the assessment of visuoperceptual (Scott & Schoenberg, 2011) or visuospatial tasks (Bauer, 2014).

In assessing visual perception, visual recognition can be assessed using tests such as Benton Visual Form Discrimination, Famous Faces (for facial recognition), or Boston Naming Test; visual organization can be examined using the Hooper Visual Organization Test, for example (see Lezak et al., 2012, for a review of specific tests). Deficits may indicate damage to ventral stream (occipitotemporal) regions such as in certain types of agnosia, however for visual recognition tests such as the Boston Naming Test, language deficits should first be ruled out.

Visuospatial functioning can be assessed with map reading, tests like Judgment of Line Orientation or Design Memory from the Wechsler Memory Scale—Fourth Edition, or visuoconstructive tasks (Lezak et al., 2012). Visuoconstructive tasks can be included under visuospatial tasks (Scott &Schoenberg, 2011), or considered a separate category (Bauer, 2014). In addition to visuospatial processing, constructional tasks include figure copying or drawing, such as clock drawing or the Rey-Osterrieth Complex Figure, two-dimensional construction such as Object Assembly from the Wechsler Adult Intelligence Scale (WAIS), or three-dimensional construction such as the WAIS Block Design subtest (see Lezak et al., 2012, for a review of specific tests). Deficits in visuoconstruction may indicate damage to parietal areas, particularly in the right hemisphere, and may correspond functionally with a constructional apraxia. Copy tasks can sometimes lateralize damage depending on whether drawings maintain the gestalt (left-hemisphere damage) or lose the gestalt (right-hemisphere damage), focusing on details (Scott & Schoenberg, 2011). It should also be noted that many of these tests also load on frontal lobe executive functioning deficits, given the planning and organization involved, as well as motor function, and, therefore, consideration for these related cognitive abilities should also be made.

The following clinical case will incorporate neuropsychological findings with assessment of visual functioning, and has the advantage of also detailing neuropsychological reassessment following oculomotor rehabilitation by OT.

Case Example Relevant to Neuropsychology

Mr. Glass is a 30-year-old male, referred by his neurologist to assess cognitive functioning following post-concussive syndrome with vestibular symptoms after he was physically assaulted and then suffered a motor vehicle accident (MVA). According to Mr. Glass, during the assault he was hit in the head, neck, back, and body, and fell down a flight of stairs; he reported that he did not lose consciousness. One month later he was involved in an MVA. He reported that his car was hit on the front driver's side and went into a spin. He reported that he did not lose consciousness but experienced neck pain afterward.

On neuropsychological evaluation 9 months following the MVA, Mr. Glass complained of headaches, neck pain, vestibular deficits with dizziness, balance problems, occasional blurred vision and light sensitivity, decreased attention and concentration, increased distractibility, changes in short-term memory, and decreased multitasking. He also reported occasional ringing in his ears. He reported an EEG of his head revealed abnormal spiking in his right temple, but this was not available for review. He was seeing a physical therapist for vestibular rehabilitation.

Mr. Glass reported that alcohol use worsened his symptoms and so he avoided drinking. He endorsed daily cannabis use, which he argued lessened the severity of his headaches. He took naproxen to lessen his pain, as well as citalopram 20 mg for headaches and dizziness. Regarding activities of daily living, he avoided driving when possible and noted that he was sleeping approximately 12–13 hours per day. Mr. Glass denied any symptoms of anxiety or depression. However, he noted increased irritability.

Mr. Glass graduated from college with a degree in computer science. His profession was that of a computer programmer, which involves extensive near point work. He continued to work as a computer programmer after the assault. However, he would frequently be sent home early because he could not concentrate and experienced headaches and dizziness due to the prolonged near point computer work. Following his MVA, he was put on short-term disability days later. He reported that he was eager to return to work as a computer programmer.

At the time of the evaluation, Mr. Glass was alert and attentive but reported that the artificial fluorescent lights gave him headaches; on subsequent test dates, he was tested in a room with large amounts of natural light. He noted that he could not focus his attention on visual tasks for extended periods without feeling dizzy and nauseous. As such, he was given multiple breaks in between short testing sessions to ensure that his effort was strong throughout. Mr. Glass's attention, concentration, and effort were best with familiar tasks that offered clear instructions, required short responses, or relied on rote memory. He appeared nervous on several occasions about his test performance, stating that he would have been able to complete tasks much more easily prior to his concussion. When confronted with a challenge or sensing failure, his confidence appeared to rapidly erode and his effort waned. During those instances, the examiner stopped testing and provided encouragement, increasing his effort.

Neuropsychological Assessment

- Results of free-standing and embedded performance validity tests indicated adequate effort (Table 10.4).

Table 10.4 Labels for psychometric data

Verbal Description	T-Score	Standard Score/IQ	Z-Score	Scaled Score	%ile Rank
Very Superior	≥ 70	≥ 130	≥ 2.0	≥ 16	≥ 98
Superior	64–69	120–129	1.34–1.9	14–15	91–97
High Average	58–63	110–119	0.67–1.33	12–13	75–90
Average	43–57	90–109	0.66 to −0.66	8–11	25–74
Low Average	37–42	80–89	−1.3 to −0.7	6–7	9–24
Borderline	30–36	70–79	−2.0 to −1.4	4–5	2–8
Impaired	< 30	< 70	< −2.0	<4	< 2

- Overall intellect was in the low average range, which was a decrease compared to estimated high average premorbid functioning. Verbal abilities were average, perceptual reasoning and working memory were low average, and processing speed was a clinically significant weakness, in the borderline range.
- Mr. Glass demonstrated difficulties with immediate attention and concentration; his verbal and visual attention fell largely in the borderline range of functioning. He showed difficulty with visual scanning (cancellation; Trails) and processing speed (Symbol Search) on visual attention tasks.
- On verbal tasks, verbal fluency emerged as a relative weakness, in the low average range, and his crystallized knowledge emerged as a relative strength. Verbal abstract reasoning was at the low end of average.
- On visuospatial tasks, his copy of a complex figure was in the borderline to low average range. Visual abstract reasoning was borderline, while three-dimensional construction was average.
- His memory appeared to be negatively affected by deficits in both attention and in executive functioning. He performed better on story learning and memory task (low average) than on a word-list task (T = 32; borderline) and was helped somewhat with cueing. Recognition on the word-list contained errors. Visual memory was impaired, and not helped with recognition.
- He demonstrated impairments in various aspects of executive functioning such as cognitive flexibility and set-shifting.
- Self-reported mood symptoms were consistent with only minimal anxiety and depression, but may have represented an overly positive representation of his current emotional functioning. He endorsed numerous concerns about his somatic symptoms and an increase in irritability.

Overall, Mr. Glass's cognitive difficulties appeared to be attributed to his multiple concussions, as impairments in information processing speed, attention, short-term memory, and executive functioning are widely reported in patients with post-concussive symptoms (Griffen & Hanks, 2014). Regarding visual and visuospatial functioning, he exhibited difficulties with visual scanning and figure copy. It was recommended that he continue vestibular physical therapy, drive only short distances, see OT for evaluation and oculomotor rehabilitation, and be retested in approximately 9–12 months.

Occupational Therapy Assessment

Mr. Glass was seen for evaluation in OT 1 year after his injury. He was seeing a physical therapist for vestibular rehabilitation and was referred for OT services by his neuropsychologist.

- Mr. Glass's OT vision evaluation revealed that he lacked a significant convergence response and he was consistently reporting diplopia at 12″ and closer. He was also significantly exophoric at 13 diopters. Both of these problems were resulting in the inability to participate in near point activities without closing one eye to eliminate ocular discomfort and diplopia. His ability to use his eyes together efficiently as a team (binocular vision) was very impaired and therefore he lacked stereopsis (fine depth perception).
- Mr. Glass's visual pursuits and saccades were significantly impaired, and he also demonstrated impairment of his vestibular-ocular reflex which helps keep an image stabilized on the retina. This was believed to be contributing to his dizziness and balance deficits. The King-Devick test of visual scanning skills could not be administered during initial evaluation due to the patient's diplopia at near point.
- Mr. Glass's memory skills were assessed 1 month after beginning his OT treatment utilizing the Rivermead Behavioral Memory Test, which evaluates a wide range of memory skills including visual memory for pictures, faces and spatial memory, verbal memory, prospective memory, and the ability to learn a new skill (Zoltan, 2007). He scored in the 2nd percentile.

His OT treatment sessions consisted of 90-minute sessions one day a week along with an extensive daily home program of exercises and activities to improve his visual skills. Approximately 2 months after starting therapy, Mr. Glass was involved in a second MVA in which he was reportedly hit head-on but suffered no loss of consciousness. He was seen in an emergency room and a CT scan revealed no acute injury, although he reported increased visual symptoms and dizziness. Many of the gains that he had begun to make in therapy were erased, and therapy had to start over.

Mr. Glass's OT sessions and his home program consisted of gaze stabilization exercises, convergence training, and scanning activities in addition to working on strategies to limit the impact of his memory deficits on his functional performance. He was also provided with yellow lenses to help him deal with his light sensitivity, particularly to florescent lighting. He worked diligently on his home program and made slow, steady progress in all areas.

At the conclusion of his therapy 9 months after his initial evaluation, Mr. Glass's convergence near point was normal at 3″ which he was able to sustain for over 1 minute. His visual scanning skills, as measured by the King-Devick test, were normal both in speed and accuracy and retesting of his memory skills utilizing the Rivermead Behavioral Memory test placed him in the 92nd percentile. He was discharged from OT. He also was discharged from vestibular physical therapy in the same month, and had resumed driving. It was recommended that he have an updated neuropsychological evaluation prior to his return to work.

Neuropsychological Reassessment

One month after discharge from therapies and 1 year since initial neuropsychological evaluation, Mr. Glass, now age 31, was re-evaluated. He reported that his concentration and memory appeared better than at previous testing. Mr. Glass denied dizziness, double vision, or sensitivity to light. He reported only occasional headaches. The pain he suffered from his second MVA was also better and he was no longer taking any medication. He reported smoking cannabis only occasionally and reported more regular sleep, of approximately 8 hours nightly. He resumed driving 1 month earlier and felt that it had been going well. Mr. Glass had only been on short-term disability and was now eager to return to work.

- Results of reassessment revealed improvements in the areas of processing speed, attention, memory, and executive functioning compared to the previous testing.
- Attention was now solidly in at least the average range, with improved processing speed and no scanning difficulty.
- Verbal fluency improved to the average to high average range.
- Visual abstract reasoning was now average. Copy of a complex design remained in the borderline range, due to poor planning.
- List-learning and memory improved to low average, and story memory improved to average. Visual memory remained poor (impaired; due to poor copy), but recognition was now average.

- Cognitive flexibility and set-shifting were now in the superior range, with no errors.
- It was recommended that he return to work. However, given his recent recovery and extended absence, it was recommended that he initially return at a reduced schedule (i.e., half-time for at least 2 weeks, etc., with a gradual increase in hours and responsibilities), in order to maximize the probability of success.
- It should be noted that in a follow-up phone call from OT to the patient made 6 months after being discharged from therapy, he stated that he was back full-time at his previous job and was feeling well.

This case demonstrates the improvements made in a patient with multiple concussions following oculomotor assessment and rehabilitation by occupational therapy and vestibular physical therapy. The multidisciplinary approach of neuropsychological assessment and OT vision evaluation both before and after treatment helped guide his rehabilitation and return to functional tasks such as driving and work.

Conclusions

Visual and visuospatial deficits are common in patients with conditions such as neurodegenerative dementias, cerebrovascular accidents, and brain injuries. Thus, neuropsychologists are often presented with visual test results for their patients that they are unable to interpret. This chapter reviewed the diagnostic testing done by the neuro-ophthalmologist, how to understand normal and abnormal values, and how impairments can impact neuropsychological assessment. In inpatient or rehabilitation settings, occupational therapists may also conduct visual assessments and treatment, and this chapter is one of the first to integrate the two perspectives of neuro-ophthalmology and occupational therapy. It is our hope that neuropsychologists gain a better understanding of visual neurodiagnostic testing as well as an appreciation for a multidisciplinary approach to working with these patients.

References

American Occupational Therapy Association. (1999). Definition of OT practice for the AOTA Model Practice Act. *American Journal of Occupational Therapy*, *53*(6), 608. https://doi.org/10.5014/ajot.53.6.608

Bauer, R. M. (2014). Visuospatial, visuoperceptual, visuoconstructional disorders. In M. W. Parsons & T. A. Hammeke (Eds.), *Clinical neuropsychology: A pocket handbook for assessment* (pp. 436–462). American Psychological Association.

Blumenfeld, H. (2021). *Neuroanatomy through clinical cases* (3rd ed.). Sinauer Associates, Oxford University Press.

Foldi, N. S., Schaefer, L. A., White, R. E., Johnson, R., Jr, Berger, J. T., Carney, M. T., & Macina, L. O. (2005). Effects of graded levels of physical similarity and density on visual selective attention in patients with Alzheimer's disease. *Neuropsychology*, *19*(1), 5.

Galetta, K. M., Liu, M., Leong, D. F., Ventura, R. E., Galetta, S. L., & Balcer, L. J. (2016). The King-Devick test of rapid number naming for concussion detection: Meta-analysis and systematic review of the literature. *Concussion (London, England)*, *1*(2), CNC8. https://doi.org/10.2217/cnc.15.8

Gallaway, M., Scheiman, M., & Mitchell, G. L. (2017). Vision therapy for post-concussion vision disorders. *Optometry and Vision Science*, *94*(1), 68–73. https://doi.org/10.1097/OPX.0000000000000935

Griffen, J., & Hanks, R. (2014). Cognitive and behavioral outcomes from traumatic brain injury. In M. Sherer & A. M. Sander (Eds.), *Handbook on the neuropsychology of traumatic brain injury* (pp. 25–45). Springer.

Holden, S. K., Van Dok, E., & Pelak, V. S. (2019). Co-occurrence of convergence insufficiency and cognitive impairment in parkinsonian disorders: A pilot study. *Frontiers in Neurology*, *10*, 864.

Kaeser, P. F., Ghika, J., & Borruat, F. X. (2015). Visual signs and symptoms in patients with the visual variant of Alzheimer disease. *BMC Ophthalmology*, *15*(1), 1–7.

Lezak, M. D., Howieson, D. B., Bigler, E.D., & Tranel, D. (2012). *Neuropsychological assessment* (5th ed.). Oxford University Press.

Mapstone, M., Stefenella, T. M., & Dufy, C. J. (2003). A visuospatial variant of mild cognitive impairment: Getting lost between aging and AD. *Neurology*, 60(5), 802–808.

Migliori, M. E., & Gladstone, G. J. (1984). Determination of the normal range of exophthalmometric values for black and white adults. *American Journal of Ophthalmology*, *98*(4), 438–442.

Moon, Y., Eah, K. S., Lee, E. J., Kang, D. W., Kwon, S. U., Kim, J. S., & Lim, H. T. (2021). Neuro-ophthalmologic features and outcomes of thalamic infarction: A single-institutional 10-year experience. *Journal of Neuro-Ophthalmology*, *41*(1), 29–36.

Morton, R. L. (2004). Visual dysfunction following traumatic brain injury. In M. J. Ashley (Ed.), *Traumatic brain injury rehabilitative treatment and case management* (pp. 183–208). CRC Press.

Murray, N. G., Szekely, B., Islas, A., Munkasy, B., Gore, R., Berryhill, M., & Reed-Jones, R. J. (2020). Smooth pursuit and saccades after sport-related concussion. *Journal of Neurotrauma*, *37*(2), 340–346.

Olds, J. J., Hills, W. L., Warner, J., Falardeau, J., Alasantro, L. H., Moster, M. L., Egan, R. A., Cornblath, W. T., Lee, A. G., Frishberg, B. M. and Turbin, R. E., Katz, D. M., Charley, J. A., & Pelak, V. S. (2020). Posterior cortical atrophy: Characteristics from a clinical data registry. *Frontiers in Neurology*, *11*, 358.

Pelak, V. S. (2022). The clinical approach to the identification of higher-order visual dysfunction in neurodegenerative disease. *Current Neurology and Neuroscience Reports*, *22*(4), 229–242. https://doi.org/10.1007/s11910-022-01186-7

Sand, K. M., Midelfart, A., Thomassen, L., Melms, A., Wilhelm, H., & Hoff, J. M. (2013). Visual impairment in stroke patients—A review. *Acta Neurologica Scandinavica*, *127*, 52–56.

Schaefer, L. A., & Pick, L. H. (2022). Considerations and suggested practices for psychological assessment and intervention when working with older adults with disabilities. *Professional Psychology: Research and Practice*, *53*(5), 446–457. https://doi.org/10.1037/pro0000472

Scheiman, M. (2002). *Understanding and managing vision deficits: A guide for occupational therapists*. Slack.

Scott, J. G., & Schoenberg, M. R. (2011). Deficits in visuospatial/visuoconstructional skills and motor praxis. In M. R. Schoenberg & J. G. Scott (Eds.), *The little black book of neuropsychology* (pp. 201–218). Springer.

Suleiman, A., Lithgow, B. J., Anssari, N., Ashiri, M., Moussavi, Z., & Mansouri, B. (2020). Correlation between ocular and vestibular abnormalities and convergence insufficiency in post-concussion syndrome. *Neuro-Ophthalmology*, *44*(3), 157–167.

Uchiyama, M., Nishio, Y., Yokoi, K., Hirayama, K., Imamura, T., Shimomura, T., & Mori, E. (2012). Pareidolias: Complex visual illusions in dementia with Lewy bodies. *Brain: A Journal of Neurology*, *135*(Pt 8), 2458–2469. https://doi.org/10.1093/brain/aws126

Ventura, R. E., Balcer, L. J., & Galetta, S. L. (2014). The neuro-ophthalmology of head trauma. *Lancet Neurology*, *13*(10), 1006–1016.

Weil, R. S., Schrag, A. E., Warren, J. D., Crutch, S. J., Lees, A. J., & Morris, H. R. (2016). Visual dysfunction in Parkinson's disease. *Brain: A Journal of Neurology*, *139*(11), 2827–2843. https://doi.org/10.1093/brain/aww175

Wu, S. Z., Masurkar, A. V., & Balcer, L. J. (2020). Afferent and efferent visual markers of Alzheimer's disease: A review and update in early stage disease. *Frontiers in Aging Neuroscience*, *12*, 572337.

Zoltan B. (2007). Visual processing skills. In B. Zoltan (Ed.), *Vision, perception, and cognition: A manual for the evaluation and treatment of adult with acquired brain injury* (4th ed., pp. 48–108). Slack.

11

CSF Biomarkers, Tissue Biopsy, and Genetic Analysis

Skyler J. Smith

Introduction

The diagnosis of complicated neurological diseases goes beyond clinical observation; it is also based on neuropsychological, neuroimaging, and laboratory diagnosis. For neuropsychologists, the integration of cerebrospinal fluid (CSF) analysis, tissue biopsy, and genetic testing findings is an essential piece of interpreting cognitive and behavioral symptoms. Several neurological disorders have overlapping clinical features, particularly in their early stages, making differential diagnosis difficult. Being familiar with biological markers can provide insight into disease mechanisms and help support guided clinical decision-making.

Biomarkers in CSF, especially, have gained a lot of importance in exploring disease-specific neurodegenerative and inflammatory processes. Markers of proteins including tau, amyloid beta, alpha synuclein, and prions could indicate the presence of Alzheimer's disease, Parkinsonian syndromes, or prion diseases, among others. These markers not only provide a biochemical perspective that complements cognitive testing but also can help in the early diagnosis—frequently before clinical symptoms have fully manifested. Other biomarker levels, such as glucose or lactate, may also be indicative of underlying metabolic or infectious processes that may impact neuropsychological functioning.

Tissue biopsies and genetic analyses provide additional tools in the neuropsychologist's diagnostic toolbox, especially for complex or atypical cases. Biopsy can confirm pathology when imaging and CSF are negative. Genetic testing, too, is becoming more important for diagnosing inherited disorders like Huntington's disease or frontotemporal dementia. For patients and families dealing with these diagnoses, genetic insights can help to guide medical planning, but also mental health support and counseling. The integrative model extends the neuropsychology domain by further specifying the systems

Skyler J. Smith, *CSF Biomarkers, Tissue Biopsy, and Genetic Analysis*. In: *Medical and Laboratory Examinations for Neuropsychologists*. Edited by: Thomas J. Farrer and R. Brock Frost, Oxford University Press.
 DOI: 10.1093/9780197619537.003.0012

for needed input/care, serving the neuropsychologists to identify the particular aspects of cognition and affected situations for the applicable intervention, depending on the pathology of underlying conditions and the cognitive presentation.

Integrating CSF Biomarkers, Tissue Biopsy, and Genetic Analysis into Neurodiagnostic Practice CSF Assessment

Importance

Cerebrospinal fluid (CSF) assessment is an integral component in the neurodiagnostic process. The examination of CSF provides invaluable insights into the physiological and pathological states of the central nervous system (CNS), assisting in the diagnosis of a wide spectrum of neurological disorders. From detecting inflammatory markers in autoimmune encephalitis to identifying oligoclonal bands in multiple sclerosis, CSF analysis is pivotal in differentiating between various neurological conditions. Its ability to reflect biochemical changes at the neuronal and synaptic levels also makes it crucial in understanding neurodegenerative diseases such as Alzheimer's and Parkinson's disease.

Physiological Parameters

The physiologic parameters used for the evaluation of CSF, including cell count, protein concentration, glucose levels, and the presence of specific biomarkers, offer a window into the metabolic and immunologic status of the brain and spinal cord. The normal range of these parameters varies, but deviations from these norms can indicate pathologies. Elevated protein levels may suggest inflammation, infection, or blood-brain barrier disruption, while abnormal glucose levels could be indicative of infections like meningitis or systemic metabolic disturbances impacting the CNS.

Opening Pressure

Values

The normal range of opening pressure in adults during a lumbar puncture is typically accepted to be between 10 and 20 cm H_2O. However, this range can slightly vary based on patient factors such as age, body habitus, and positional changes

(Norager et al., 2021; Wright et al., 2012). In pediatric populations, the normal range can be higher (Avery et al., 2010). Various studies, electronic medical records, and medical systems will have slightly differing reference values. Be sure to review the normal reference values as they are continually updated to reflect new data.

Values exceeding 25 cm H_2O in adults and 28 cm H_2O in pediatric patients are generally considered elevated. Elevated opening pressures are commonly associated with increased intracranial pressure, which can be due to various causes such as brain tumors, intracranial hemorrhages, or hydrocephalus. Symptoms can include headaches behind the eyes, vision changes, nausea, dizziness, and papilledema (diagnostic). In contrast, opening pressures below 10 cm H_2O are considered low. While less common, low CSF pressures can be observed in cases of spontaneous intracranial hypotension (SIH), often resulting from a CSF leak (Wright et al., 2012). SIH could be associated with cranial trauma or spinal injury. Clinically, common symptoms associated with decreased CSF pressure include an occipital headache which worsens when standing and can be associated with changes in visual acuity. Other symptoms, less specific, include nausea, dizziness, changes in cognitive ability, and tinnitus. SIH can appear similar to Chiari I malformation on imaging due to the presence of low-lying cerebellar tonsils in both, so care should be taken to differentiate between the two pathologic conditions (Bin Wan Hassan et al., 2023).

Patient-related factors play a significant role in the accuracy of opening pressure readings. Anxiety, pain, or discomfort can lead to involuntary muscular contractions or Valsalva maneuvers, which can transiently increase intrathecal pressure (Wright et al., 2012). In pediatric patients, crying can similarly elevate CSF pressure, complicating the interpretation of the readings (Wiegand & Richards, 2007). Patient comfort and calmness during the procedure is necessary to ensure accuracy of the results, and patient behaviors during the procedure should be noted.

There are other considerations when evaluating the utility of a lumbar puncture. BMI is associated with a mild relative increase in CSF pressure (Whiteley et al., 2006). Hydration status is another crucial consideration. Dehydration can lead to a lower CSF pressure, while overhydration may result in elevated pressures (Burger et al., 2013; Jabbari et al., 2013). The impact of medications, including diuretics, also cannot be overlooked as they significantly influence fluid balance and CSF dynamics. Medications such as corticosteroids and vitamin A infusions should also be considered, as they can increase intracranial pressure and CSF volume respectively (Mokri, 2013; Wile, 2012). A thorough review of the patient's medication list is necessary to provide the optimal predictive power for a lumbar puncture. Reviewing, if available, the opening pressures for past evaluations could also be helpful when differentiating between physiological or pathological CSF pressures. There could be a physiological variance around 6 cm H_20 in adults and significantly decreased variance in pediatric patients who are less than 1 year old (Sakka et al., 2011).

Technique

When performing CSF lumbar puncture, the patient is positioned in either the lateral recumbent or the sitting position, depending on clinical indications and patient comfort. The lateral recumbent position is often preferred for measuring opening pressure, as it reduces the impact of gravity. The target site is usually between the L3 and L4, or L4 and L5 lumbar vertebrae, chosen to avoid damage to the spinal cord, which typically ends at the L1–L2 level in adults. The area is cleaned with antiseptic solution, and local anesthesia is administered to minimize discomfort. A spinal needle is inserted in the midline, between the spinous processes of the chosen vertebrae. The practitioner applies steady, controlled pressure to advance the needle while monitoring patient feedback. Loss of resistance is often felt as the needle penetrates the dura mater, signaling entry into the subarachnoid space. Once the needle is correctly positioned, CSF may spontaneously flow from the needle, or a manometer can be attached to measure the opening pressure. The fluid is collected in sterile tubes for various analyses, including cell count, protein and glucose levels, and microbiological cultures (Wright et al., 2012).

Closing Pressure

Closing pressure, typically measured at the end of a lumbar puncture procedure, should be similar to the opening pressure. Significant deviations from this alignment could be indicative of procedural complications or alterations in CSF dynamics during the procedure. An elevated closing pressure compared to opening pressure may suggest an obstruction in CSF flow due to surgical error or an increase in intracranial pressure due to a medication or other cause. A decrease in closing pressure might be attributed to excessive removal of CSF or potential leakage, leading to a drop in CSF volume and pressure and is often associated with post-lumbar puncture headaches (Wright et al., 2012). A recent study has suggested that removal of CSF for hydrocephalus is not clinically useful due to replacement of the CSF within 7 hours, but might be a therapeutic option, according to another study, for certain conditions such as acute intracranial hypertension secondary to malignancy and exacerbated by pregnancy (Godoy et al., 2022; Mollan et al., 2023).

Biochemical Markers in CSF Analysis

Importance

CSF analysis is a powerful tool in diagnosing and monitoring neurological disorders. Biochemical markers in CSF provide valuable insights into the pathophysiological mechanisms underlying various neurologic diseases. These markers can indicate neuronal or axonal injury, inflammation, infection, and neurodegenerative processes.

Values

Normal reference values for CSF components are essential for interpretation. Variations from these norms can indicate pathology. For example, elevated protein levels might suggest inflammation, while abnormal glucose levels could indicate infection or malignancy. The values have to be taken in context with the clinical presentation of the patient for the best "fit" for the neurodegenerative condition.

Protein

Total protein content in CSF is a key indicator of the integrity and permeability of the blood-CSF barrier. It is a nonspecific marker but crucial for differential diagnosis. Elevated protein levels may indicate a breach in the blood-brain barrier due to inflammation, bacterial or viral infection, or trauma. Low levels can occur with increased CSF production associated with cranial or spinal leakage. Previously, normal ranges were accepted to be between 15–45 mg/dL (Aluise et al., 2008), but recent systematic reviews suggested that the upper range should be incrementally increased based on age, gender-assigned at birth, and other pathologic conditions in order to reduce the risk of over-diagnosis (Breiner et al., 2019; Fautsch et al., 2023).

Prions

Prion proteins are unique in that they are infectious agents composed of protein in a misfolded form that are resistant to protease digestion. These abnormal proteins can induce normal protein in the host to also misfold, creating a cascading effect that causes accumulation of these proteins within cells and the body. Prions can cause rapidly progressive neurodegenerative disorders. Prion-related neurodegenerative conditions can be transmitted by exposure to infected soil, animals, or other infected humans. They can also be passed on to offspring. In CSF analysis, the detection of these proteins, particularly the 14-3-3 protein, is a significant biomarker for prion diseases like Creutzfeldt-Jakob disease (CJD). Sensitivity and specificity of CSF tests for prions are continually improving, aiding in early and accurate diagnosis (A. G. B. Thompson & Mead, 2019). It is suggested to avoid exposure/contact to contaminated medical equipment or body fluids from known or suspected cases of CJD as the disease can be spread through these fomites.

Alpha-Synuclein

Alpha-synuclein, a protein predominantly expressed in the brain, has gained significant attention as a biomarker in the CSF for the diagnosis and monitoring of neurodegenerative disorders, particularly Parkinson's disease and other synucleinopathies (i.e., Lewy body disease, multiple system atrophy) (McCann et al., 2014). Alpha-synuclein is a presynaptic neuronal protein implicated in synaptic vesicle regulation. Pathologically, it is known for its propensity to form insoluble fibrils, which accumulate in neurons such as Lewy bodies and Lewy neurites. Lewy bodies are the hallmark lesions in PD and other Lewy body diseases. The abnormal aggregation of alpha-synuclein is believed to contribute to neurodegeneration, possibly through mechanisms that disrupt cellular homeostasis and neuronal communication (Visanji et al., 2019).

The quantification of total alpha-synuclein levels in CSF is being explored as a biomarker for these diseases. However, its specificity and sensitivity are still under investigation, as levels can be influenced by various factors including blood contamination and rapid clearance from CSF. For instance, lower levels of alpha-synuclein in CSF have been reported in PD compared to Alzheimer's disease (AD) and healthy controls (Wennström et al., 2013). One of the major challenges in using CSF alpha-synuclein as a biomarker is the variability in measurement techniques and lack of standardized assays. Additionally, overlapping levels of alpha-synuclein across different neurodegenerative diseases and in normal aging populations can limit its utility as a standalone diagnostic marker (Hansson, 2021). Ongoing research is focused on identifying post-translationally modified forms of alpha-synuclein, such as phosphorylated or oligomeric alpha-synuclein, which might offer higher specificity for synucleinopathies (Brembati et al., 2023). Although currently not a definitive diagnostic tool at this time, CSF alpha-synuclein levels can complement other diagnostic modalities in neurodegenerative diseases.

Tau Proteins

Tau proteins are integral to the stability and function of neurons. They are primarily associated with microtubules, helping with their assembly and stabilization, and the MAPT gene (Leveille et al., 2021). In the context of neurodegenerative diseases, especially AD, tau proteins undergo pathological alterations that significantly impact their function and form.

Total Tau

In CSF analysis, total tau (t-tau) serves as a marker for neuronal damage and degeneration (Pillai et al., 2019). Under normal conditions, tau proteins are relatively contained within cells and only a small fraction is released into the CSF. In neurodegenerative conditions, neuronal damage results in the breakdown of cell membrane integrity, leading to an increased release of t-tau into the CSF. Elevated levels of t-tau are not specific to AD but are observed in a variety of neurological conditions that involve neuronal death or injury, such as traumatic brain injury, stroke, and Creutzfeldt-Jakob disease. The quantification of t-tau in CSF provides a measure of the degree of neuronal damage and can be particularly useful in differential diagnosis and monitoring disease progression (Hansson, 2021).

P-Tau (Phosphorylated Tau)

The phosphorylation state of tau is crucial in its ability to bind to and stabilize microtubules. In AD and certain other neurodegenerative diseases, tau proteins become hyperphosphorylated at one of more than 40 different positions (Hansson, 2021). This hyperphosphorylation reduces tau's affinity for microtubules, leading to microtubule destabilization and the subsequent formation of intracellular neurofibrillary tangles, one of the pathological hallmarks of AD (Li et al., 2007). Measuring specific phosphorylated isoforms of tau, such as p-tau181 or p-tau217, in the CSF can increase the specificity for AD. Elevated levels of p-tau are more closely associated with the AD pathology compared to t-tau, as they are indicative of the disease-specific process of tau hyperphosphorylation. P-tau thus offers the potential for the early detection and differentiation of AD from other forms of dementia (Hansson, 2021; Zhang et al., 2024).

The combined analysis of t-tau and p-tau levels in CSF provides valuable information in the context of neurodegenerative diseases. Serum p-tau181 and p-tau217 are observed in autosomal dominant AD, and both CSF and serum p-tau181 and p-tau217 begin to increase years before dementia onset. High levels of t-tau indicate nonspecific neuronal damage, while elevated p-tau levels suggest AD-specific pathological processes. This distinction is particularly important in clinical settings for the differential diagnosis of AD from other neurodegenerative diseases (Hansson, 2021). It is of note that looking for serum biomarkers is the preferred diagnostic methodology due to it being a less invasive procedure (Zhang et al., 2024).

Amyloid-Beta (Aβ) Peptides

Amyloid-beta (Aβ) peptides, particularly Aβ42, are central to the pathophysiology of Alzheimer's disease. These peptides are produced from the amyloid precursor protein (APP) through sequential cleavages by β- and γ-secretases. The Aβ peptides, especially Aβ42, have a propensity to aggregate, forming insoluble fibrils and amyloid plaques in the brain. The process of plaque formation is believed to initiate a cascade of pathological events, including tau pathology, inflammation, and neuronal death, which contribute to the clinical manifestations of AD. The peptides, especially the longer sequences with their increased hydrophobicity, are able to change cellular metabolism and induce insulin resistance in neurons. They congregate to lipid rafts and can create pores in neurons. The influx of Ca+2 due to the pore formation is another mechanism that can lead to neuronal death (Arbor et al., 2016).

The measurement of Aβ42 in CSF is a useful biomarker in an early AD diagnosis. Typically, patients with AD show significantly reduced levels of Aβ42 in the CSF. This reduction is thought to reflect sequestration of Aβ42 in brain plaques, thereby decreasing its presence in the CSF (Grimmer et al., 2009). Decreased CSF Aβ42 levels can be observed years before the onset of clinical symptoms, making it a valuable biomarker for early detection of AD (Arbor et al., 2016). Aβ peptides in CSF are also important in the context of development of therapeutic interventions. Monitoring Aβ levels can be useful in assessing the efficacy of treatments that target amyloid pathology, such as β-secretase inhibitors or immunotherapies directed against Aβ. There are many modalities currently in clinical trial; however, the benefit for the prevention of the progression of AD is still uncertain (Jeremic et al., 2021).

Aβ Ratios

To improve diagnostic accuracy, the ratio of plasma Aβ42 to other Aβ peptides, such as Aβ40, is often used. This ratio adjusts for individual variability in overall Aβ production. A decreased plasma Aβ42/Aβ40 ratio is more specific for AD pathology as it reflects the selective deposition of Aβ42 in plaques (Fandos et al., 2017).

CSF Aβ42 levels, along with the presence of tau proteins, form part of the core biomarkers in the diagnostic criteria for AD. These biomarkers help in differentiating AD from other forms of dementia and in predicting the progression from mild cognitive impairment to AD (Hansson, 2021).

Oligoclonal Bands (OCBs)

Oligoclonal bands (OCBs) are distinct immunoglobulin bands seen in CSF electrophoresis, representing localized production of immunoglobulins within the CNS. These bands are detected using isoelectric focusing followed by immunofixation. The presence of OCBs in CSF, but not in serum, indicates intrathecal synthesis of immunoglobulins and suggests an immune process within the CNS. They are markers for inflammatory conditions of the CNS, especially multiple sclerosis (MS). OCBs are produced when activated B-lymphocytes within the CNS synthesize immunoglobulins. OCBs are found in approximately 90%–95% of MS patients. Their presence supports the diagnosis of MS, especially when other clinical and radiological findings are consistent with the disease (Kennedy et al., 2024). The presence and number of OCBs in CSF may correlate with disease activity and progression in MS. Indeed, some studies suggest that a higher number of bands might be associated with a more aggressive disease course, although this is not universally accepted (Jin et al., 2023). While OCBs are strongly associated with MS, they are not specific to this disease process (Jin et al., 2023). They can also be detected in other inflammatory neurological disorders such as neurosarcoidosis, autoimmune encephalitis, and CNS infections like neurosyphilis or Lyme disease. In infectious conditions, OCBs reflect an immune response to the pathogen (Libbey et al., 2014).

The diagnostic value of OCBs lies in their ability to support an inflammatory or immune-mediated etiology of neurological symptoms. In terms of prognosis, while OCBs have been studied, their role in predicting disease progression in MS and other conditions is still an area of ongoing research. OCBs are useful to provide further evidence of an accurate diagnosis of a pathologic process. Interpretation of the presence of OCBs should be taken in the context of the patient presentation and other laboratory findings. Additionally, a negative OCB result does not rule out MS or other inflammatory CNS diseases (Jin et al., 2023).

Big Picture

Protein biomarkers, when evaluated in conjunction with clinical findings and other diagnostic tests, significantly enhance the diagnostic accuracy and management of various neurological diseases. Understanding the nuances of these proteins in CSF is beneficial to providing optimal care for patients experiencing neuropathological disorders.

Glucose

Glucose in the CSF is a critical marker for various neurological conditions. Normally, CSF glucose levels are about 60%–70% of plasma glucose levels, with standard reference ranges typically between 50 and 80 mg/dL. These values can be influenced by systemic glucose levels, so it is necessary to compare CSF glucose with concurrent blood glucose levels (Shahan et al., 2021).

Hypoglycorrhachia, or low CSF glucose, is found in various neurological conditions. The most common cause of reduced CSF glucose varies depending on patient factors, such as neurosurgical and infectious disease history. For example, a retrospective analysis of 89 adults with hypoglycorrhachia conducted by Chow and Troy found that while bacterial meningitis, where glucose utilization is increased by bacteria and the host's immune response, was the most common cause of hypoglycorrhachia in the sample, this was mediated by the high rates in neurosurgical patients. When the sample was subdivided by history, fungal meningitis was most common in individuals with HIV and noninfectious etiologies (e.g., cerebrovascular events and malignancies) were the commonest etiologies in individuals without neurosurgical or HIV histories (Chow & Troy, 2014). Viral meningitis often does not significantly affect glucose levels. Other causes of low CSF glucose include fungal and tuberculous meningitis, CNS sarcoidosis, malignancies, and subarachnoid hemorrhage (Shahan et al., 2021). GLUT1 deficiency syndrome, while rare, can also present with hypoglycorrhachia if the CSF is assessed in a stepwise age-related fashion (Leen et al., 2013). Elevated CSF glucose levels are rare but could be seen in conditions where blood glucose levels are extremely high, as in diabetic ketoacidosis (Tigchelaar et al., 2022).

Lactate

Lactate in CSF serves in helping differentiate between bacterial and viral meningitis. Normal CSF lactate levels range from 1.1 to 2.4 mmol/L. In bacterial meningitis, lactate levels are typically elevated (>4 mmol/L), reflecting anaerobic metabolism due to bacterial proliferation and the host immune response. Aseptic meningitis usually does not cause a significant increase in CSF lactate (Huy et al., 2010).

Elevated lactate levels are also seen in conditions that result in impaired oxidative metabolism within the CNS, such as mitochondrial disorders, cerebral hypoxia, and ischemia. In such conditions, cells switch to anaerobic metabolism, leading to lactate accumulation. Elevated lactate can also be observed in neoplastic diseases involving the CNS, due to the high metabolic rate of tumor cells and the Warburg effect (Huy et al., 2010).

It is important to note that while elevated CSF lactate is a useful marker in differentiating bacterial from viral meningitis, it is not infallible. Some viral infections, particularly those causing encephalitis or severe systemic illness, can also lead to elevated lactate levels due to inflammatory-induced anaerobic metabolism. Several other CNS diseases such as stroke, seizure, and cerebral trauma can also cause elevated CSF lactate. It is therefore important to evaluate the patient and use the CSF lactate in conjunction with symptom presentation and other laboratory findings (Huy et al., 2010).

Evaluation

Ratios

The analysis of protein ratios in CSF is a powerful tool in neuropsychology, aiding in the diagnosis and differentiation of major neuropathic illnesses. These ratios, when used in conjunction with other diagnostic information, can significantly improve the accuracy of clinical assessments and inform treatment strategies. However, their interpretation requires careful consideration of the entire clinical context and an understanding of the limitations of current testing methodologies.

Tau/Amyloid-Beta Ratios

The tau/amyloid-beta (Aβ) ratio, specifically the ratio of total tau (t-tau) and phosphorylated tau (p-tau) to Aβ peptides in the CSF, is an established diagnostic tool for Alzheimer's disease (AD) and other neurodegenerative conditions. In AD, the characteristic accumulation of amyloid-beta plaques and tau neurofibrillary tangles in the brain, manifesting as decreased Aβ42 and increased t-tau and p-tau levels in CSF, is central to the disease's pathogenesis. The elevated tau (t-tau or p-tau)/Aβ42 ratio in CSF, indicative of both neuronal injury/death and amyloid pathology, is useful for enhancing diagnostic accuracy for AD (Bouwman et al., 2022).

This ratio's utility extends beyond diagnosis; it serves as a valuable tool in differentiating AD from other forms of dementia that present with similar clinical symptoms but diverge in underlying pathologies. It also provides the ability to detect changes in tau and Aβ1-42 levels, prior to the onset of clinical symptoms of AD. Providing a potential picture of when mild cognitive impairment will occur and how aggressive the disease may present in a patient (Bouwman et al., 2022; Irwin et al., 2013).

Prognostically, higher tau/Aβ1-42 ratios in AD correlate with faster disease progression and greater cognitive decline. This ratio also facilitates monitoring disease progression and evaluating treatment efficacy targeting amyloid-beta and tau pathologies. It should be noted that the utility of this biomarker is subject to the variability in laboratory methods for measuring tau and Aβ1-42, indicating further assay standardization is needed for consistent interpretation (Bouwman et al., 2022).

Understanding the tau/Aβ1-42 ratio's dynamics in different dementia types can improve clinical efficiency. In AD, a high tau/Aβ1-42 ratio is typical, reflecting decreased Aβ1-42 amyloid plaques and increased t-tau and p-tau (Bouwman et al., 2022). Vascular dementia typically shows a lower ratio, with mildly elevated tau levels and variable amyloid-beta levels, suggesting no pathologic variations of amyloid are present in neurons. Frontotemporal dementia often presents with normal or slightly elevated tau levels and normal amyloid-beta levels, resulting in a lower tau/Aβ1-42 ratio (Irwin et al., 2013). Other conditions, like Creutzfeldt-Jakob disease (CJD), show distinct tau/Aβ1-42 ratio patterns. With CJD, for instance, patients might display markedly high levels of t-tau but relatively unaffected amyloid-beta levels (Randall et al., 2013).

Alpha-Synuclein/Tau Ratio in Parkinson's Disease (PD) and Lewy Body Dementia (LBD)

PD and LBD are characterized by the pathological accumulation of alpha-synuclein in the brain, forming Lewy bodies and Lewy neurites. This contrasts starkly with the tau pathology predominant in AD. Research suggests that in PD and LBD, CSF levels of alpha-synuclein are generally reduced, likely reflecting its sequestration in neuronal aggregates. In contrast, tau levels in these conditions may not be as markedly affected as in AD (Bouwman et al., 2022; Moussaud et al., 2014). In clinical practice, the alpha-synuclein/tau ratio augments the differential diagnosis of dementia and parkinsonian syndromes. It adds a layer of specificity to standard diagnostic procedures and potentially plays a role in monitoring disease progression and therapeutic response. However, the applicability of this ratio in tracking disease trajectory and response to treatment is still a subject of ongoing research and should be used with other diagnostic tools for decision-making (Moussaud et al., 2014).

IgG Index in Cerebrospinal Fluid (CSF) Analysis

The IgG index is a valuable tool in the assessment of neuroinflammatory and infectious processes within the CNS. It quantifies the production of IgG within the CNS and is a measure of intrathecal IgG synthesis. IgG Index is calculated as the ratio of (CSF IgG/serum IgG) to (CSF albumin/serum albumin). A normal IgG index typically is less than 0.70. An elevated IgG index (>0.71) suggests increased production of IgG within the CNS, indicative of an inflammatory or immune process (Simonsen et al., 2020).

In multiple sclerosis (MS), the IgG index is often elevated, reflecting the intrathecal production of IgG as part of the autoimmune process underlying the disease. It is one of the supportive laboratory findings used in the diagnosis of MS, according to the McDonald criteria (A. J. Thompson et al., 2018). An elevated IgG index in MS can be associated with the presence of oligoclonal bands, another marker of CNS-specific immune activity (Magliozzi & Cross, 2020).

Besides MS, an increased IgG index can be seen in other neuroinflammatory conditions such as neurosarcoidosis, CNS lupus, and some infectious diseases like neuroborreliosis (Lyme disease). It is important to note that an elevated IgG index is not just specific to MS and should be interpreted in the broader context of clinical presentation and other diagnostic findings (Lepennetier et al., 2019).

14-3-3 Proteins in the Evaluation of Neuropathic Conditions

14-3-3 proteins, a group of regulatory molecules involved in various cellular processes, have gained prominence in the diagnostic evaluation of certain neuropathic conditions, most notably CJD, a rapidly progressive and fatal neurodegenerative disorder. In the context of CJD, 14-3-3 proteins are released into the CSF because of the extensive neuronal damage characteristic of the disease. Their presence in the CSF serves as a marker of neuronal death and is indicative of the underlying pathology of CJD. The detection of 14-3-3 proteins in CSF is part of the diagnostic criteria for sporadic CJD, particularly when accompanied by typical clinical symptoms and electroencephalogram or MRI findings (Figgie & Appleby, 2021).

The presence of 14-3-3 proteins in the CSF can support the diagnosis of CJD in the appropriate clinical context, but it is not specific to CJD and can be elevated in other acute or rapidly progressive neurological conditions, such as acute stroke, encephalitis, or hypoxic brain injury. The test for 14-3-3 proteins is most useful when there is a high clinical suspicion of CJD, and it should be interpreted alongside other diagnostic investigations. If available, RT-QuIC or MRI are preferred diagnostic tools (Figgie & Appleby, 2021). For neuropsychologists, the interpretation of these markers should be integrated with comprehensive clinical and neuropsychological evaluations to aid in diagnosis, patient management, and supportive care for individuals with rapidly progressive neurodegenerative diseases.

Neurofilament Light Chain Concentrations

Neurofilament light chain (NfL) is released with neuronal damage or degeneration to the axon and has a very high sensitivity in patients with CJD, even in patients with more slowly progressive forms (Thompson & Mead, 2019). NfL is useful for differentiating atypical rapid neurodegenerative dementias from atypical prion disease. However, in similar neurological pathologies, clinical usefulness of NfL is reduced (Figgie & Appleby, 2021).

RT-QuIC Assay in Neuropathic Condition Evaluation

The Real-Time Quaking-Induced Conversion (RT-QuIC) assay has emerged as a significant advancement in the diagnostic evaluation of prion diseases, including CJD, and is gaining attention for its potential application in other protein misfolding neuropathologies. RT-QuIC amplifies minute amounts of misfolded prion proteins (PrPSc) present in biological samples, typically CSF or nasal brushings. The assay induces the misfolded proteins to convert normal prion proteins (PrPC) into the pathological form, which can then be detected and quantified due to amyloid formation. The assay's high sensitivity (>90%) and specificity (>98.5%) reduce the need for more invasive procedures like brain biopsy, historically used in the definitive diagnosis of CJD. The assay requires specialized equipment and expertise, limiting its availability to certain specialized centers (Figgie & Appleby, 2021).

Cellular Markers

Cellular components in CSF provide critical insights into various neurological conditions. The presence and type of cells in CSF can indicate inflammation, infection, hemorrhage, or malignancies within the CNS.

White Blood Cells (WBCs)

Normally, CSF contains around 5 WBCs/μL, predominantly lymphocytes and monocytes. Elevated WBC counts in CSF, known as pleocytosis, indicate inflammation or infection. The differential count of WBCs can help in diagnosing the type of infection: predominantly lymphocytic pleocytosis suggests viral infections, malignancy, or chronic inflammatory conditions like multiple sclerosis, whereas neutrophilic pleocytosis is more indicative of bacterial meningitis or early stages of viral meningitis (de Graaf et al., 2011; Jerrard et al., 2001).

Red Blood Cells (RBCs)

RBCs are usually absent from CSF. Their presence often indicates a hemorrhage into the subarachnoid space or a traumatic lumbar puncture (i.e., blood contamination during the procedure). In subarachnoid hemorrhage, xanthochromia—a yellow discoloration caused by the breakdown of RBCs—can also be observed in the CSF (Jerrard et al., 2001).

Eosinophils

Eosinophils are typically not present in normal CSF. Their presence, eosinophilic pleocytosis, can be seen in parasitic infections, fungal meningitis, and allergic reactions. They are also occasionally observed in conditions like eosinophilic meningitis, which can be caused by medications or malignancies.

Malignant Cells
The detection of malignant cells in CSF is indicative of primary CNS tumors, such as gliomas and medulloblastomas, or metastases from systemic cancers. The identification of specific cell types through cytological examination can guide the diagnosis and treatment of these conditions.

Other Cellular Components
Occasionally, other cellular components like macrophages and plasma cells can be found in CSF. Macrophages are typically seen following hemorrhage or chronic inflammation, while plasma cells might be observed in infectious or autoimmune conditions after a breach of the blood-brain barrier.

Tissue Biopsies

Tissue biopsies are a useful diagnostic tool for the evaluation of neurodegenerative and neuromuscular conditions if other less invasive tools do not reveal a definitive answer as to the underlying pathology associated with the patient's condition. The American Academy of Neurology's guidelines should be followed before biopsy is recommended (American Academy of Neurology, 2024).

Central Nervous System

Biopsies can be performed in various locations in the CNS depending on the suspected pathology. Common sites include the frontal and temporal lobes, which are more easily accessible and carry a lower risk of complications. The choice of location is guided by imaging studies that pinpoint areas of abnormality, such as tumors, lesions, or areas of inflammation. CNS biopsies are typically indicated when imaging and noninvasive testing fail to provide a definitive diagnosis. Common indications include unexplained neurological symptoms, suspected tumors, inflammatory diseases, and infectious lesions. Counter-indications include high-risk locations (such as the brainstem), severe coagulopathy, uncontrolled hypertension, and in cases where surgery poses greater risk than the potential diagnostic benefit. The risks associated with brain biopsies include hemorrhage, infection, neurological deficits (e.g., weakness, language impairment), and, in rare cases, death. The risk varies based on the biopsy location and the patient's overall health status (Mathon et al., 2020).

CNS biopsies are effective in diagnosing and understanding major neurological diseases. For instance, in neuro-oncology, biopsies can differentiate between primary brain tumors (like gliomas) and metastatic lesions, influencing treatment regimens significantly (Mathon et al., 2020). In neuroinflammatory conditions, such as MS or vasculitis, biopsies can reveal the characteristic pathological features, helping to confirm the diagnosis—although a biopsy as a diagnostic tool in MS is rarely needed (Kuhlmann et al., 2008). Finally, in infectious diseases like progressive multifocal leukoencephalopathy or brain abscesses, biopsies can identify specific pathogens, guiding targeted antimicrobial therapy (Mathon et al., 2020).

Peripheral Nervous System

PNS biopsies often involve the sural nerve in the lower leg, as it is sensory, and its biopsy is less likely to cause significant motor disability. Other nerves, like the radial cutaneous nerve, may also be biopsied depending on the clinical scenario. Indications for nerve biopsy include unexplained peripheral neuropathies, particularly when they are asymmetrical or rapidly progressive, and suspected cases of vasculitis or amyloidosis. Counter-indications include situations where a less invasive test can provide the same information, or in cases where the biopsy is unlikely to affect clinical management (Nathani et al., 2021). Nerve biopsies are crucial in diagnosing diseases like amyloid neuropathy, where amyloid deposits are seen in nerve tissue, or in vasculitis-based neuropathies, where inflammation of small vessels can be directly observed. They are also essential in diagnosing certain infectious neuropathies, such as leprosy or Lyme disease, where the causative organisms can be identified in nerve tissue. In hereditary neuropathies, such as Charcot-Marie-Tooth disease, nerve biopsies can reveal characteristic structural abnormalities. Risks include pain, infection, sensory deficits at the biopsy site, and, rarely, neuroma formation. The severity of these complications varies based on the nerve biopsied and the technique used.

Genetic Analysis

Genetic studies have become an indispensable part of neuropathology, providing insights into the etiology and pathogenesis of various neurological disorders. The integration of genetic data with clinical and pathological findings facilitates a more accurate diagnosis and personalized treatment approaches. Further, it provides guidance for patients to be able to discuss family planning in light of the risk of inheritable diseases.

Huntington's Disease and CAG Repeats

Huntington's disease (HD) is a paradigmatic example of a neurogenetic disorder where neuropathology and genetics intersect significantly. It is an autosomal dominant, progressive neurodegenerative disorder characterized by a triad of motor, cognitive, and psychiatric disturbances. Huntington's disease is an anticipatory condition that can worsen with each new generation (Reiner et al., 2011). The genetic mutation responsible for HD involves an abnormal expansion of CAG triplet repeats in the HTT gene on chromosome 4. The CAG repeat in the HTT gene encodes a polyglutamine (polyQ) tract in the huntingtin protein. The expanded polyQ tract leads to the production of a mutant huntingtin protein, which accumulates and forms aggregates in neurons, particularly affecting the striatum and cerebral cortex (Podvin et al., 2019).

The age of onset and the rate of disease progression in HD are inversely correlated with the number of CAG repeats. A higher number of repeats is associated with an earlier onset and more rapid progression. This information is vital for prognostic counseling and planning long-term care. Normal individuals typically have fewer than 36 repeats, while individuals with HD usually have 40 or more repeats. Those with repeat numbers in the intermediate range (35–39) may or may not develop HD. Patients with more than 60 repeats will experience juvenile HD (Podvin et al., 2019).

Neuropathologically, HD is characterized by selective neuronal loss and atrophy predominantly in the striatum (caudate nucleus and putamen), leading to characteristic choreiform movements. However, as the disease progresses, more widespread brain atrophy is observed, correlating with the diverse range of clinical symptoms (Reiner et al., 2011), including neuropsychiatric (Paoli et al., 2017).

Genetic testing not only confirms the diagnosis in symptomatic individuals but also allows for predictive testing in asymptomatic at-risk individuals. Genetic testing for HD involves quantifying the number of CAG repeats in the HTT gene. Given the direct correlation between repeat length and disease manifestation, as well as implications for offspring and family members, pre- and post-test genetic counseling is essential but should be done according to the patients' wishes (Craufurd et al., 2015).

Understanding the genetic basis of HD has opened avenues for research into targeted therapies aimed at reducing the production or aggregation of mutant huntingtin protein. While there is currently no cure for HD, ongoing research in gene therapy and molecular-targeted treatments offers hope for altering the disease course (Kim et al., 2021).

Alzheimer's Disease and APOE Status

Alzheimer's disease, the most common form of dementia, has a complex etiology with both genetic and environmental factors playing roles in its pathogenesis. Among the genetic factors, the apolipoprotein E (APOE) gene is one of the most significant. The APOE gene has three major alleles: ε2, ε3, and ε4. The ε4 allele is associated with an increased risk of developing AD and is also linked to an earlier age of onset. The ε2 allele is thought to have a protective effect against the development of AD. The ε3 allele is neutral and is the most common (Arbor et al., 2016).

Neuropathologically, AD is characterized by amyloid plaques and neurofibrillary tangles. APOE ε4 has been linked to a higher burden of amyloid plaques in the brain. This allele influences amyloid-beta aggregation and clearance, thereby accelerating plaque formation. Additionally, APOE ε4 is associated with increased tau pathology, another key feature of AD (Arbor et al., 2016).

While genetic testing for APOE status can provide information about an individual's risk for AD, it is not determinative. The presence of an APOE ε4 allele does not guarantee that a person will develop AD, nor does the absence of ε4 preclude it. Therefore, genetic counseling is crucial when interpreting these results, especially considering the psychological impact and the current lack of curative treatments (Roses, 1997). Knowledge of a patient's APOE status can inform the clinical management of AD. Individuals with an ε4 allele may benefit from more aggressive risk factor management (e.g., controlling hypertension, diabetes, and hyperlipidemia) and might be monitored more closely for cognitive decline. It also has implications for clinical trials, as APOE ε4 carriers might respond differently to therapeutic interventions (Ji et al., 2024).

The APOE ε4 allele has become a significant target in AD research, with efforts directed at understanding how it contributes to the pathogenesis of AD and developing therapies that could mitigate its effects. This includes research into therapies that can modify amyloid-beta aggregation or enhance its clearance.

Presenilin is another gene that impacts the production of amyloid-beta and can increase the formation of Aβ42 variants, which are more likely to lead to pathological aggregation. Mutations in presenilin 1 and 2 (PSEN1/2) play significant roles in familial forms of AD, and are, therefore, targets of genetic studies. As with APOE, identification of mutations in PSEN 1 and 2 have implications for onset and course of disease, possible treatment targets, and family planning (An et al., 2015; Kelleher & Shen, 2017).

Parkinson's Disease

Parkinson's disease, predominantly characterized by motor symptoms such as bradykinesia, rigidity, and tremor, also includes non-motor features and has a multifactorial etiology, including genetic factors. Several genes have been implicated in PD, with mutations in LRRK2, SNCA (which encodes alpha-synuclein), and PARK genes being the most common. LRRK2 mutations are among the most frequent causes of familial and sporadic PD cases. Mutations in SNCA lead to the accumulation of alpha-synuclein in Lewy bodies, a neuropathological hallmark of PD (Cook et al., 2021).

Genetic testing for PD-related mutations can be informative, particularly in familial cases or early-onset PD. However, the penetrance of these mutations can vary, and not all mutation carriers will develop PD. Genetic counseling is useful for interpreting test results and discussing the implications with patients and their families in a way that prevents unnecessary worry or concern (Schulte & Gasser, 2011).

The identification of specific genetic mutations can have implications for the clinical management of PD. For example, patients with LRRK2 mutations may exhibit a clinical course and response to dopaminergic therapy similar to idiopathic PD, while those with PARK mutations may have a more aggressive disease progression. Understanding the genetic underpinnings of PD has significant implications for research and development of targeted therapies. For instance, therapies aimed at reducing alpha-synuclein accumulation are under investigation. Genetic testing of patients with clinical suspicion for PD could create avenues for the patients to be involved in clinical trials (Cook et al., 2021).

As our understanding of the genetic basis of PD evolves, it holds promise for improved diagnostic, therapeutic, and prognostic approaches in the management of this complex neurodegenerative disorder.

Amyotrophic Lateral Sclerosis

Amyotrophic lateral sclerosis, also known as Lou Gehrig's disease, is a progressive neurodegenerative disorder characterized by the loss of motor neurons in the spinal cord, brainstem, and motor cortex. While most ALS cases are sporadic, approximately 10% are familial (FALS), highlighting the role of genetic factors in the disease. Several genes have been identified in association with ALS. Mutations in the SOD1 (superoxide dismutase 1) gene were the first to be linked to FALS. The C9orf72 gene mutation is currently recognized as the most common genetic cause of both FALS and sporadic ALS cases. Other genes, including TARDBP (which encodes TDP-43) and FUS, are also implicated in ALS (Andersen & Al-Chalabi, 2011).

The main pathologic process of ALS includes the degeneration of motor neurons and the presence of protein inclusions in the surviving neurons. TDP-43 inclusions are found in the majority of ALS cases. The specific genetic mutations can influence the type of protein inclusions and the pattern of neuronal degeneration (Mejzini et al., 2019).

Genetic testing in ALS can confirm a suspected familial case and guide family members to understand their own risk. It is important for genetic counseling to accompany testing, given the implications for both the patient and their family, particularly as some of these genetic forms of ALS may have a more aggressive course or earlier onset. The identification of specific genetic mutations in ALS patients can provide insights into the likely course of the disease and potential responses to therapy. For example, ALS patients with SOD1 mutations may have a different disease trajectory compared to those with C9orf72 mutations. Even with the potential benefits of genetic testing, there is a lack of consensus on the application of genetic counseling and use of genetic testing in ALS (Vajda et al., 2017).

Frontotemporal Dementia (FTD) and Genetic Factors

Frontotemporal dementia, a heterogeneous group of disorders characterized by progressive atrophy of the frontal and temporal lobes of the brain, is notable for its genetic links and distinctive neuropathology.

Genetic factors play a significant role in FTD, with key mutations identified in several genes. The C9orf72 gene mutation is the most common genetic cause of both familial FTD and ALS, underscoring the genetic and clinical overlap between these two conditions. Mutations in the MAPT gene lead to the accumulation of abnormal tau proteins. Mutations in the GRN gene, leading to progranulin deficiency, are another notable cause of familial FTD and in TDP-43 pathologies. These genetic abnormalities contribute to the diverse clinical and neuropathological manifestations of FTD (Greaves & Rohrer, 2019).

Genetic testing for FTD is essential, particularly in familial cases, and should be accompanied by comprehensive genetic counseling due to the implications for the patient and their relatives. Identifying a specific genetic mutation can provide insights into the likely disease course, potential comorbidities, and implications for other family members. Genetic testing can also further facilitate accurate diagnosis of the neuropathic condition (Koriath et al., 2021).

Quick Tips for Genetic Analysis in Neuropathology

- Always interpret genetic findings in conjunction with clinical presentation, neuropathological findings, and family history. Genetic mutations should be considered as part of a broader diagnostic picture.
- Understand the implications of genotype-phenotype correlations. Certain genetic mutations are associated with specific clinical manifestations and neuropathological features.
- Genetic counseling is essential before and after testing. It helps patients and families understand the implications of genetic findings, including risk for other family members, potential for disease progression, and therapeutic options.
- Address ethical considerations, such as the impact of genetic information on insurability and employment, and the decision-making process for asymptomatic at-risk family members.
- Use genetic testing to confirm or rule out suspected genetic disorders, especially in atypical or early-onset cases.
- Genes like APOE and PSEN1/2 in Alzheimer's disease or C9orf72 in ALS and FTD can provide prognostic information, influencing clinical management and patient counseling.
- Tailor treatment strategies based on genetic findings. For instance, certain genetic forms of Parkinson's disease might respond differently to conventional dopaminergic therapies.
- Stay informed about emerging gene-targeted therapies. As our understanding of neurogenetics expands, so does the potential for targeted treatments.
- Genetic information can guide patient selection for clinical trials, ensuring appropriate targeting of novel therapies.
- Participate in or refer patients to registries and research studies. This can help in understanding the natural history of genetic neurological disorders and in developing new treatments.
- Keep abreast of advancements in genetic testing technologies, such as next-generation sequencing, which can simultaneously screen multiple genetic abnormalities.
- Understand the limitations of current genetic tests and the potential for false negatives or uncertain significance results.

Genetic analysis in neuropathology is a dynamic and integral aspect of patient care, offering profound insights into diagnosis, prognosis, and treatment. A nuanced understanding of genetic data, coupled with a comprehensive, patient-centered approach, is essential for harnessing the full potential of genetic testing in clinical practice.

Conclusion

In the diagnostic workup for genetic and neurodegenerative illnesses, biomarkers, genetic testing, and tissue biopsies have become an important step toward enhancing the assessment and management of patients with neurologic disorders. These tools afford much deeper and more accurate insights into the biological processes behind cognitive and behavioral signs/symptoms, thus enhancing the accuracy and clinical relevance of neuropsychological assessment. The combination of laboratory data with cognitive profiles will therefore allow clinicians to be more precise in their diagnoses of neuropsychological disorders, thereby improving the formulation of treatment plans and prognosis counseling. The multidimensional approach is of vital importance to the holistic care aspect.

References

Aluise, C. D., Sowell, R. A., & Butterfield, D. A. (2008). Peptides and proteins in plasma and cerebrospinal fluid as biomarkers for the prediction, diagnosis, and monitoring of therapeutic efficacy of Alzheimer's disease. *Biochimica et Biophysica Acta, 1782*(10), 549–558. https://doi.org/10.1016/j.bbadis.2008.07.008

American Academy of Neurology. (2024, November 24). *Clinical guidelines*. American Academy of Neurology Clinical Guidelines. https://www.aan.com/practice/guidelines

An, S. S., Cai, Y., & Kim, S. (2015). Mutations in presenilin 2 and its implications in Alzheimer's disease and other dementia-associated disorders. *Clinical Interventions in Aging*, 1163. https://doi.org/10.2147/CIA.S85808

Andersen, P. M., & Al-Chalabi, A. (2011). Clinical genetics of amyotrophic lateral sclerosis: What do we really know? *Nature Reviews Neurology, 7*(11), Article 11. https://doi.org/10.1038/nrneurol.2011.150

Arbor, S. C., LaFontaine, M., & Cumbay, M. (2016). Focus: The aging brain: Amyloid-beta Alzheimer targets—Protein processing, lipid rafts, and amyloid-beta pores. *Yale Journal of Biology and Medicine, 89*(1), 5.

Avery, R. A., Shah, S. S., Licht, D. J., Seiden, J. A., Huh, J. W., Boswinkel, J., Ruppe, M. D., Chew, A., Mistry, R. D., & Liu, G. T. (2010). Reference range for cerebrospinal fluid opening pressure in children. *New England Journal of Medicine, 363*(9), 891–893. https://doi.org/10.1056/NEJMc1004957

Bin Wan Hassan, W. M. N., Mistretta, F., Molinaro, S., Russo, R., Bosco, G., Gambino, A., & Bergui, M. (2023). Overview of spontaneous intracranial hypotension and differential diagnosis with Chiari I malformation. *Journal of Clinical Medicine, 12*(9), Article 9. https://doi.org/10.3390/jcm12093287

Bouwman, F. H., Frisoni, G. B., Johnson, S. C., Chen, X., Engelborghs, S., Ikeuchi, T., Paquet, C., Ritchie, C., Bozeat, S., Quevenco, F.-C., & Teunissen, C. (2022). Clinical application of CSF biomarkers for Alzheimer's disease: From rationale to ratios. *Alzheimer's & Dementia: Diagnosis, Assessment & Disease Monitoring, 14*(1), e12314. https://doi.org/10.1002/dad2.12314

Breiner, A., Moher, D., Brooks, J., Cheng, W., Hegen, H., Deisenhammer, F., McCudden, C. R., & Bourque, P. R. (2019). Adult CSF total protein upper reference limits should be age-partitioned and significantly higher than 0.45 g/L: A systematic review. *Journal of Neurology, 266*(3), 616–624. https://doi.org/10.1007/s00415-018-09174-z

Brembati, V., Faustini, G., Longhena, F., & Bellucci, A. (2023). Alpha synuclein post translational modifications: Potential targets for Parkinson's disease therapy? *Frontiers in Molecular Neuroscience, 16*. https://www.frontiersin.org/articles/10.3389/fnmol.2023.1197853

Burger, B. M., Chavis, P. S., & Purvin, V. (2013). A weed by any other name. *Survey of Ophthalmology, 58*(2), 176–183. https://doi.org/10.1016/j.survophthal.2012.02.005

Chow, E., & Troy, S. B. (2014). The differential diagnosis of hypoglycorrhachia in adult patients. *American Journal of the Medical Sciences, 348*(3), 186–190. https://doi.org/10.1097/MAJ.0000000000000217

Cook, L., Schulze, J., Naito, A., & Alcalay, R. N. (2021). The role of genetic testing for Parkinson's disease. *Current Neurology and Neuroscience Reports, 21*(4), 17. https://doi.org/10.1007/s11910-021-01100-7

Craufurd, D., MacLeod, R., Frontali, M., Quarrell, O., Bijlsma, E. K., Davis, M., Hjermind, L. E., Lahiri, N., Mandich, P., Martinez, A., Tibben, A., & Roos, R. A. (2015). Diagnostic genetic testing for Huntington's disease. *Practical Neurology, 15*(1), 80–84. https://doi.org/10.1136/practneurol-2013-000790

de Graaf, M. T., de Jongste, A. H. C., Kraan, J., Boonstra, J. G., Smitt, P. A. E. S., & Gratama, J. W. (2011). Flow cytometric characterization of cerebrospinal fluid cells. *Cytometry Part B: Clinical Cytometry, 80B*(5), 271–281. https://doi.org/10.1002/cyto.b.20603

Fandos, N., Pérez-Grijalba, V., Pesini, P., Olmos, S., Bossa, M., Villemagne, V. L., Doecke, J., Fowler, C., Masters, C. L., & Sarasa, M. (2017). Plasma amyloid β 42/40 ratios as biomarkers for amyloid β cerebral deposition in cognitively normal individuals. *Alzheimer's & Dementia: Diagnosis, Assessment & Disease Monitoring, 8*, 179–187. https://doi.org/10.1016/j.dadm.2017.07.004

Fautsch, K. J., Block, D. R., Graff-Radford, J., Wang, F., Craver, E. C., Hodge, D. O., Cutsforth-Gregory, J. K., Kilgore, K. P., Petersen, R. C., Knopman, D. S., Flanagan, E., Toledano, M., Mielke, M. M., Bhatti, M. T., & Chen, J. J. (2023). Population-based evaluation of total protein in cerebrospinal fluid. *Mayo Clinic Proceedings, 98*(2), 239–251. https://doi.org/10.1016/j.mayocp.2022.10.012

Figgie, M. P., & Appleby, B. S. (2021). Clinical use of improved diagnostic testing for detection of prion disease. *Viruses, 13*(5), Article 5. https://doi.org/10.3390/v13050789

Godoy, D. A., Robba, C., Paiva, W. S., & Rabinstein, A. A. (2022). Acute intracranial hypertension during pregnancy: Special considerations and management adjustments. *Neurocritical Care, 36*(1), 302–316. https://doi.org/10.1007/s12028-021-01333-x

Greaves, C. V., & Rohrer, J. D. (2019). An update on genetic frontotemporal dementia. *Journal of Neurology, 266*(8), 2075–2086. https://doi.org/10.1007/s00415-019-09363-4

Grimmer, T., Riemenschneider, M., Förstl, H., Henriksen, G., Klunk, W. E., Mathis, C. A., Shiga, T., Wester, H.-J., Kurz, A., & Drzezga, A. (2009). Beta amyloid in Alzheimer's disease: Increased deposition in brain is reflected in reduced concentration in cerebrospinal fluid. *Biological Psychiatry, 65*(11), 927–934. https://doi.org/10.1016/j.biopsych.2009.01.027

Hansson, O. (2021). Biomarkers for neurodegenerative diseases. *Nature Medicine, 27*(6), Article 6. https://doi.org/10.1038/s41591-021-01382-x

Huy, N. T., Thao, N. T., Diep, D. T., Kikuchi, M., Zamora, J., & Hirayama, K. (2010). Cerebrospinal fluid lactate concentration to distinguish bacterial from aseptic meningitis: A systemic review and meta-analysis. *Critical Care, 14*(6), R240. https://doi.org/10.1186/cc9395

Irwin, D., Trojanowski, J., & Grossman, M. (2013). Cerebrospinal fluid biomarkers for differentiation of frontotemporal lobar degeneration from Alzheimer's disease. *Frontiers in Aging Neuroscience, 5*. https://www.frontiersin.org/articles/10.3389/fnagi.2013.00006

Jabbari, A., Alijanpour, E., Mir, M., Bani Hashem, N., Rabiea, S. M., & Rupani, M. A. (2013). Post spinal puncture headache, an old problem and new concepts: Review of articles about predisposing factors. *Caspian Journal of Internal Medicine, 4*(1), 595–602.

Jeremic, D., Jiménez-Díaz, L., & Navarro-López, J. D. (2021). Past, present and future of therapeutic strategies against amyloid-β peptides in Alzheimer's disease: A systematic review. *Ageing Research Reviews, 72*, 101496. https://doi.org/10.1016/j.arr.2021.101496

Jerrard, D. A., Hanna, J. R., & Schindelheim, G. L. (2001). Cerebrospinal fluid 1 1 Clinical Laboratory in Emergency Medicine is coordinated by Jonathan S. Olshaker, MD, of the University of Maryland, Baltimore, Maryland. *Journal of Emergency Medicine, 21*(2), 171–178. https://doi.org/10.1016/S0736-4679(01)00360-2

Ji, X.-Y., Peng, X.-Y., Tang, H.-L., Pan, H., Wang, W.-T., Wu, J., Chen, J., & Wei, N.-L. (2024). Alzheimer's disease phenotype based upon the carrier status of the apolipoprotein E ε4 allele. *Brain Pathology, 34*(1), e13208. https://doi.org/10.1111/bpa.13208

Jin, H., Lu, Q., Gao, F., & Hao, H. (2023). Application of oligoclonal bands and other cerebrospinal fluid variables in multiple sclerosis and other neuroimmunological diseases: A narrative review. *Annals of Translational Medicine, 11*(7), 282. https://doi.org/10.21037/atm-21-3073

Kelleher, R. J., & Shen, J. (2017). Presenilin-1 mutations and Alzheimer's disease. *Proceedings of the National Academy of Sciences, 114*(4), 629–631. https://doi.org/10.1073/pnas.1619574114

Kennedy, P. G. E., George, W., & Yu, X. (2024). The elusive nature of the oligoclonal bands in multiple sclerosis. *Journal of Neurology, 271*(1), 116–124. https://doi.org/10.1007/s00415-023-12081-7

Kim, A., Lalonde, K., Truesdell, A., Gomes Welter, P., Brocardo, P. S., Rosenstock, T. R., & Gil-Mohapel, J. (2021). New avenues for the treatment of Huntington's disease. *International Journal of Molecular Sciences, 22*(16), 8363. https://doi.org/10.3390/ijms22168363

Koriath, C. A. M., Kenny, J., Ryan, N. S., Rohrer, J. D., Schott, J. M., Houlden, H., Fox, N. C., Tabrizi, S. J., & Mead, S. (2021). Genetic testing in dementia—Utility and clinical strategies. *Nature Reviews Neurology, 17*(1), Article 1. https://doi.org/10.1038/s41582-020-00416-1

Kuhlmann, T., Lassmann, H., & Brück, W. (2008). Diagnosis of inflammatory demyelination in biopsy specimens: A practical approach. *Acta Neuropathologica, 115*(3), 275–287. https://doi.org/10.1007/s00401-007-0320-8

Leen, W. G., Wevers, R. A., Kamsteeg, E.-J., Scheffer, H., Verbeek, M. M., & Willemsen, M. A. (2013). Cerebrospinal fluid analysis in the workup of GLUT1 deficiency syndrome: A systematic review. *JAMA Neurology, 70*(11), 1440–1444. https://doi.org/10.1001/jamaneurol.2013.3090

Lepennetier, G., Hracsko, Z., Unger, M., Van Griensven, M., Grummel, V., Krumbholz, M., Berthele, A., Hemmer, B., & Kowarik, M. C. (2019). Cytokine and immune cell profiling in the cerebrospinal fluid of patients with neuro-inflammatory diseases. *Journal of Neuroinflammation, 16*(1), 219. https://doi.org/10.1186/s12974-019-1601-6

Leveille, E., Ross, O. A., & Gan-Or, Z. (2021). Tau and MAPT genetics in tauopathies and synucleinopathies. *Parkinsonism & Related Disorders, 90*, 142–154. https://doi.org/10.1016/j.parkreldis.2021.09.008

Li, B., Chohan, M. O., Grundke-Iqbal, I., & Iqbal, K. (2007). Disruption of microtubule network by Alzheimer abnormally hyperphosphorylated tau. *Acta Neuropathologica, 113*(5), 501–511. https://doi.org/10.1007/s00401-007-0207-8

Libbey, J. E., Cusick, M. F., & Fujinami, R. S. (2014). Role of pathogens in multiple sclerosis. *International Reviews of Immunology, 33*(4), 266–283. https://doi.org/10.3109/08830185.2013.823422

Magliozzi, R., & Cross, A. (2020). Can CSF biomarkers predict future MS disease activity and severity? *Sage Journals, 26*(5). https://doi.org/10.1177/1352458519871818

Mathon, B., Le Joncour, A., Bielle, F., Mokhtari, K., Boch, A.-L., Peyre, M., Amoura, Z., Cacoub, P., Younan, N., Demeret, S., Shotar, E., Burrel, S., Fekkar, A., Robert, J., Amelot, A., Pineton De Chambrun, M., Carpentier, A., Capelle, L., Navarro, S., . . . Navarro, V. (2020). Neurological diseases of unknown etiology: Brain-biopsy diagnostic yields and safety. *European Journal of Internal Medicine, 80*, 78–85. https://doi.org/10.1016/j.ejim.2020.05.029

McCann, H., Stevens, C. H., Cartwright, H., & Halliday, G. M. (2014). α-Synucleinopathy phenotypes. *Parkinsonism & Related Disorders, 20*, S62–S67. https://doi.org/10.1016/S1353-8020(13)70017-8

Mejzini, R., Flynn, L. L., Pitout, I. L., Fletcher, S., Wilton, S. D., & Akkari, P. A. (2019). ALS genetics, mechanisms, and therapeutics: Where are we now? *Frontiers in Neuroscience, 13*. https://www.frontiersin.org/articles/10.3389/fnins.2019.01310

Mokri, B. (2013). Spontaneous low pressure, low CSF volume headaches: Spontaneous CSF leaks. *Headache: The Journal of Head and Face Pain, 53*(7), 1034–1053. https://doi.org/10.1111/head.12149

Mollan, S. P., Momin, S. N., Khatkar, P. S., Grech, O., Sinclair, A. J., & Tsermoulas, G. (2023). A neuro-ophthalmologist's guide to advances in intracranial pressure measurements. *Eye and Brain, 15*, 113–124. https://doi.org/10.2147/EB.S404642

Moussaud, S., Jones, D. R., Moussaud-Lamodière, E. L., Delenclos, M., Ross, O. A., & McLean, P. J. (2014). Alpha-synuclein and tau: Teammates in neurodegeneration? *Molecular Neurodegeneration, 9*(1), 43. https://doi.org/10.1186/1750-1326-9-43

Nathani, D., Spies, J., Barnett, M. H., Pollard, J., Wang, M., Sommer, C., & Kiernan, M. C. (2021). Nerve biopsy: Current indications and decision tools. *Muscle & Nerve, 64*(2), 125–139. https://doi.org/10.1002/mus.27201

Norager, N. H., Olsen, M. H., Pedersen, S. H., Riedel, C. S., Czosnyka, M., & Juhler, M. (2021). Reference values for intracranial pressure and lumbar cerebrospinal fluid pressure: A systematic review. *Fluids and Barriers of the CNS, 18*(1), 19. https://doi.org/10.1186/s12987-021-00253-4

Paoli, R., Botturi, A., Ciammola, A., Silani, V., Prunas, C., Lucchiari, C., Zugno, E., & Caletti, E. (2017). Neuropsychiatric burden in Huntington's disease. *Brain Sciences, 7*(6), 67. https://doi.org/10.3390/brainsci7060067

Pillai, J. A., Bonner-Jackson, A., Bekris, L. M., Safar, J., Bena, J., & Leverenz, J. B. (2019). Highly elevated cerebrospinal fluid total tau level reflects higher likelihood of non-amnestic subtype of Alzheimer's disease. *Journal of Alzheimer's Disease*, *70*(4), 1051–1058. https://doi.org/10.3233/JAD-190519

Podvin, S., Reardon, H. T., Yin, K., Mosier, C., & Hook, V. (2019). Multiple clinical features of Huntington's disease correlate with mutant HTT gene CAG repeat lengths and neurodegeneration. *Journal of Neurology*, *266*(3), 551–564. https://doi.org/10.1007/s00415-018-8940-6

Randall, C., Mosconi, L., de Leon, M., & Glodzik, L. (2013). Cerebrospinal fluid biomarkers of Alzheimer's disease in cognitively healthy elderly. *Frontiers in Bioscience (Landmark Edition)*, *18*, 1150–1173.

Reiner, A., Dragatsis, I., & Dietrich, P. (2011). Genetics and neuropathology of Huntington's disease. In J. Brotchie, E. Bezard, & P. Jenner (Eds.), *International review of neurobiology* (Vol. 98, pp. 325–372). Academic Press. https://doi.org/10.1016/B978-0-12-381328-2.00014-6

Roses, A. D. (1997). Genetic testing for Alzheimer disease: Practical and ethical issues. *Archives of Neurology*, *54*(10), 1226–1229. https://doi.org/10.1001/archneur.1997.00550220036011

Sakka, L., Coll, G., & Chazal, J. (2011). Anatomy and physiology of cerebrospinal fluid. *European Annals of Otorhinolaryngology, Head and Neck Diseases*, *128*(6), 309–316. https://doi.org/10.1016/j.anorl.2011.03.002

Schulte, C., & Gasser, T. (2011). Genetic basis of Parkinson's disease: Inheritance, penetrance, and expression. *Application of Clinical Genetics*, *4*, 67–80. https://doi.org/10.2147/TACG.S11639

Shahan, B., Choi, E. Y., & Nieves, G. (2021). Cerebrospinal fluid analysis. *American Family Physician*, *103*(7), 422–428.

Simonsen, C. S., Flemmen, H. Ø., Lauritzen, T., Berg-Hansen, P., Moen, S. M., & Celius, E. G. (2020). The diagnostic value of IgG index versus oligoclonal bands in cerebrospinal fluid of patients with multiple sclerosis. *Multiple Sclerosis Journal—Experimental, Translational and Clinical*, *6*(1), 2055217319901291. https://doi.org/10.1177/2055217319901291

Thompson, A. G. B., & Mead, S. H. (2019). Review: Fluid biomarkers in the human prion diseases. *Molecular and Cellular Neuroscience*, *97*, 81–92. https://doi.org/10.1016/j.mcn.2018.12.003

Thompson, A. J., Banwell, B. L., Barkhof, F., Carroll, W. M., Coetzee, T., Comi, G., Correale, J., Fazekas, F., Filippi, M., Freedman, M. S., Fujihara, K., Galetta, S. L., Hartung, H. P., Kappos, L., Lublin, F. D., Marrie, R. A., Miller, A. E., Miller, D. H, Montalban, X., & Cohen, J. A. (2018). Diagnosis of multiple sclerosis: 2017 revisions of the McDonald criteria. *The Lancet. Neurology*, *17*(2), 162–173. https://doi.org/10.1016/S1474-4422(17)30470-2

Tigchelaar, C., Zuylen, M. L. van, Hulst, A. H., Preckel, B., Beek, A. P. van, Kema, I. P., Hermanides, J., & Absalom, A. R. (2022). Elevated cerebrospinal fluid glucose levels and diabetes mellitus are associated with activation of the neurotoxic polyol pathway. *Diabetologia*, *65*(7), 1098. https://doi.org/10.1007/s00125-022-05693-7

Vajda, A., McLaughlin, R. L., Heverin, M., Thorpe, O., Abrahams, S., Al-Chalabi, A., & Hardiman, O. (2017). Genetic testing in ALS. *Neurology*, *88*(10), 991–999. https://doi.org/10.1212/WNL.0000000000003686

Visanji, N. P., Lang, A. E., & Kovacs, G. G. (2019). Beyond the synucleinopathies: Alpha synuclein as a driving force in neurodegenerative comorbidities. *Translational Neurodegeneration*, *8*(1), 28. https://doi.org/10.1186/s40035-019-0172-x

Wennström, M., Surova, Y., Hall, S., Nilsson, C., Minthon, L., Boström, F., Hansson, O., & Nielsen, H. M. (2013). Low CSF levels of both α-synuclein and the α-synuclein cleaving enzyme neurosin in patients with synucleinopathy. *PloS One*, *8*(1), e53250. https://doi.org/10.1371/journal.pone.0053250

Whiteley, W., Al-Shahi, R., Warlow, C. P., Zeidler, M., & Lueck, C. J. (2006). CSF opening pressure: Reference interval and the effect of body mass index. *Neurology*, *67*(9), 1690–1691. https://doi.org/10.1212/01.wnl.0000242704.60275.e9

Wiegand, C., & Richards, P. (2007). Measurement of intracranial pressure in children: A critical review of current methods. *Developmental Medicine & Child Neurology*, *49*(12), 935–941. https://doi.org/10.1111/j.1469-8749.2007.00935.x

Wile, D. (2012). Diuretics: A review. *Annals of Clinical Biochemistry*, *49*(5), 419–431. https://doi.org/10.1258/acb.2011.011281

Wright, B. L. C., Lai, J. T. F., & Sinclair, A. J. (2012). Cerebrospinal fluid and lumbar puncture: A practical review. *Journal of Neurology*, *259*(8), 1530–1545. https://doi.org/10.1007/s00415-012-6413-x

Zhang, X., Wang, J., Zhang, Z., & Ye, K. (2024). Tau in neurodegenerative diseases: Molecular mechanisms, biomarkers, and therapeutic strategies. *Translational Neurodegeneration*, *13*(1), 40. https://doi.org/10.1186/s40035-024-00429-6

Index

For the benefit of digital users, indexed terms that span two pages (e.g., 52–53) may, on occasion, appear on only one of those pages.

Note: Tables and figures are indicated by an italic *t* and *f* following the page number.